Studies in Systems, Decision and Control

Volume 339

The series "Studies in Systems, Decision and Control" (SSDC) covers both new developments and advances, as well as the state of the art, in the various areas of broadly perceived systems, decision making and control–quickly, up to date and with a high quality. The intent is to cover the theory, applications, and perspectives on the state of the art and future developments relevant to systems, decision making, control, complex processes and related areas, as embedded in the fields of engineering, computer science, physics, economics, social and life sciences, as well as the paradigms and methodologies behind them. The series contains monographs, textbooks, lecture notes and edited volumes in systems, decision making and control spanning the areas of Cyber-Physical Systems, Autonomous Systems, Sensor Networks, Control Systems, Energy Systems, Automotive Systems, Biological Systems, Vehicular Networking and Connected Vehicles, Aerospace Systems, Automation, Manufacturing, Smart Grids, Nonlinear Systems, Power Systems, Robotics, Social Systems, Economic Systems and other. Of particular value to both the contributors and the readership are the short publication timeframe and the world-wide distribution and exposure which enable both a wide and rapid dissemination of research output.

Indexed by SCOPUS, DBLP, WTI Frankfurt eG, zbMATH, SCImago.

All books published in the series are submitted for consideration in Web of Science.

More information about this series at http://www.springer.com/series/13304

Ali Ismail Awad · Steven Furnell ·
Marcin Paprzycki · Sudhir Kumar Sharma
Editors

Security in Cyber-Physical Systems

Foundations and Applications

Springer

Editors
Ali Ismail Awad
Department of Computer Science
Electrical and Space Engineering
Luleå University of Technology
Luleå, Sweden

Faculty of Engineering
Al-Azhar University
Qena, Egypt

Marcin Paprzycki
Systems Research Institute
Polish Academy of Sciences
Warszawa, Poland

Steven Furnell
School of Computer Science
University of Nottingham
Nottingham, UK

Sudhir Kumar Sharma
Institute of Information Technology
and Management
New Delhi, India

ISSN 2198-4182 ISSN 2198-4190 (electronic)
Studies in Systems, Decision and Control
ISBN 978-3-030-67360-4 ISBN 978-3-030-67361-1 (eBook)
https://doi.org/10.1007/978-3-030-67361-1

This Springer imprint is published by the registered company Springer Nature Switzerland AG
The registered company address is: Gewerbestrasse 11, 6330 Cham, Switzerland

Preface

Cyber-Physical Systems (CPS) are characterized by the intrinsic combination of software and physical components. Inherent elements often include wired or wireless data communication, sensor devices, real-time operation, and automated control of physical elements. Typical examples of associated application areas include, among others, industrial control systems, smart grids, autonomous vehicles and avionics, medial monitoring, and robotics. The incarnation of CPS can therefore range from considering individual Internet-of-Things devices through to large-scale infrastructures.

Against this backdrop, it is easy to recognize an inherent requirement for security, protection, and trust. CPS will readily exhibit the standard security requirements that are established elsewhere—the need for confidentiality of data, integrity of data and operation, and the overall availability of the data and services being provided. They also require resilience against both failure and attack. However, the context in which they are operating often means that the importance of these aspects is amplified. The impact of a breach may be more significant, particularly in the scenario where failure or compromise of CPS could lead to loss of life. Moreover, the nature and importance of the systems themselves can increase their desirability as targets of attack. Here, it is important to note that as CPS are ingrained into every aspect of our lives, public trust in them becomes of utmost importance. Only if we can trust the CPS, we will be able to reap all potential benefits that they can bring. However, trust starts with assuring highest possible levels of security and protection.

As a consequence, the protection of Cyber-Physical Systems has become a recognized theme within the domain of cybersecurity, with a growing body of knowledge being established around the topic. However, as with cybersecurity in other contexts, the issue is far from solved, and CPS have essentially broadened the landscape in which ongoing attention will now be needed. As such, issues such as recognizing risk, designing resilient architecture, and combatting attacks remain fertile areas for new research and further contribution. It should be also stressed that, on the meta-level, CPS are a coherent whole, but when it comes to individual application areas, domain-specific security and protection requirements have to be formulated and dealt with.

The purpose of this book is to fill in the gaps in the security of Cyber-Physical Systems literature, by providing cutting-edge research findings. The book covers both strategic security research such as security frameworks and risk assessment, and technical security research like intrusion detection, where different case studies are presented. The volume introduces a collection of ten chapters written by experts in the fields that cover strategic and technical security domains. Furthermore, the material has been prepared in a way that makes each chapter independently readable from the others, while still contributing a collective overall insight into the topic area.

The book begins with a chapter authored by *Haque et al.*, which introduces a conceptual analysis of cybersecurity frameworks for improving the knowledge and understanding of the defense-in-depth security architectures. In doing so, it provides a meta-level state-of-the-art reflection on the area. Specifically, the authors conducted a comprehensive analysis of security and resilience frameworks proposed by different governing bodies. Besides, the realization techniques for the most common resilience frameworks and security practices are proposed. According to the NIST Cybersecurity Framework, risk assessment is the first phase toward building any security perimeter. It is also demanded to integrate best practices and standards to manage emerged security risks. The chapter gives sufficient information for understanding the available frameworks, which, if being considered, should facilitate the interoperability between existing and developed attack countermeasures.

The resilience of CPS against security attacks can be partially achieved through the common Confidentiality, Integrity, and Availability (CIA) security triad. Deploying cryptographic algorithms in resource-constrained systems is still a challenge that needs further research. The key establishment can also be a problem or weakness that faces any cryptographic mechanism, where security failure may result from a successful attack compromising the security key. Chapter "Key-Establishment Protocols for Constrained Cyber-Physical Systems", written by*Lara-Nino, Diaz-Perez, and Morales-Sandoval*, tackles the key establishment problem by introducing three alternative protocols for resource-constrained CPSs, where Wireless Sensor Networks (WSNs) are used as an application domain. Two Elliptic Curve Cryptography-based (ECC) protocols are reviewed, analyzed, and improved. As a result, a novel key-establishment protocol that uses the implemented enhancements of ECC-based solutions, but employs quantum-safe cryptographic algorithms, is proposed. The proposed protocol is theoretically valid, however, further empirical research such as security testing, cost calculation, and under attack evaluations need to be carried out.

In connection with the WSNs, the Internet-of-Things (IoT) plays a major role in building any CPS, such as autonomous vehicles, smart homes, and industrial systems. The wide diversity and the nonhomogeneity of the IoT devices make the IoT traffic different from the normal network one. Chapter "Empirical Characterization of Network Traffic for Reliable Communication in IoT Devices", authored by *Bebortta and Senapati*, tries to reveal the myth, by studying IoT traffic at the packet and flow levels, toward efficient IoT traffic management. Of course, understanding the characteristics of IoT traffic offers better control over IoT networks and precise identification of IoT devices via, for example, device profiling. It also opens doors for using Artificial Intelligence (AI) in developing proactive security mechanisms via detecting well-known attack patterns and predicting unknown ones.

Security analytics is a timely trend in Cybersecurity that uses AI tools for analyzing data to produce proactive security measures. Network traffic is a type of data that can be fed to AI or Machine Learning (ML) algorithms for attack detection and forecasting. Chapter "Machine Learning for Fostering Security in Cyber-Physical Systems", written by *Dhiman, Gupta, and Sharma,* discusses the usability of ML techniques for enhancing CPS security. A general overview of ML notions is provided, alongside applications of ML in risk assessment and the use of ML for detecting anomalous behavior. The autonomous vehicle system is used to reflect the role of ML in providing security in the device layer, network layer, and applications layer of CPS. The autonomous vehicle system is used to reflect the role of ML in foresting security in the device layer, network layer, and applications layer of CPS. The chapter serves as a good overview of ML applications in CPS security, which can be taken forward toward developing ML-based security measures.

Proceeding further with security measures, we find Chapter "A Model for Auditing Smart Intrusion Detection Systems (IDSs) and Log Analyzers in Cyber-Physical Systems (CPSs)", authored by *Nehinbe,* which introduces a model for auditing the smart Intrusion Detection Systems (IDSs) and log analyzers, for improving the efficiency of IDSs in terms of, for example, identifying false alarms. Furthermore, the auditing process should help detect any misbehavior of IDSs, such as disabled rules and policies, which can happen intentionally, or as a consequence of a successful attack of the IDS itself. The introduced audit model serves as a guide to human elements in CPSs for auditing IDSs. The chapter covers an important security concept where security measures should not only be deployed, tested, and operated but also should be continuously monitored for detecting any self-suspicious behavior.

Given the complex interactions between the physical and cyber components, the detection of attacks in CPS has been approached in various ways. Chapter "Model-Based CPS Attack Detection Techniques: Strengths and Limitations", authored by *Athalye, Ahmed, and Zhou,* compares different attack detection mechanisms, and evaluates them using a defined set of metrics. Model-based attack detection methods comprise statistical change monitoring (CUSUM and bad-data detectors) and a device fingerprinting technique. Several types of attacks were simulated, to experimentally analyze the performance of the detection methods. Here, two real-world case studies, namely a smart water treatment plant and a water distribution plant, have been used to facilitate realistic environments to assess the security measures.

The remaining four chapters of the book deal with security in individual application domains. Reading them illustrates how different, while similar, key security concerns are when CPS materializes in environment monitoring, smart grid, gas pipelines, and autonomous vehicles. Here, the first three areas are deployed on a daily basis across the world, while the latter one is likely to become of extreme importance in the near future.

The first CPS case, as considered by *Hajder, Hajder, and Nycz* in Chapter "Security of Cyber-Physical Monitoring and Warning Systems for Natural and Technological Threats", facilitates regional environmental monitoring. It is a resident-focused application, which delivers alerts on the basis of information gathered from an area surrounding a small city. In the considered scenario, a mixture of "state-owned" and "private" sensors (weather stations, in particular) is used. Material presented in the chapter outlines issues related to security and protection in a mixed public-private ecosystem and proposes the functional organization and architecture of a trustworthy environmental monitoring system. In this context, methods and means of counteracting security and accessibility threats are discussed and evaluated. Presented work is based on an implementation of an actual monitoring system, which has been installed in a small town adjacent to a larger agglomeration.

The second application of interest is the functionalization of CPS in Smart Grids (SG). Recently, to facilitate uninterrupted and reliable transmission, generation, and distribution of electricity, two trends can be observed: First, the increased use of renewable energy, which requires more intelligence in the grid; Second, responding to this and other needs, the conventional power grid is being upgraded to become a smart grid CPS. Here, Chapter "Risk Identification and Risk Assessment of Communication Networks in Smart Grid Cyber-Physical Systems", written by *Jha et al.*, is focused on the identification of risk factors for communication networks in smart grid CPS. Further, risk assessment strategies for applications that are being developed for such environments are formulated, with a detailed discussion. Presented work is grounded in real-world examples for multiple, different, smart grid CPS applications.

The third application domain is somewhat similar to the second one. It is also focused on a critical infrastructure, but this time, instead of the smart grid, natural gas networks are considered. The operation of natural gas pipelines relies heavily on Industrial Control Systems and Supervisory Control and Data Acquisition Systems. These systems introduce additional vulnerabilities, which materialize only in their presence. Interestingly, work presented by *Wang, Zhao, and Blum* in Chapter "An Overview of Cybersecurity for Natural Gas Networks: Attacks, Attack Assessment and Attack Detection" is one of the few that addresses vulnerabilities in gas pipeline infrastructures. Here, the authors describe the gas pipelines, approached from the point of view of CPS. They also provide an overview of models, theories, and representative detection approaches. Using real-life-based examples, the damage caused by three cyber-physical attacks on natural gas systems is analyzed. The discussion also illustrates the performance of representative attack detection approaches.

The final chapter presents the forward-looking context of Connected and Automated Vehicles (CAVs), seen as a large class of CPS. CAVs have recently emerged as an interesting approach to (limited) car autonomy, which aims to improve safety, fuel consumption, road throughput, and driving comfort, by forming so-called vehicle platoons. However, their actual viability in the long time remains to be established. Nevertheless, *Basiri, Azad, and Fischmeister* introduce a Secure Distributed Nonlinear Model Predictive Control (Secure-DNMPC) algorithm, which allows the secure control of vehicle platoons. The proposed approach is investigated in different communication topologies, under Denial of Service (DoS) attacks. Moreover, the security of platoons, where arbitrary vehicles perform cut-in and/or cut-out maneuvres, is considered. In this work, DoS attacks are conceptualized by introducing varying time delay into the data transmission, and simulation results then demonstrate the fruitfulness of the proposed method.

The book as a whole clearly evidences the security challenges of cyber-physical systems, as well as proposals for how to address them. Of course, the highly heterogeneous nature of CPS themselves means that a one-size-fits-all solution does not exist (and is not likely to be created in the future). However, the individual approaches and findings will nonetheless highlight ideas and directions that can help in various circumstances, as well as support the understanding of the CPS security landscape as a whole. As such, we hope that readers will find the book interesting and relevant as a contribution to the body of literature in this important area.

November 2020

Ali Ismail Awad
Luleå University of Technology
Luleå, Sweden

Al-Azhar University
Qena, Egypt

Steven Furnell
University of Nottingham
Nottingham, UK

Marcin Paprzycki
Systems Research Institute
Polish Academy of Sciences
Warsaw, Poland

Sudhir Kumar Sharma
Institute of Information Technology
and Management
New Delhi, India

Contents

About the Editors

Ali Ismail Awad is currently an Associate Professor (Docent) with the Department of Computer Science, Electrical and Space Engineering, Luleå University of Technology, Luleå, Sweden. He is an Associate Professor with the Electrical Engineering Department, Faculty of Engineering, Al-Azhar University at Qena, Qena, Egypt. He is also a Visiting Researcher with University of Plymouth, United Kingdom. His research interests include Cybersecurity, Internet-of-Things security, image analysis with applications in biometrics and medical imaging, and network security. He has edited or co-edited five books and authored or co-authored several journal articles and conference papers in these areas. He is an Editorial Board Member of Future Generation Computer Systems Journal, Computers & Security Journal, Internet of Things; Engineering Cyber-Physical Human Systems Journal, IET Image Processing Journal, and Health Information Science and Systems Journal. Dr. Awad is currently an IEEE senior member.

Steven Furnell is a professor of cybersecurity at University of Nottingham. He is also an Adjunct Professor with Edith Cowan University in Western Australia and an Honorary Professor with Nelson Mandela University in South Africa. His research interests include usability of security and privacy, security management and culture, and technologies for user authentication and intrusion detection. He has authored over 330 papers in refereed international journals and conference proceedings, as well as books including *Cybercrime: Vandalizing the Information Society* and *Computer Insecurity: Risking the System.* Professor Furnell is the current Chair of Technical Committee 11 (security and privacy) within the International Federation for Information Processing, and a member of related working groups on security management, security education, and human aspects of security. He is the Editor-in-Chief of *Information and Computer Security*, as well as an Associate Editor for various other journals including *Computers & Security* and *The Computer Journal.* His activities also include extensive contributions to international conferences in the security field, including keynote talks, event chairing, and programme committee memberships. In terms of professional affiliations, Prof. Furnell is a senior member of the IEEE and the ACM, and a fellow of BCS, the Chartered Institute for IT. He is also a Fellow and Board Member of the Chartered Institute of Information Security and chairs the academic partnership committee.

Marcin Paprzycki Systems Research Institute Polish Academy of Sciences. He has an MS degree from Adam Mickiewicz University in Poznań, Poland, a Ph.D. from Southern Methodist University in Dallas, Texas, USA, and a Doctor of Science degree from Bulgarian Academy of Sciences, Sofia, Bulgaria. He is a Senior Member of IEEE, a Senior Member of ACM, a Senior Fulbright Lecturer, and was an IEEE CS Distinguished Visitor. His original research interests were in the area of high-performance computing/parallel computing/computational mathematics. Over time they shifted toward intelligent systems, software agents and agent systems, and application of semantic technologies, among others. Currently, he serves as the Vice Chair of IEEE Poland Section. He has contributed to more than 500 publications and was invited to the programme committees of over 800 international conferences. He is on the editorial boards of 12 journals.

Sudhir Kumar Sharma is currently a Professor and Head of the Department of Computer Science, Institute of Information Technology & Management affiliated to GGSIPU, New Delhi, India. He has extensive experience for over 21 years in the field of Computer Science and Engineering. He obtained his Ph.D. degree in Information Technology in 2013 from USICT, Guru Gobind Singh Indraprastha University, New Delhi, India. Dr. Sharma obtained his M. Tech degree in Computer Science & Engineering in 1999 from Guru Jambheshwar University, Hisar, India, and an M.Sc. degree in Physics from University of Roorkee (now IIT Roorkee), Roorkee, in 1997. His research interests include Machine Learning, Data Mining, and Security. He has published more than 50 research papers in various prestigious International Journals and International Conferences. He is a life member of CSI and IETE. Dr. Sharma is the lead guest editor of the special issue in Multimedia Tools and Applications, Springer. He was a convener and Volume Editor of ICETIT-2019 and ICRIHE-2020. He authored and edited 6 Computer Science books in the field of Internet of Things, WSN, Blockchain, Elsevier, CRC Press, USA. He was selected as a reviewer/editorial board member for several reputed international journals. He has also served as a speaker, session chair or co-chair at various national and international conferences.

Realizing Cyber-Physical Systems Resilience Frameworks and Security Practices

Md Ariful Haque, Sachin Shetty, Kimberly Gold, and Bheshaj Krishnappa

Abstract Cyber-Physical Systems (CPSs) are complex systems that evolve from the integrations of components dealing with real-time computations and physical processes, along with networking. CPSs often incorporate approaches merging from different scientific fields such as embedded systems, control systems, operational technology, information technology systems (ITS), and cybernetics. Major cybersecurity concerns are rising around CPSs because of their expanding uses in the modern world today. Often the security concerns are limited to deriving risk analytics and security assessment. Others focus on the development of intrusion detection and prevention systems. To make the CPSs resilient, it needs a thorough understanding of the current cybersecurity frameworks proposed by different governing bodies in this domain. It is also imperative to realize how these frameworks are applying established security practices. To address the gap in understanding the defense-in-depth security architectures and achieving them within the CPS domain, we analyze the cybersecurity frameworks and the challenges in applying them. To give some background information, we start a discussion of the differences between ITS and CPS. We then present a state-of-the-art review of some of the existing cybersecurity frameworks for risk and resilience management. Finally, we propose formal techniques to realize the frameworks and security practices in the CPS domain by providing quantitative resilience analytics.

M. A. Haque (✉) · S. Shetty
Computational Modeling and Simulation Engineering, Old Dominion University,
5115 Hampton Blvd, Norfolk, VA 23529, USA
e-mail: mhaqu001@odu.edu

S. Shetty
e-mail: sshetty@odu.edu

K. Gold
Naval Surface Warfare Center, Crane Division, Crane, IN 47522, USA
e-mail: kimberly.gold@navy.mil

B. Krishnappa
Risk Analysis and Mitigation, ReliabilityFirst Corporation, 3 Summit Park Drive, Suite 600,
Cleveland, OH 44131, USA
e-mail: bheshaj.krishnappa@rfirst.org

A. I. Awad et al. (eds.), *Security in Cyber-Physical Systems*, Studies in Systems, Decision and Control 339, https://doi.org/10.1007/978-3-030-67361-1_1

Keywords Cyber-Physical Systems · Cybersecurity frameworks · Security practices · Criticality assessment · Resilience metrics · Graphical modeling · Analytical hierarchy process · TOPSIS

1 Introduction

In the modern world today, we observe a steep increase in the usage of Cyber-Physical Systems (CPSs). For example, critical infrastructures (i.e., energy delivery systems, oil and gas industry, healthcare systems, transportation systems), industrial manufacturing plants, autonomous vehicles, smart cities, etc. profoundly use CPSs. CPS is a class of complex systems of systems that integrate cyber operations with the physical processes. In CPSs, we use computing and networking devices to perform computation and communication. The networked devices also control the underlying instrumental processes. We need the communication network to monitor and control the physical devices' operations and performances in real-time, some of which may base on remote field locations.

In the broad sense, CPSs consists of the cyber domain (or, the information technology systems (ITS)), and the physical domain (or, the operational technology (OT) network). The cyber section consists of servers and hosting devices for organization-wide communications. On the other hand, the physical domain contains the field devices and the industrial control systems (ICS), which again comprise sensors, actuators, control functions, feedback systems, etc. We need the OT network for handling the production processes and the ITS for the business communications. The advancements in monitoring and controlling the production processes bring the risk of malicious cyber attacks on these systems. The increment in risk comes from the integrated interconnection between the cyber components and the physical elements, more precisely, the amalgamation of ITS and ICS.

The CPS's security concerns are often addressed by focusing on the development of intrusion detection and prevention system (IDS/IPS) and generating security metrics for risk assessment. While the IDS and IPS are necessary for mitigating the attack impact and quick recovery of the system, we cannot overlook the concern regarding systems' resilience and reliability. The resilience posture indicates the overall system security and guides network administrators for developing effective and optimized mitigation strategies and remediation plans.

To make the CPS cyber resilient, regulatory bodies and researchers propose several frameworks and provide essential instructions in standards. The standard bodies that we are referring are the National Institute of Standards and Technology (NIST), the North American Electric Reliability Corporation critical infrastructure protection (NERC-CIP), and the Industrial Control Systems Cyber Emergency Response Team (ICS-CERT). The open question is how to apply those frameworks and security practices as comprehensively as possible without affecting the regular business operations.

In this chapter, we address the implementation challenges of the theoretical frameworks. We discuss the threats and vulnerabilities that CPS is facing today. We also cover how the security frameworks, recommended defense architectures, and standard practices can help design and develop resilient CPS. This chapter aims to mathematically realize the frameworks and security practices using established theoretical analysis methodologies. Significant contributions of the chapter are:

- A detailed discussion on the CPS threats, vulnerabilities, and cyber resilience
- A comprehensive review of cybersecurity and resilience frameworks and recommended defense-in-depth security practices for CPS
- A proposed qualitative approach for quantifying cyber resilience using analytical hierarchy process (AHP)
- A quantitative realization of defense-in-depth security architectures using a multi-level directed acyclic graph modeling technique
- Critical and cyber vulnerable assets identification using the vulnerability graph model
- A concise discussion on the mapping of CPS security, resilience, and operational domains.

We organize the rest of the chapter as follows. Section 2 presents a brief description of CPS and components of CPS (i.e., IT network, supervisory control and data acquisition (SCADA), and OT network), and cyber resilience. Section 3 provides a state-of-the-art review of the cybersecurity and resilience frameworks. Section 4 discusses some of the critical security guidelines presented by the standard bodies. Section 5 proposes different mathematical techniques for the realization of the frameworks. Section 6 highlights the challenges in mapping CPS security, resilience, and operational domains. Finally, Sect. 7 concludes the chapter with some significant takeaways.

2 Cyber-Physical Systems

Cyber-Physical Systems (CPSs) represent a composite class of engineered systems consisting of physical processes and computational resources. The National Institute of Standards and Technology (NIST) CPS Public Working Group (CPS PWG) defines CPS as "smart systems that include engineered interacting networks of physical and computational components" (Griffor et al. [1]). CPS technologies continue helping to transform people's approaches to interact with engineered systems. Advances in CPS bring extended capability, adaptability, and usability, making them crucial in many industries. Today we observe CPS are in use to implement most modern technologies such as the Internet of Things (IoT), industrial internet, Industrial Control Systems (ICS), smart devices, etc. In this chapter, we sometimes use the phrases CPS and ICS interchangeably to mean the same systems.

We present a conceptual representation of the Cyber-Physical Systems in Fig. 1. We divide the discussion area into feedback systems, application domains, system security, and system challenges. CPS consists of control and feedback systems, which are highly interconnected and heterogeneous. The control systems are either networked or distributed and include physical processes such as sensors and actuators, which operate in real-time. There may be human and environmental interactions involved in the process.

We illustrate the CPS here by using the example of the power systems as researchers consider the power systems as cyber-physical power systems (CPPS) [2, 3]. The power system's physical domain consists of the generation and distribution devices such as generators, transformers, electric buses, etc. The physical part also comprises ICS devices. There are different ICS devices in use based on requirements such as the phasor measurement units (PMU), intelligent electronic devices (IED), the programmable logic controllers (PLC), and remote terminal units (RTU), etc. To monitor and control the field devices' performances, we need the supervisory control and data acquisition (SCADA) systems. As we know, SCADA is the central control system used to monitor and control the equipment in the industrial production systems. In general, SCADA contains the master terminal unit (MTU), human-machine interface (HMI), and input/output (I/O) devices, etc. The field ICS devices such as RTU sends real-time system performance data to MTU. The operators in SCADA observe the performance measures, compare those values with desired values, and, if necessary, issue control commands through HMI. The commands issued from HMI control the system to function at the desired service level (Macaulay and Singer [4]).

Due to the complicated operational requirements, the CPS itself has challenges such as modeling the underlying physical processes and real-time behavior, modeling interconnectivity, and interoperability in the heterogeneous SoS, secure integration of different components of CPS, etc. The CPS needs to handle the analysis of specification, design methodologies, scalability and complexity, and overall verification and validation of the systems from the modeling & simulation perspective. On the other hand, because of the amalgamation of the IT and OT domains, CPS needs to handle many cyber threats. Thus, understanding the proposed cyber frameworks and applying the recommended practices in developing a resilient system are integral parts of CPS security analysis.

We start with a short discussion on the primary differences between ITS and CPS in the next subsections. We then gradually proceed to CPS threats, vulnerabilities, and cyber resilience to smooth transition to the cyber framework analysis.

2.1 Primary Differences Between CPS and ITS Security

Today, the extensive access of ITS devices into the control systems makes CPS vulnerable to cyberattacks. Cyberattacks are different than physical attacks on several points. In the physical attacks, the defenders are aware of the system units under the target, the impact is immediate, and there are policies to handle such attacks. On the other hand, cyberattacks are remote, repeatable, and can occur over extended

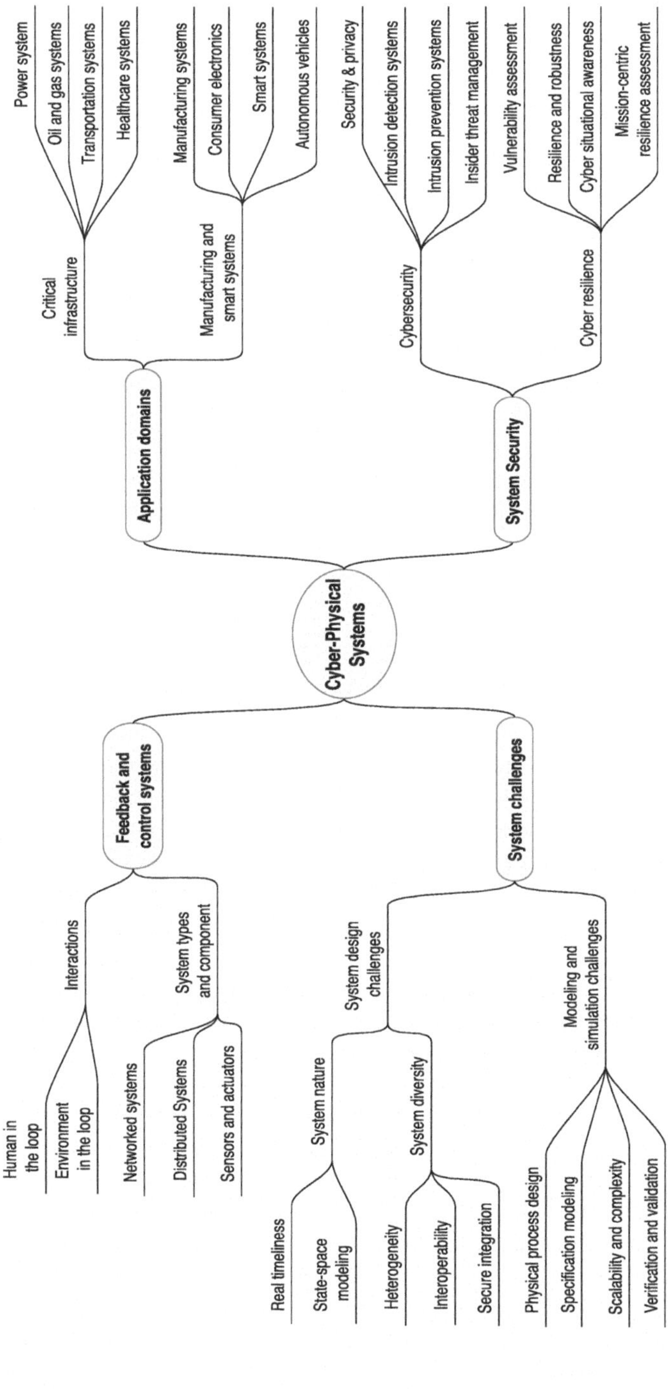

Fig. 1 Cyber-Physical Systems concept map

periods. Cyber intruders can execute cyberattacks, with the objective of a long-term intrusion and identification of potential attractive targets (e.g., advanced persistent threat). The impact can be less intense for the time being but can lead to disastrous consequences in the long run. We provide a summary of the fundamental differences between ITS and ICS/CPS security in Table 1.

Overseeing cybersecurity in the CPS domain is far more daunting than controlling the same in the information technology context. The reason lies on the ground that CPS has unique operational requirements than ITS. Firstly, for CPS, real-time availability and operational continuity are of utmost importance. But for ITS, data confidentiality and integrity are crucial. Momentary downtime in ITS does not hamper any production processes (see Macaulay and Singer [4]). Secondly, it is easy to apply patching through anti-malware and anti-virus software in ITS, and they often automatically download and install the necessary security patches or updates. But ICSs are generally old proprietary technologies intended for functionality (not focus on security issues). ICSs have limited memory and other processing capacities. These hardware-level limitations make it hard to install anti-malware or anti-virus solutions, which consume a lot of memory for automatic updates and delay monitoring and controlling the production process. Thirdly, ICS operates in diverse fields such as in the oil, gas, and electric industries. So the application of security measures should be adapted to fit the structure of these sectors.

2.2 CPS Threats and Vulnerabilities

This section starts with a brief definition of vulnerability and threat, as we find in the literature to facilitate the audience with the necessary information for the next discussion. In information systems, a vulnerability is a flaw in the software program or system that an intruder may exploit to gain unauthorized access to a cyber asset. NIST defines vulnerability as "weakness in an information system, system security procedures, internal controls, or implementation that could be exploited or triggered by a threat source" (see Johnson et al. [7]). On the other hand, a threat is anything that "can exploit a vulnerability, intentionally or accidentally, and obtain, damage, or destroy an asset" (Cyware [8]). The Joint Task Force Transformation Initiative defines threat as "threat is any circumstance or event with the potential to adversely impact organizational operations and assets, individuals, other organizations, or the Nation through an information system via unauthorized access, destruction, disclosure, or modification of information, and/or denial of service" (Blank [9]). Lewis [10] defines vulnerability and threat as "the probability that a component or asset will fail when attacked" and "the probability that an attack will happen", respectively.

We have already highlighted that CPS consist of physical, control, and communication layers. CPS threat vectors can come from adversaries in any of those layers. In the physical layer, the availability of the field devices' services and functionalities are of utmost concern. There is the risk of information alteration by modifying the physical device codes (e.g., PLC logic codes). In the control layer, most attacks occur in

Table 1 Primary differences in operations and security in ITS and CPSs/ICS

Category	Information Technology Systems (ITS)	Cyber-Physical Systems (CPSs/ICS)
Performance constraints[a]	• High throughput demanded • Non real-time response is ok • High delay and jitter are tolerable	• Modest throughput is allowable • Real-time response in essential • Delay and jitter over certain threshold are not tolerable
Resource constraints[b]	• Updated hardware and software products are used • Systems have enough memory and processing capabilities • Regular security updates are maintained through patching	• Old and less secured proprietary products are used • Products are designed with low memory and processing capabilities • Often security updates and patches are not implemented to avoid system unavailability due to reboot requirements after configuration changes
Confidentiality, integrity, and availability	• Data confidentiality and integrity are critical • Temporary unavailability is tolerable	• Confidentiality and integrity is not important • High availability is required. Momentary downtime may not be acceptable
Communication protocol	Standard communication protocols (i.e., TCP, UDP, etc.)	Proprietary protocols (i.e., MODBUS, DNP3, etc.)
Patch and change management[a]	• Software updates and patching are applied regularly according to the organization's security policy • Rebooting the system to re-initialize the hardware or software devices is acceptable	• Any configuration changes need to test, and deploy in test mode before committing the changes to live system to avoid unexpected outages • Unplanned rebooting of the system is not acceptable
Password and authentication[b]	• Multi-factor authentication is possible to deploy • Passwords need to change after certain time • Security is enhanced through encryption mechanisms	• Sometimes lack of any sort of authentication requirement • Passwords are hard-wired in legacy ICS and cannot be changed • Lack of encryption mechanisms in message communication
Component lifetime and technical support[a]	• Lifetime generally spans from 3 to 5 years • Ample technical support available from either own IT experts or diversified managed services	• Lifetime varies between 15 and 20 years • Support solely vendor dependent. Some product supports may be ceased by the vendor due to lifetime expiry
Operational command and control	Mostly central monitoring	Distributed field operations, but central monitoring through SCADA

[a]Stouffer et al. [5]
[b]Colbert et al. [6]

the form of distributed denial of service (DDoS), eavesdropping (man-in-the-middle attack), jamming, selective forwarding, etc. Threats in the communication layer can lead to leaking of confidentiality, stealing credentials, unauthorized access to the system, social engineering, etc. Based on the type of threats, we classify them in the discussion below, as pointed out by Haque et al. [11].

External Threats: By external threats, we mean any cyberattack coming from outside of the organization. External threats arise from different rival groups, including nation sponsored hackers, terrorist organizations, or industry competitors. Cyber intruders may launch an advanced persistent threat attack, where the goal is to theft crucial data and login information (e.g., password) on the network's assets without getting caught. One such example is the Stuxnet attack on the Iranian nuclear centrifuges in the year 2010 (see Chen and Abu-Nimeh [12]).

Internal Threats: The internal threat comes from either within the organization or from the affiliated parties. Today, the industry's operating processes are segmented and done by third-party vendors or contractors. Thus organizations need to share system information with outside business partners to some extent. Sharing the network information (e.g., network design documents) makes the CPS/ICS system vulnerable to potential cyber threats. There is also the risk of insider attacks from the organization's employees as some employees have authorized access to the ICS network for managing the network operations. This type of insider threat falls in the category of credentialed ICS insider attack [13, 14].

Technology Threats: Even today, most ICS systems run on old technologies, where the primary concern is the matching of protocol-level message communications among different ICS products from other vendors. Thus, many ICSs lack strong authentication and encryption mechanism (see Laing [15]). Some ICS use authentication procedures, but the weak security mechanisms (e.g., insecure password, default user accounts, and inadequate password policies) are not enough to protect the system from intelligent adversaries [13, 14].

Integration and Inter-connectivity Threats: In the enterprise networks, business units are interconnected. Due to the interconnection of the corporate network with the control system network, ICS devices become vulnerable to cyberattacks. This vulnerability arises because part of the corporate network is open for communication over the internet, and ITS hosts and servers contain vulnerabilities. Merely putting the ICS devices behind the firewalls do not necessarily protect the ICS components [13].

Next, we discuss the cyber resilience from the CPS perspective to understand how to protect the CPS form the threats discussed above.

2.3 Cyber Resilience: What Does It Mean for CPS?

Some of the early definitions of resiliency had concentrated on disaster resiliency. From the disaster resilience perspective, Bruneau et al. [16] had proposed a conceptual framework to define seismic resilience. In another work, Tierney and Bruneau [17] later introduced the R4 framework for disaster resilience. The R4 model [17] comprises of four metrics: robustness, redundancy, resourcefulness, and rapidity. 'Robustness' means systems' ability to function and provide services even under degraded performance, probably with reduced quality of services. 'Redundancy' means identifying substitute elements that satisfy functional requirements in the event of significant performance degradation or service disruption. 'Resourcefulness' is to initiate solutions by identifying the required resources based on the consequence, nature, or depth of degradation by prioritizing problems that need to solve. 'Rapidity' indicates the ability to restore functions within the required time-stamp.

The National Academy of Science (NAS) defines resilience as "the ability to prepare and plan for, absorb, respond, recover from, and more successfully adapt to adverse events" (see National Research Council [18]). The National Institute of Standards and Technology (NIST) defines the information system resilience as follows. Resilience is "the ability of an information system to continue to: (i) operate under adverse conditions or stress, even if in a degraded or debilitated state while maintaining essential operational capabilities; and (ii) recover to an effective operational posture in a time frame consistent with mission needs" (see Ross [19]).

A lot of research works are going on the cyber resiliency study of CPS. We mention a few of them here which deal with frameworks and security guidelines. The NIST provides a framework (Sedgewick [20]) for improving the cybersecurity and resilience of critical infrastructures that support both ITS and ICS. NIST provides another framework specifically for Cyber-Physical Systems (Griffor et al. [1]). We elaborate on the frameworks in Sect. 3.2. Haque et al. [21] illustrate the gap in resilience analysis and propose a cyber resilience framework to quantify resilience metrics. The framework considers the physical, technical, and organizational aspects of cyber operations to assess ICS's cyber resilience. Haque et al. also introduce a qualitative cyber resilience assessment tool [22] based on the framework.

In the ICS domain, Stouffer et al. [5] provide detailed guidelines for ICS system security. The policies cover secure ICS architecture and the methods for applying the security controls to the ICS environment. Barker et al. [23] propose resilience analytics for social networks that depends on each other. The metrics describe how risk analysis can help in the modeling and quantification of systems resiliency. DiMase et al. [24] present a systems engineering framework for Cyber-Physical Systems security and resiliency. The paper focuses on CPS security and relates to resiliency to handle integrated and targeted security measures and policies. We would cover some of the frameworks in Sect. 3 and thus omit the detailed discussion here to avoid repetition.

In the modeling context, Haque et al. [25] highlight ways of modeling resilience in CPS by considering the criticality of the cyber asset. Haque et al. [26] present cyber modeling techniques by utilizing the critical system functionality for energy delivery systems specifically. In the resilience analytics, Clark and Zonouz [27] present intrusion resilience metrics for Cyber-Physical Systems by segregating the cyber and control layers of CPS. In another work, Haque et al. [14] explain the challenges in resilience assessment in CPS and discuss ways to develop a simulation platform for resilience assessment.

Wei and Ji [28] discuss a model named the resilient industrial control system (RICS). The authors mentioned the following characteristics of resilient ICS:

- Capability to reduce the unexpected consequence or impact of a cyber incidence to as minimum as possible
- Capability to mitigate a major portion of undesirable events
- Capability to recover normal operations within an expected time frame.

The R4 metrics [17] presented above are in line with the resilient characteristics provided by Wei and Ji [28]. Most of the above works address resilience by developing security frameworks and deriving quantitative analytic for the CPS or ICS. In this chapter, we want to focus on understanding the cybersecurity frameworks and standard practices proposed by the governing bodies; That discussion follows in Sect. 3 and Sect. 4, respectively.

3 State-of-the-Art Review of Cybersecurity Frameworks

In this section, we cover four crucial cybersecurity frameworks applicable to CPS. These are (1) NIST framework for improving critical infrastructure cybersecurity, (2) NIST framework for Cyber-Physical Systems, (3) NIST risk management framework for information systems cybersecurity, and (4) cyber resiliency engineering framework of MITRE Corporation. We consider these frameworks for our analysis as researchers consider these frameworks as mostly adopted frameworks in the cybersecurity domain. We also provide a comparative analysis of several other cybersecurity frameworks in Sect. 3.5.

3.1 NIST Framework for Improving Critical Infrastructure Cybersecurity

The NIST cybersecurity framework (Sedgewick [20]) version 1.0 provides broad guidelines to manage cybersecurity risk and resilience. It has three main sections: core, implementation, tiers, and profiles. It can also help to identify operations needed to reduce risks and enhance resilience. The NIST framework identifies and proposes five security functions. These functions help managing systems cybersecurity, as we illustrate in Fig. 2.

The core piece of the framework provides actions to achieve specific results in the cybersecurity area. The element "Functions" organize necessary cybersecurity activities at the uppermost level. These functions are to identify, protect, detect, respond, and recover, respectively. These functions help organizations in managing cybersecurity risk and resilience. Here, the 'identify' function implies developing an

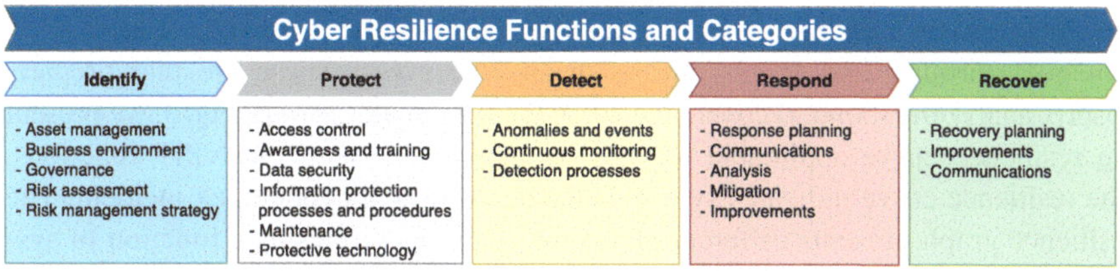

Fig. 2 NIST cybersecurity framework core functions and categories. We present here only the core functions and categories adapting from the initially proposed framework by Sedgewick [20] to explain the essential ideas

understanding of system risks and managing assets, data, capabilities, skills, etc. The 'protect' function deals with developing necessary defensive measures and implementing those to ensure the continuity of services. Detect realizes the capability to capture the occurrence of a cyberattack incident. The function 'respond' refers to taking actions regarding a detected cyber breach incident. Lastly, the recovery means restoring any damaged capabilities or services due to a cyberattack incident.

The framework presents a high-level risk and resilience assessment. It guides what to do during a cyber attack event. However, the model framework lacks pointing on how to implement those actions. Also, the model needs to consider system differences among different critical infrastructures. For example, if the same attack happens in the energy and water sectors, the methodologies and actions to be taken, as mentioned, are the same, which may not consider the system differences.

We adapt the resilience curve presented by Wei and Ji [28] and map the graph with the five functions offered by the NIST framework. The curve is similar to

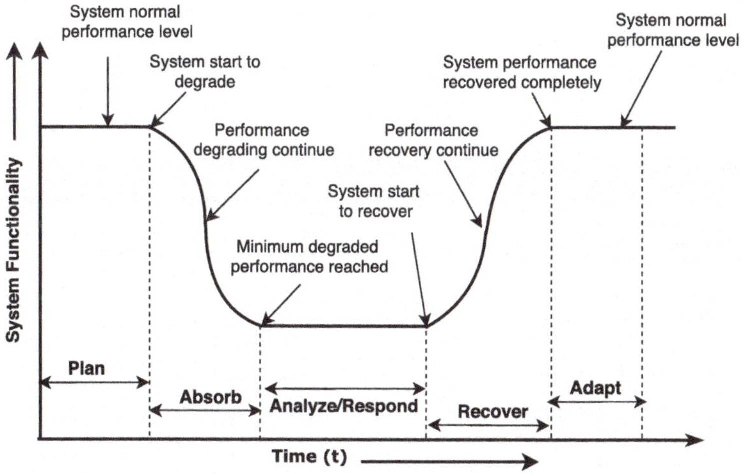

Fig. 3 CPS cyber resilience graph with different phases of action. We adjust the graph from the original graph presented in RICS model by Wei and Ji [28] to incorporate the resilience phases

the duck curve in energy systems reliability analysis. In general, a resilient system goes through five stages during an adverse event. These are plan/prepare, absorb, analyze/respond, recover, and adapt (Linkov et al. [29]). In Fig. 3, we present the resilience curve applicable to CPS by mapping it with the NIST functions. The resilience curve indicates system behaviors during a cyberattack incident. The resilience graph presents different phases of cyber operations as a function of system functionality over time. The five stages complete the resilience cycle, and the area formed by the enclosed curve is the quantitative measure of the system's cyber resilience.

3.2 NIST Framework for Cyber-Physical Systems

Griffor et al. [1] propose a framework for CPS that captures the generic CPS functionalities. The framework focuses on the activities required to support conceptualization, realization, and assurance of CPS. The framework requires identifying CPS domains, facets, aspects, concerns, activities, and artifacts [1]. Here, 'domains' represent the CPS application areas; 'concerns' are concepts that drive the CPS framework methodology. Activities within the facets address the 'aspects.' And 'aspects' consist of a group of related concerns. There are nine defined aspects. These are functional, business, human, trustworthiness, timing, data, boundaries, composition, and lifecycle (see Griffor et al. [1]). 'Facets' encompass identified activities to perform in the systems engineering process within the CPS. Each facet contains a set of well-defined activities and artifacts (i.e., outputs) for addressing the concerns. In Fig. 4, the middle rectangular box layer shows what to do at each facet step. The bottom parallelograms indicate the outcomes (i.e., artifacts) of the facet steps.

In Fig. 4, we observe the three identified facets: conceptualization, realization, and assurance. 'Conceptualization' means things to perform. These are the group of actions that constitute a CPS model. 'Realization' means how things are to make

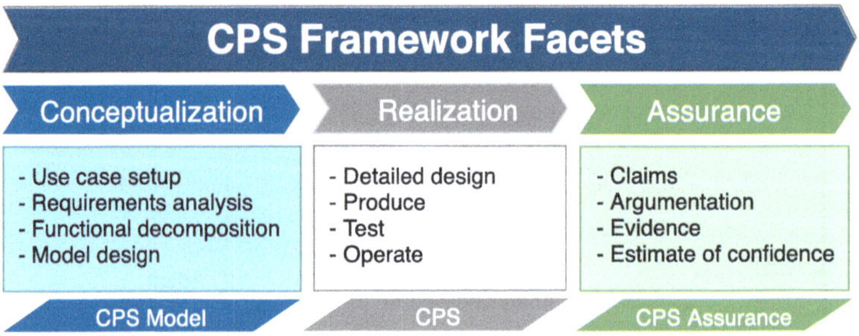

Fig. 4 Main facets of the NIST framework for Cyber-Physical Systems [1]. We have adapted the figure from the original framework to focus only on the important ideas. Here, conceptualization, realization, and assurance are the three facets, as proposed in the framework

and operate. Realization encompasses the group of measures that create, deploy, and manage a CPS. 'Assurance' is to achieve the desired level of confidence that the system will work as planned. This facet includes the group of actions that provide the belief that CPSs work as intended.

The CPS framework is still in first draft format and yet not fully established. The CPS framework's primary goal is to be actionable. From the critics' point of view, the framework is nothing but a systematic approach for realizing CPS's process. The three main facets on which the framework is sitting upon require activities that depend solely on the expertise from subject matter experts. The framework hardly illustrates how to handle cyber and physical challenges from design, modeling, and security perspectives. Defining the CPS aspects and updating the facet activities and artifacts would differ from domain to domain. They would require gathering a vast amount of data and expertise from system administrators or subject matter experts. Overall, the framework does not explain how to handle the security, reliability, and resilience issues of the complex CPS.

3.3 NIST Risk Management Framework for Information Systems Cybersecurity

In collaboration with the US Department of Defense, the Office of the Director of National Intelligence, and the Committee on National Security Systems, NIST has developed the Risk Management Framework (RMF) [30]. The RMF has conceived to improve information and data security in a networked environment. The RMF encourages sharing data and information among organizations and strengthens risk and resilience management processes. The RMF has considered a three-layered pyramid-shaped approach to handle and manage risks within the organization. The bottom layer is the information systems layer. The middle layer deals with business or mission processes. Finally, the topmost layer handles the organizational processes. Here we only analyze the core processes as proposed in the framework.

Figure 5 illustrates the RMF steps in the risk management process flow. We explain here the seven steps involved in the comprehensive risk assessment in brief.

1. **Prepare**: The preparation step incorporates essential tasks at all the three levels of the enterprise network. The 'prepare' step is to keep the organization ready to manage risks associated with its security and privacy. The three levels that we are referring are the organizational level, mission and business process level, and information systems level.
2. **Categorize**: The categorization step classifies the system based on the impact analysis. Here the classification of the categories considers the study of the amount of information processed by the system. The categorization also takes into consideration the volume of data stored and transmitted by the system.
3. **Select**: This step guides the organization to choose an initial set of baseline security controls. The security controls come from the analysis of the security

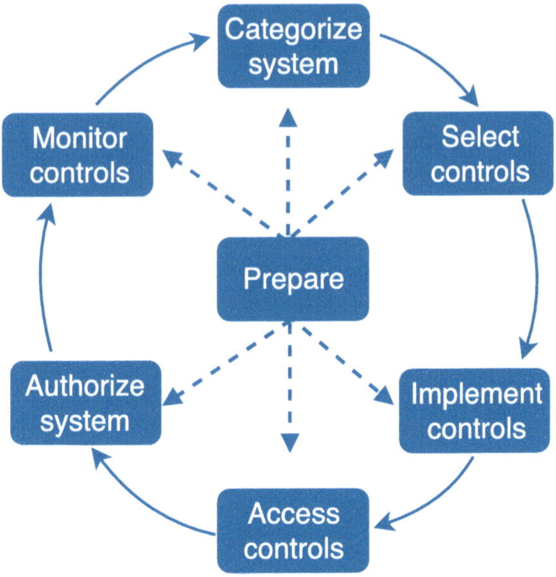

Fig. 5 Steps in NIST risk management framework for information systems cybersecurity [30]. It consists of seven steps: (1) prepare, (2) categorize system, (3) select controls, (4) implement controls, (5) access controls, (6) authorize system, and (7) monitor controls. The steps need are to follow sequentially, although the preparation phase needs to consider the constraints in other stages. The figure is adapted from the proposed framework [30] to help in realizing the discussion

categorization. This 'select' step handles managing and rectification of security controls standards as needed. The baseline standard comes from the study of the organization's risk conditions assessment.

4. **Implement**: This implementation step emphasizes on the execution of the security controls from the operational perspective. The step also incorporates documentation of the controls.
5. **Assess**: This step is to assess the security controls using the right measures to estimate the extent of the correctness of the implemented controls. It considers whether the system is operating as planned. The step also evaluates if the implemented actions produce the expected outcome considering the established security requirements.
6. **Authorize**: This authorization step is to authorize system operations when there is an identified risk. This risk is directly related to organizational assets and operations. The operation of the system goes on if the assessment outcome finds that the risk is acceptable.
7. **Monitor**: The last step is to monitor and assess the implemented security measures regularly. This monitoring includes evaluating the effectiveness of the implemented security control and documenting any operational environment changes. Other tasks, such as conducting impact analyses of the security alterations and reporting to appropriate personnel, are part of the monitoring process.

This framework is one of the advanced cybersecurity frameworks existing today to address cybersecurity and cyber resiliency concerns. The seven sequential steps, as previously mentioned, are possible to tailor depending on the CPS domain areas with the help of subject matter experts. One of the framework's primary focus is to

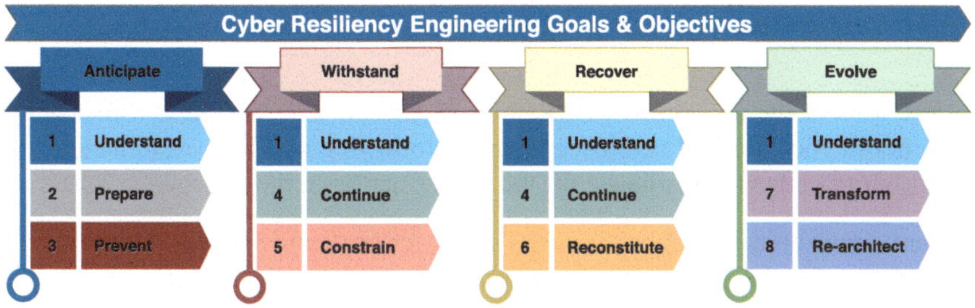

Fig. 6 MITRE cyber resiliency engineering framework [31]. We only present here goals and objectives, adapting from the proposed framework to help in realizing the essential ideas

monitor and assess the security control mechanisms and evaluate the impact of any cyberattack incident. We think the framework addresses that concern conclusively and comprehensively.

3.4 MITRE Cyber Resiliency Engineering Framework

MITRE Corporation has proposed a cyber resiliency engineering framework (see Bodeau and Graubart [31]). The framework consists of cyber resiliency goals, objectives, and cyber resiliency practices. It also incorporates threat models associated with cyber risk and resiliency. The framework focuses on characterizing cyber resilience metrics. Figure 6 illustrates the framework. The elements of cyber resiliency consist of four goals: (1) anticipate, (2) withstand, (3) recover, and (4) evolve [31]. There are eight objectives: (1) understand, (2) prepare, (3) prevent, (4) continue, (5) constrain, (6) reconstitute, (7) transform, and (8) re-architect. The framework consists of fourteen practices that intend to maximize cyber resiliency. These are (1) adaptive response, (2) privilege restriction, (3) deception, (4) diversity, (5) substantiated integrity, (6) coordinated defense, (7) analytic monitoring, (8) non-persistence, (9) dynamic positioning, (10) redundancy, (11) segmentation, (12) unpredictability, (13) dynamic representation, and (14) realignment. In this framework, the different goals, objectives, and practices may work together or operate separately.

Although the NIST frameworks presented earlier deal with cybersecurity in broad, the MITRE framework focuses specifically on the cyber resilience engineering and assessment. The goals and objectives guide us to take the correct action under each step of the resilience management cycle. The proposed goals align with the NAS resilience definition, which includes the plan, absorb, recover, adapt [18]. The framework offers several practices which, with careful consideration, apply to ICS/CPS domain by adjusting the rules considering the system constraints and design methodologies.

3.5 Comparison of the Frameworks

A close look at the above frameworks reveals that the frameworks consider management of cyber risk and resilience from the following perspectives to handle the cybersecurity and cyber resilience for the infrastructure or the systems.

- Plans, goals, objectives, practices, and strategies (*risk and resilience perspective*)
- Identify, protect, detect, respond, recover, and adapt (*resilience perspective*)
- Anticipate, recover, withstand, and evolve (*resilience perspective*)

In Table 2, we present a structured comparison among a couple of crucial cybersecurity and cyber resilience frameworks proposed by different standard bodies and research organizations. If we look in-depth, we find that most of the frameworks discuss some common areas. These are identifying critical assets, securing the network through multi-level access controls, and assessing cyber risks on the business and organization as a whole. Finally, the frameworks propose techniques to safeguard the critical system functions or services by developing mitigation plans and strategies. What is a lack in those frameworks is to formalize those guidances using established mathematical methods. In this work, we understand the need for formal approaches, and we address that need to develop mathematical techniques for risk and resilience assessment in detail in Sect. 5.

4 Cyber Standards and Recommended Practices for CPS

In this section, we briefly discuss control system specific recommendations suitable for ICS or CPS. The ICS-CERT provides the following critical control system specific cyber recommendations that give a solid baseline regarding what to do and how to prevent cyberattacks in ICS.

- ***Developing cyber forensics plans for control systems***: Developing a cyber forensics program is challenging for control systems environments. The challenges arise because of the system limitations, such as nonstandard protocols, old designs, and irregular proprietary technologies. Cornelius and Fabro [32] address the challenges of traditional forensics to ICS and provides detailed guidance to develop a cyber forensics program through identifying system environment and uniqueness, defining context-specific requirements, and identifying and collection of system data.
- ***Applying defense-in-depth strategies to improve industrial control systems cybersecurity***: The ICS defense-in-depth strategies (see Fabro et al. [33]) provide comprehensive guidance for improving cybersecurity in control systems such as CPS/ICS.
- ***ICS security incident response plan***: The standard [35] primarily focuses on the preparation and response mechanisms for a cyberattack incident on the ICSs network. The policy has four segments. The first segment concentrates on planning

Table 2 Major cybersecurity and cyber resilience frameworks proposed by standard bodies and research organizations

Framework	Publishing organization	System	Intended use	Major functions, processes, and/or metrics	Year	Document version
Risk management framework for information systems cybersecurity	Joint Task Force	ITS	Managing security and privacy risk for organization-wide information systems	Prepare, categorize, select, implement, access, authorize, and monitor	2018	NIST SP 800-37 Rev. 2[a]
Framework for Cyber-Physical Systems	National Institute of Standards and Technology (NIST)	Mainly CPS	Broad design and security guidelines for CPS	Conceptualization, realization, and assurance	2017	NIST SP 1500-201[b]
Framework for improving critical infrastructure cybersecurity	National Institute of Standards and Technology (NIST)	Critical Infrastructures (CI)	Managing risk and resilience of CI	Identify, protect, detect, respond, and recover	2014	Version 1.0[c]
Conceptual Framework for developing resilience metrics for the electricity, oil and gas sectors in the United States	Sandia National Laboratory	Mainly energy and oil and gas sector. Also covers ICS, CPS, SCADA	Developing cyber resilience analytics for energy, and oil and gas sectors	Define goals and metrics, characterize threats, apply system model, evaluate and incorporate improvements	2014	Version not specified[d]
Cyber resiliency engineering framework	MITRE Corporation	ITS	Developing cyber resilience goals, objectives, and practices for ITS	Anticipate, withstand, recover, and evolve	2011	Version not specified [e]
R4 resilience framework	Multidisciplinary Center for Earthquake Engineering Research (MCEER)	Critical Infrastructure (CI)	Resilience assessment for CI using quantitative security metrics	Robustness, redundancy, resourcefulness, and rapidity	2007	Version not specified[f]

[a]JOINT TASK FORCE [30]
[b]Griffor et al. [1]
[c]Sedgewick [20]
[d]Watson et al. [34]
[e]Bodeau and Graubart [31]
[f]Tierney and Bruneau [17]

for a potential cyber event. This part also incorporates establishing a response team and setting up a response plan for cyber incidents. The plan should include policies, procedures, and personnel as per the organization's established standards. The second segment focuses on incident prevention. The third segment is incident management, which again subdivides into four operations: (1) detection of potential threats; (2) containment of the event (e.g., quarantine malware installed on the servers); (3) remediation including the eradication of the risk (e.g., malware); and finally (4) recovering from the event and restoring the system to its full-service capability. The fourth segment deals with the post-event analysis. This analysis includes determining the root cause, access path, vulnerability, and other necessary information to understand the incident better. The review would help to prevent the system in the future, including cyber forensics and data preservation.
- ***Patch management for control systems***: There is no "one size fits all" solution that adequately addresses the patch management processes of IT and OT networks. There are some differences in implementing the patches in information technology systems and industrial control systems, as discussed earlier in Table 1. The recommended practices (see Tom et al. [36]) provide a detailed explanation of the patch management program (e.g., backup, testing of a patch, disaster recovery, etc.), patching analysis (e.g., vulnerability analysis), and deployment in the control systems environment.
- ***Updating Antivirus in industrial control systems***: Antivirus has widely used in information technology than the ICS. The application of antivirus software is to comply with the defense-in-depth strategy in ICS. Thus antivirus software and patches need to keep updated periodically in ICS. These recommendations [37] guide how to update the Antivirus in the control system environment without impacting the OT production systems.

Again, most of these standards are very generic and may vary from system to system, depending on the area of applications. In this chapter, we want to provide quantifiable resilience assessment methodologies that would help make informed decisions by incorporating the security guidelines.

5 Formal Approaches for Realizing CPS Resilience

One way to realize the frameworks and security practices within the CPS domain is to provide quantitative cyber resilience analytics. The quantitative cyber resilience analytics could help network administrators and operators in two ways: (1) It can help in assessing systems and evaluating the weak areas and (2) assist in developing optimal mitigation strategies. Researchers utilize both qualitative and quantitative modeling approaches for deriving quantitative cyber resilience metrics. In this section, we present formal mathematical methods and procedures to quantify cyber resilience for the CPS. We first offer a subjective approach for quantifying cyber resilience utilizing the analytical hierarchy process (AHP) in Sect. 5.1. We then propose a quantitative resilience assessment approach using a multi-level vulnerability graph model utilizing the graph properties in Sect. 5.2. Next, we offer a plan for critical cyber asset identification utilizing the technique for order of preference by similarity to ideal solution (TOPSIS) method in Sect. 5.3.

We know that we need to choose only specific aspects from the frameworks for formal modeling within this chapter's context. That is why we model here network criticality, system functionality, and cyber resilience analytics for the CPS utilizing the system's vulnerabilities. We also provide methods to rank critical assets.

5.1 Cyber Resilience Quantification by Subjective Evaluation Using Analytical Hierarchy Process (AHP)

The Analytic Hierarchy Process (AHP) is an organized technique for analyzing complex decisions based on mathematical and psychological comparison (Saaty [38]). AHP has been in use in the cybersecurity domain to assess security metrics for a long time because of its ability to combine mathematical objectivity with the psychological subjectivity to evaluate information and help make decisions [39, 40]. We use AHP to quantify cyber resilience analytics using the subjective evaluation method based on specific questionnaires. First, in the next paragraphs, we present the mathematical process involved in AHP. Then we discuss a case study to assess the robustness metrics for a hypothetical ICS network in Sect. 5.1.1.

In AHP, we form the hierarchy by setting a goal to evaluate, criteria to meet that goal, and available possibilities or options or alternatives. Here we illustrate the AHP procedures for the cyber resilience analytics following Haque et al. [21]. We collect subjective judgment data from N subject matter experts (SME). We compare m criteria pairwise and form a comparison matrix P of dimension $m \times m$. An element P_{ij} in P represents the subjective comparison between the two criteria P_i and P_j. We provide the pairwise comparison matrix P in Eq. (1) below where $P_{ij} = \frac{1}{P_{ji}}$.

$$P = \begin{bmatrix} 1 & P_{12} & \cdots & P_{1m} \\ \frac{1}{P_{12}} & 1 & \cdots & P_{2m} \\ \cdots & \cdots & \cdots & \cdots \\ \frac{1}{P_{1m}} & \frac{1}{P_{2m}} & \cdots & 1 \end{bmatrix} \quad (1)$$

We then derive the normalized comparison matrix N_{CMP} from the original comparison matrix P above where $N_{CMP}(i, j) = \frac{P_{ij}}{\sum_{i=1}^{m} P_{ij}}$.

$$N_{CMP} = \begin{bmatrix} N_{CMP}(1,1) & N_{CMP}(1,2) & \cdots & N_{CMP}(1,m) \\ N_{CMP}(2,1) & N_{CMP}(2,2) & \cdots & N_{CMP}(2,m) \\ \cdots & \cdots & \cdots & \cdots \\ N_{CMP}(m,1) & N_{CMP}(m,2) & \cdots & N_{CMP}(m,m) \end{bmatrix} \quad (2)$$

Each criterion has a weight. We compute the weights of the criteria using the normalized matrix N_{CMP}. The weights are none other than the normalized right eigenvector of the pairwise comparison matrix P.

$$W = \begin{bmatrix} W_1 \\ W_2 \\ \dots \\ W_m \end{bmatrix} \tag{3}$$

where, $W_i = \frac{1}{m}\left(\sum_{j=1}^{m} N_{CMP}(ij)\right)$. We also need to check the consistency of the pairwise comparison. We can do that by computing the consistency ratio, CR by using the expression $CR = \frac{CI}{RI}$, where RI is the random index, and CI is the consistency index. We calculate CI by utilizing the principle eigenvalue λ_{max}, as given in Eq. (4).

$$CI = \frac{\lambda_{max} - 1}{m - 1} \tag{4}$$

Here, we compute λ_{max} by

$$\lambda_{max} = \sum_{j=1}^{m}\left(\sum_{i=1}^{m} P_{ij}\right) * W_j \tag{5}$$

We find the value of random index RI from Table 6 of the article by Saaty [38]. We accept the comparison if the consistency ratio $CR \leq 0.1$ (this means that out of 10 sample responses 9 responses are consistent to each other). Table 3 provides default RI values for the corresponding m values for cases $m < 10$. Next, in Sect. 5.1.1, we present an illustration of assessing robustness metric.

5.1.1 A Hypothetical ICS Network 'Robustness' Assessment Using AHP

To explain how the mechanism of the AHP applies in cyber resilience assessment, we provide here an illustration using the 'robustness' metric (one of the broad four resilience metrics of R4 model [17]). Let us consider a hypothetical ICS network, and our goal is to evaluate the cyber robustness metric for that ICS network quantitatively. Here, we utilize the robustness metric's decomposition, as illustrated by Haque et al.

Table 3 Values of the random index (RI) for small problems ($m < 10$)

m-factor[a]	2	3	4	5	6	7	8
Random Index (RI)	0.00	0.58	0.9	1.12	1.24	1.32	1.41

[a]See Table 6 of Satty [38]

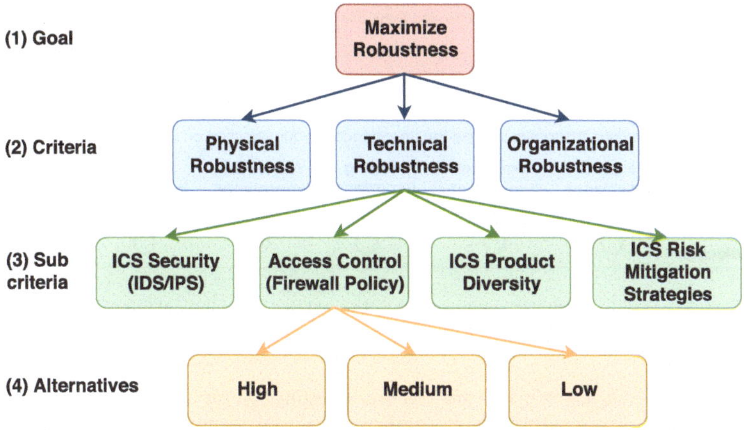

Fig. 7 Decomposition of 'robustness' metric for ICS using the AHP process hierarchy. Robustness is one of the four broad categories of cyber resilience metrics in R4 model

Table 4 List of possible values for each of the sub-criteria

Alternative values	Interpretation of the options
High (H)	Specialized security measures are already implemented in the ICS for the associated sub category
Medium (M)	Some or partial security measures are implemented in the ICS for the associated sub criteria
Low (L)	No or very few measures are implemented in the ICS for the corresponding sub criteria

[22]. As shown in Fig. 7, the robustness metric has three assessment criteria: physical robustness (C_1), technical robustness (C_2), and organizational robustness (C_3). Each criterion has four sub-criteria SC_1, SC_2, SC_3, and SC_4. SC_1 is ICS security (e.g., using IDS/IPS or physical security), SC_2 is access control (e.g., using firewall policy), SC_3 is ICS product diversity, and finally, SC_4 is ICS risk mitigation strategies. Finally, to be aligned with the Common Vulnerability Scoring System (CVSS) (Mell et al. [41]), we design each sub-criteria to take values among three alternative options: high (H), medium (M), and low (L). We present the meaning of high, medium, and low in Table 4.

We first use the Likert scale to set up the pairwise comparison. In comparison, having equal importance is the lowest parameter with a numeric value of 1, and having extreme importance is the highest-ranked parameter with a numerical value of 9. We present a sample assessment question in Fig. 8. We prefer to use the same Likert scale numerical scores to align with the Likert range used by Satty [38].

We then utilize the subject matter experts to assess the questionnaires. We have conducted the survey and collected a total of $N = 15$ sample data sets from which we exclude $N' = 5$ because of inconsistency (CR < 0.1) in responses. Table 5 con-

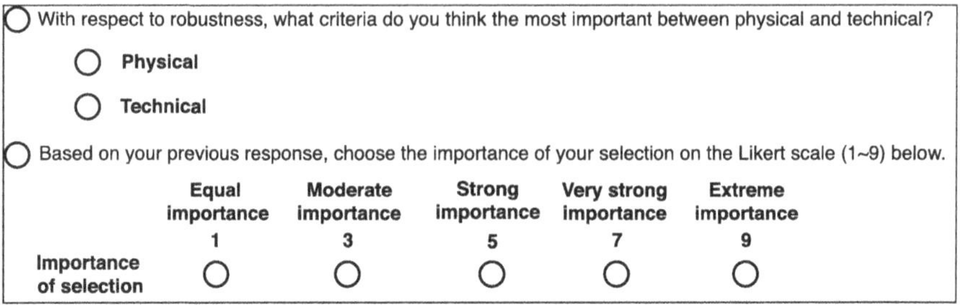

Fig. 8 Sample pairwise comparison between the physical and the technical criteria for the 'Robustness' metric using Likert scale

Table 5 Pairwise comparison matrix and eigenvector estimation for maximizing robustness with respect to the three considered criteria: physical, technical, and organizational

	Pairwise comparison			Evaluated score
Criteria	Physical (C_1)	Technical (C_2)	Organizational (C_3)	Normalized eigenvector
Physical (C_1)	1	0.11	0.20	0.0578
Technical (C_2)	8.95	1	5.101	0.7383
Organizational (C_3)	4.89	0.20	1	0.2039

tains the aggregated pairwise comparison matrix that we have computed from the consistent data set for the three criteria C_1, C_2, and C_3. From the normalized right eigenvector of Table 5, we find the robustness as a function of the physical (C_1), technical (C_2), and organizational (C_3) criteria as given in the Eq. (6):

$$\left.\begin{aligned} &Robustness = 0.06 \times Physical + 0.74 \times Technical + 0.20 \times Organizational \\ &or, \\ &R_1 = 0.06 \times C_1 + 0.74 \times C_2 + 0.20 \times C_3 \end{aligned}\right\} \tag{6}$$

Similarly, we have made pairwise comparisons for sub-criteria and alternatives. We derive the weights of each criterion from the normalized eigenvector corresponding to the options at the lower level of the hierarchy in the AHP model. Figure 9 shows a sample pairwise comparison illustration between options high (H) and medium (M). Equation 7 provides the numerical values that we have obtained for the weights and the normalized eigenvector for the four sub-criteria and three alternatives in the matrix form.

$$\begin{bmatrix} ICS\ Security \\ Access\ Control \\ ICS\ Product\ Diversity \\ ICS\ Risk\ Management \end{bmatrix} = \begin{bmatrix} 0.4 \\ 0.2 \\ 0.1 \\ 0.3 \end{bmatrix}, \ and \ \begin{bmatrix} High\ (H) \\ Medium\ (M) \\ Low\ (L) \end{bmatrix} = \begin{bmatrix} 0.7 \\ 0.2 \\ 0.1 \end{bmatrix} \tag{7}$$

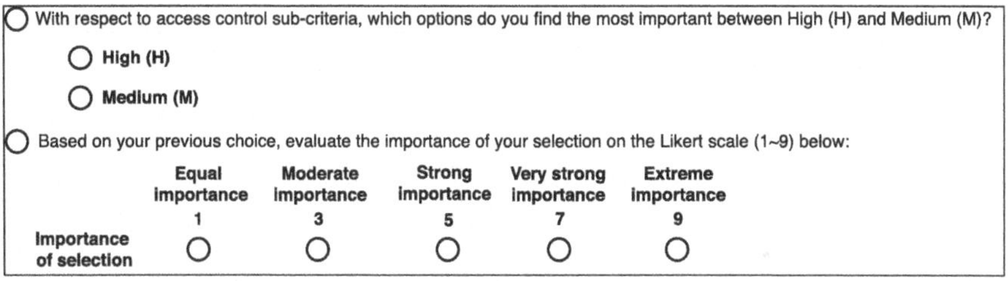

Fig. 9 Sample pairwise comparison between the options high (H) and medium (M) for the access control sub-criteria using Likert scale

This way, we can assess the four broad cyber resilience metrics one at a time using the AHP, and then combine the assessment to reach a consolidated value for the resilience metric.

5.2 *Cyber Resilience Assessment Using Multi-level Directed Acyclic Vulnerability Graph Model*

As we have already stated, our goal within the context of this chapter is to apply the frameworks and security practices to evaluate the quantitative cyber resilience metric. In this section, we describe a multi-level vulnerability graph model to assess the cyber resilience quantitatively. We find graph-theoretic security analytics is one of the most common methods to address cyber risk and resilience. Next, we present the background information necessary to understand the cyber resilience modeling approach.

5.2.1 Background Information for Graph-Theoretic Modeling and Analysis

This section discusses the graph-based modeling approach to provide the readers with the necessary background information about our resilience quantification methodology. Some of the definitions we have taken from one of our recent works [26]. We frequently refer to SCADA systems for illustration purposes as we formulate the mathematical models by keeping in mind the energy delivery systems (EDS) as an example of CPS. Readers may consider SCADA as a monitoring and control system for the physical field devices. We find that researchers commonly refer to cyber-physical power systems (CPPS) [2, 3] when it comes to the discussion of energy systems cybersecurity. That is why we take the power systems' case to illustrate the model that applies equally to other CPS.

(1) ***Vulnerability graph***: We define a vulnerability graph as a directed acyclic graph (DAG). In general, a vulnerability graph is a type of attack graph. Mathematically, we represent the vulnerability graph as $G = (N, E, W)$, where N is the set of vertices; E is the set of edges where $E \subseteq N \times N$; and W is the weight matrix of the graph. If there exist and edge $e = (i, j)$ between vertex i and j, then the vertex i and j are adjacent to each other. An adjacency matrix A of a graph $G = (N, E, W)$ with $|N| = n$ is an $n \times n$ matrix, where $A_{ij} = W_{ij}$, if $(i, j) \in E$ and $A_{ij} = 0$ otherwise. The weight value W_{ij} between the edge (i, j) is coming from the CVSS vulnerability base score (see Mell et al. [41]) of the node j. The multi-level vulnerability graph is the same as the vulnerability graph, but here different layers (as per the defense in depth security strategy) model themselves as separate graphs. There can be single or multiple perimeter devices between the layers, such as a firewall that connects the two consecutive layers. We encourage readers to explore more about the multi-level vulnerability graph in the article by Haque [42].

(2) ***Network topology***: In a CPS network, the network design follows specific system architecture and security policies (e.g., firewall rule-sets). In the CPS, as per the NIST guidelines (see Stouffer et al. [5]), ICS firewalls control the allowed protocols or message communications among the field devices through the rule-sets or policies. We consider that the adjacency matrix is sufficient to represent the network connectivity in the vulnerability graph.

(3) ***Control function***: We consider a control function a logical connection that carries (or transmit) the data from the field devices to SCADA and controls commands from SCADA to the field devices. These functions perform specific tasks such as voltage regulation adjustment, etc. Formally, we define a control function $CF(i, j)$ between node i & j as $\{CF(i, j) = e(i, j) \mid \exists\, e(i, j) \in E,\ A_{ij} \neq 0 \;\&\; W_{ij} > 0\}$, and thus, basically, the edges represents the control functions in the vulnerability graph. As we utilize the CVSS base scores, this edge weight or importance indicates the possible exploitability and impact of exploiting the particular control function. We do not consider the degree of operability of the control functions in this model as described in FDNA [43], because that brings a different research question of modeling and incorporating the functional dependencies in the cyber resilience assessment.

(4) ***CVSS base, exploitability, and impact scores***: CVSS [41] defines the exploitability and impact metrics for every known vulnerability. The national vulnerability database [44] provides the CVSS scores for all the reported (i.e., known) vulnerabilities. The exploitability metric comprises three base metrics: access vector A_V, access complexity A_C, and access authentication A_U. Similarly, the impact metric is also composed of three base metrics: confidentiality impact I_C, integrity impact I_I, and availability impact I_A. CVSS computes the exploitability E_i and impact I_i of a vulnerability i using Eq. (8).

$$\left.\begin{aligned} E_i &= 20 \times A_V^i \times A_C^i \times A_U^i \\ I_i &= 10.41 \times (1 - (1 - I_C^i)(1 - I_I^i)(1 - I_A^i)) \end{aligned}\right\} \tag{8}$$

The measurement of exploitability, impact, and base scores are on a scale of 0–10. The higher the value, the higher the exploit capability or consequences. To define the base score, CVSS define an impact function as given below:

$$f(I_i) = \begin{cases} 0 & \text{if } I_i = 0 \\ 1.176 & \text{otherwise} \end{cases} \tag{9}$$

Finally, CVSS computes the base score (BS) of vulnerability i using the below equation (see [41]):

$$BS_i = \text{roundTo1Decimal}\left(\left(\left(0.6 \times I_i\right) + \left(0.4 \times E_i\right) - 1.5\right) \times f(I_i)\right) \tag{10}$$

(5) ***Multi-edge to single edge transformation***: In a network, if a node has multiple vulnerabilities, the graph becomes a multi-digraph. The number of paths from source to destination increases exponentially and creates scalability problems for large networks. To avoid this, we transform the multi-edged directed vulnerability graph to a single-edged directed graph (simple graph) using the composite exploitability score. As the severity of the exploitability and impact are different for different vulnerabilities, we use a severity-based weight approach (see Table 3 of [25]) to incorporate the severity level of the vulnerability. The composite exploitability score (ES), impact score (IS), and base score (BS) for node j, having vulnerabilities $i = 1 \sim n$ is defined in Eqs. (11), (12), and (13).

$$ES_j = \frac{\sum_{i=1}^{n} w_i^j \times E_i^j}{\sum_{i=1}^{n} w_i^j} \tag{11}$$

$$IS_j = \frac{\sum_{i=1}^{n} w_i^j \times I_i^j}{\sum_{i=1}^{n} w_i^j} \tag{12}$$

$$BS_j = \frac{\sum_{i=1}^{n} w_i^j \times BS_i^j}{\sum_{i=1}^{n} w_i^j} \tag{13}$$

Here, w_i^j, E_i^j, I_i^j, and BS_i^j are the severity weights, exploitability score, impact score, and base score of vulnerability i of node j. We find BS_i^j from NVD database [44] or using Eq. (10), and we compute BS_j using Eq. (13) which refers to the composite base score of node j.

(6) ***Computation of edge weight***: We utilize CVSS base scores in computing the edge weights using Eq. (13). This way, we consider both the exploitability and impact of a vulnerability in our edge weight. The weight matrix is as follows.

$$W_{ij} = \begin{cases} BS_j & \text{if } (i, j) \in E \\ 0 & \text{otherwise, i.e., if } (i, j) \notin E \end{cases} \tag{14}$$

(7) ***Betweenness Centrality (BC)***: Betweenness Centrality (BC) is a graph-theoretic metric that measures the number of times a node acts as a bridge along the shortest paths between two other nodes. If we translate a network into a graph-theoretic model, then the BC of a node indicates the possibility of attack progression through that node. Mathematically, BC of node n (i.e., B_n) is as follows:

$$B_n = \sum_{s \neq n \neq t} \frac{\sigma_{st}(n)}{\sigma_{st}} \tag{15}$$

Here, σ_{st} = total number of shortest paths from source node s to target node t and $\sigma_{st}(n)$ = number of paths that pass-through node n among those shortest paths.

(8) ***Katz Centrality (KC)***: Katz Centrality (KC) is another graph-theoretic parameter that gives the importance of the node considering the network structure and node position in the network. KC quantifies the number of nodes connected through a path, while we penalize the contributions of distant nodes. Mathematically, we define KC of node i as given in Eq. (16), where β is an attenuation factor and $0 \leq \beta \leq 1$.

$$C_{Katz}(i) = \sum_{p=1}^{\infty} \sum_{m=1}^{n} \beta^p (A^p)_{mi} \tag{16}$$

The following subsections present the derivation of system critical functionality and resilience metrics, as shown by Haque et al. [26]. We utilize network criticality to formulate system functionality.

5.2.2 Critical System Functionality (CSF)

System critical functionality is the level of minimum functionality maintained by a system during any adverse scenario. It depicts the extent to which the system's typical performance can degrade. While discussing resiliency, Arghandeh et al. [45] illustrate resilience as a multi-dimensional property, which requires managing disturbances of the network performance. This disturbance may originate either from physical or cyber devices malfunctions or failures or due to a cyberattack incident. Arghandeh et al. also describe critical system functionality as maintaining the system's minimal required services in the presence of unexpected extreme disturbances. In another study, Bharali and Baruah [46] define average network functionality using the network criticality metric. Bharali and Baruah consider random network failures while determining network functionality using a graph-theoretic approach. We extend the analysis of Bharali and Baruah [46] for the case of random cyberattacks on the CPS. We think removing an edge in the vulnerability graph makes a service unavailable or deactivates a control function due to disconnecting the logical connection. Here we consider the same average network functionality metric as the system's critical functionality. This is the level of functionality maintained by the CPS under a cyberattack.

Let us denote the original graph before any attack incident happens by $G_o = G$, and the graph obtained by removing the edge e during an attack incident by $G_e = G \backslash e$. Let us also consider τ and τ_e be the network criticality of the graphs G_o and G_e. Then we define the critical system functionality by considering the effect of the edges removed from the original graph as given by Eq. (17).

$$\eta = 1 - \frac{1}{m} \sum_{e \in E} \left[I^{+}(\tau_e - \tau)\frac{\tau}{\tau_e} + I^{-}(\tau_e - \tau)\frac{\tau}{\tau_e + \frac{2n}{\mu}} \right] \tag{17}$$

where m denotes the number of edges in G_o, μ is the smallest non-zero eigenvalue of G_o, $I^{+}(x) = 1$ if $x \geq 0$ and 0 otherwise, and $I^{-}(x) = 1$ if $x < 0$ and 0 otherwise. For a connected graph G_o, $\mu = \mu_1$ is the algebraic connectivity of G_o. Here, $0 \leq \eta \leq 1$. Thus, η indicates the system functionality of the CPS under cyber attack events, i.e., the functionality or services available during the attack event considering the impacts on the links. A higher value of η means a higher degree of system functionality is maintained. We discuss the computation process of the network criticality τ in Sect. 5.2.4.

5.2.3 Cyber Resilience Metric

Deriving resilience analytics requires understanding and incorporating system behavior (linear or non-linear) during the recovery phase. It also needs to incorporate critical system functionality while generating resilience metrics. Roberson et al. [47] define resilience from the bulk power system perspective, where the authors consider that the safeguarding and restoration of the system functionality subject to perturbations are key elements of resilience. We compute the CPS's cyber resilience by utilizing the system performance or recovery curve, as given in Fig. 10 incorporating the critical system functionality metric. Typically, during an adverse event, the recovery behavior of a system is non-linear. This recovery is a function of the system (S) under consideration, duration of recovery (T), recovery rate (r), time (t), and the functionality level (η). Zobel [48] addresses the power system recovery behavior from disaster resilience and proposes several functional forms to model the recovery over time. In this work, we utilize the *inverted exponential* form of the recovery curve from Zobel [48], which considers the non-linearity and suitable to model the resilience for the CPS. We model the time-dependent system recovery behavior $Q_r(t)$ by following the Eq. (6) of Zobel [48] to demonstrate the quantitative resilience metric under any adverse event. Here the impact is equivalent to the loss of system performance or $1 - \eta$ where $0 \leq \eta \leq 1$.

$$Q_r(t) = (1 - \eta)\left(1 - \mathrm{e}^{\left(-\frac{\left(T-(t-{t_i}^{ri})\right)ln(n)}{T}\right)} + \frac{(T - (t - {t_i}^{ri}))}{nT}\right) \tag{18}$$

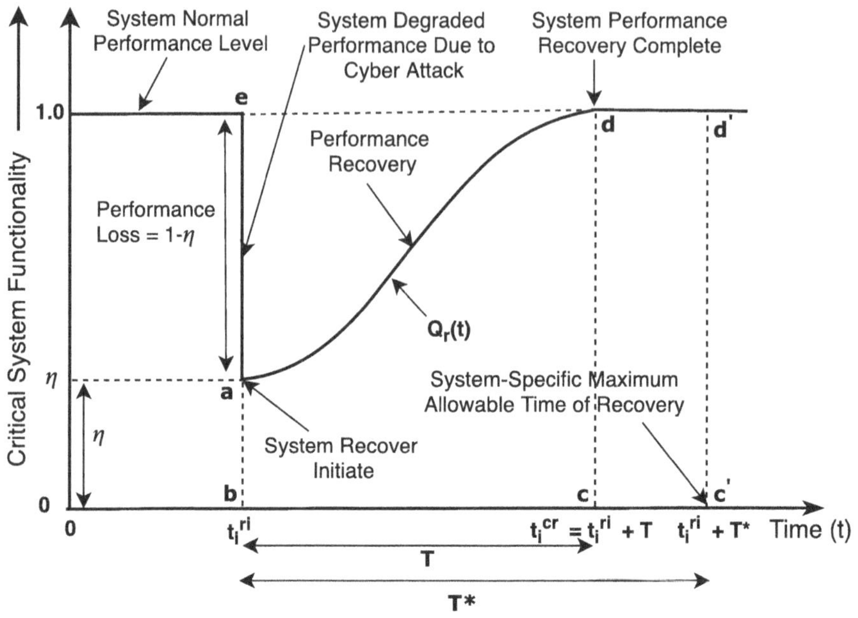

Fig. 10 System performance recovery curve during a cyberattack incident i on the CPS. We use the graph from Haque et al. [26], which is a modified form of the resilience graph presented in Wei and Ji [28]

Table 6 Notations used for resilience modeling

Notations	Explanation of notations
$Q_r(t)$	Time-dependent system recovery behavior
t_i^{ri}	Time instance of initiating system recovery for incident i
t_i^{cr}	Time instance of complete recovery of system functions for attack incident i
$T = t_i^{cr} - t_i^{ri}$	The period of recovery
T^*	System-dependent maximum allowable time for the recovery
n (in Eq. 18)	The level of concavity of the inverted exponential curve

We provide the notations used in Eq. (18) in Fig. 10 and Table 6. Here, T^* is the system-dependent maximum allowable time to recover. Typically, system administrators or designers select T^* as the maximum acceptable time for the system to recover. The area under the points e, a, d represents the amount of losses in system functionality over time due to the cyberattack incident i. Thus, the area enclosed by the marks a-b-c′-d′ is the area of system resilience. To compute the resilience metrics, we first calculate the area enclosed by the points e-a-d. We then can compute the area covered by the points e-a-d as follows:

$$A_{\text{e-a-d}} = (1-\eta)\int_{t_i^{ri}}^{t_i^{ri}+T}\left(1-\mathrm{e}^{\left(-\frac{\left(T-(t-t_i{}^{ri})\right)ln(n)}{T}\right)}+\frac{\left(T-(t-t_i{}^{ri})\right)}{nT}\right)dt \tag{19}$$

Simplifying the above equation, we find the following reduced form as in Eq. (20).

$$A_{e-a-d} = (1-\eta)T\left[1-\frac{n-1}{nln(n)}+\frac{1}{2n}\right] \tag{20}$$

From Fig. 10, e-b-c′-d′ is $1 * T^* = T^*$ and the area of e-a-d is defined by Eq. (20). Thus, the cyber resilience of the CPS system is the area under the curve enclosed by the points a-b-c′-d′ over period T^* as given in Eq. (21).

$$\xi = \frac{1}{T^*}\left[T^* - (1-\eta)T\left(1-\frac{n-1}{nln(n)}+\frac{1}{2n}\right)\right] \tag{21}$$

The term $\left(1-\frac{n-1}{nln(n)}+\frac{1}{2n}\right)$ is a constant term for specific n, and is denoted by γ. Thus, Eq. (21) becomes $\xi = \frac{1}{T^*}\left[T^* - (1-\eta)T\gamma\right]$.

5.2.4 Network Criticality

As we have seen earlier, to compute the CSF, we need the criticality metric. Bharali and Baruah [46], and Tizghadam and Garcia [49] proposed a graph-based network criticality metric. We apply the same here to measure the criticality of the overall CPS network. We use the Moore-Penrose inverse of the Laplacian matrix L to compute the network criticality τ. As we are using the directed weighted graph, we define the Laplacian matrix L as per Chung [50] as given in Eq. (22). In Eq. (22), P is the graph transition matrix, Ψ is a matrix with the Perron vector of P in the diagonal and zeros in all other matrix elements.

$$L = I - \left(\Psi^{\frac{1}{2}}P\Psi^{\frac{-1}{2}} + \Psi^{\frac{-1}{2}}P^T\Psi^{\frac{1}{2}}\right)/2 \tag{22}$$

We can also derive L is by using the normalized graph Laplacians L_{sym} and random walk Laplacian L_{rw}, as below.

$$\left.\begin{aligned} L_{sym} &= D^{\frac{-1}{2}}LD^{\frac{-1}{2}} = I - D^{\frac{-1}{2}}WD^{\frac{-1}{2}} \\ L_{rw} &= D^{\frac{-1}{2}}L_{sym}D^{\frac{1}{2}} \end{aligned}\right\} \tag{23}$$

D is a diagonal matrix formed by the degree of the nodes. We define it as $D = diag(d_1, d_2, \ldots, d_m)$. Here $d_i = \sum_{j=1}^{m} W_{ij}$. We use Bernstein [51] to compute the Moore-Penrose inverse of the Laplacian matrix (L), i.e., L^+ as we provide in Eq. (24).

$$L^+ = \left(L + \frac{J}{n}\right)^{-1} - \frac{J}{n} \tag{24}$$

where J is an $n \times n$ matrix whose entries are all equal to 1. We then define the network criticality metric τ by Eq. (25).

$$\tau = 2n * trace(L^+) \tag{25}$$

Here, n is the number of nodes, and $trace(L^+) = \sum_{i=1}^{n}(L^+)_{ii}$. The larger the value of τ means the network is more vulnerable from the exploitability perspective.

We can apply the above vulnerability graph-based resilience analytics derivation approaches in the CPS context to assess the overall network functionality, criticality, and resiliency. Next, we present a process to identify and rank the critical cyber assets using the TOPSIS method in Sect. 5.3. We consider the ranking an essential step towards realizing the security guidelines as identifying critical assets of the network is among the criteria in the recommended defense-in-depth security measures.

5.3 Ranking Critical Assets Using TOPSIS Method

Determining criticality for the network devices is a multi-attribute decision analysis (MADA) problem. Haque et al. [25] have identified some of the crucial parameters for ranking the critical devices in the power system network from a cyberattack perspective using the vulnerability graph model. Here we apply the TOPSIS method as a MADM (Multiple-Attribute Decision Making) technique to rank the critical devices in a CPS network.

Assessment Parameters: Here, we consider four parameters to assess each device's criticality, although it is possible to take N parameters into the decision-making process. The parameters are (1) device's asset value represented by Katz centrality, (2) device's briding capability, which we model using betweenness centrality, (3) attack occurrence exploitability, and (4) potential attack impact. One can compute the attack exploitability and attack result on a device using Eqs. (11) and (12). Also, we find the BC and KC using Eqs. (15) and (16). For details on the meaning of asset value, exploitability, and attack impact, we encourage readers to check the article by Haque et al. [25], which we omit here to narrow down our focus to the specific problem under consideration.

TOPSIS Method for Device Criticality Assessment: The Technique for Order of Preference by Similarity to Ideal Solution (TOPSIS) is a MADM technique. It builds on the concept that the chosen alternative should have the shortest geometric distance from the positive ideal solution and the longest geometric distance from the ideal negative solution (see Hwang et al. [52]). Kim and Kang [53] use and illustrates TOPSIS to determine the device criticality. Here, we briefly present the TOPSIS method's steps for facilitating an understanding of the ranking process.

Step I: At first, we form an $m \times n$ matrix with m criteria (i.e., parameters) and n alternatives (i.e., nodes/devices), with the intersection of each criteria and alternative contains a value y_{ij}, where $Criteria_i$ and $Alternative_j$ are the ith criteria and jth alternative.

$$Y_{m\times n} = \begin{array}{c} \\ Criteria_1 \\ Criteria_2 \\ \cdots \\ Criteria_m \end{array} \begin{array}{c} \begin{array}{cccc} Alternative_1 & Alternative_2 & \cdots & Alternative_n \end{array} \\ \begin{bmatrix} y_{11} & y_{12} & \cdots & y_{1n} \\ y_{21} & y_{22} & \cdots & y_{2n} \\ \cdots & \cdots & \cdots & y_{3n} \\ y_{m1} & y_{m2} & \cdots & y_{mn} \end{bmatrix} \end{array} \tag{26}$$

Step II: In this step, we normalize the matrix $Y_{m\times n}$ to form a normalization matrix $R_{m\times n} = (R_{ij})_{m\times n}$ using the below equation.

$$R_{ij} = \frac{y_{ij}}{\sqrt{\sum_{i=1}^{m}(y_{ij})^2}} \tag{27}$$

Step III: Here, we calculate the weighted normalized decision matrix T as below. We need to compute the weights using AHP. We illustrate an example in the Sect. 5.3.1.

$$T = (t_{ij})_{m\times n} = (W_i R_{ij})_{m\times n},\ j = 1, 2, \ldots, n \tag{28}$$

Step IV: We determine the worst alternative A_w and the best alternative A_b.

$$\begin{aligned} A_w = \{\langle \max(t_{ij}|j = 1, 2, \ldots, n|i \in I_-\rangle, \\ \langle \min(t_{ij}|j = 1, 2, \ldots, n|i \in I_+\rangle\} = \{t_{wi}|i = 1, 2, \ldots, m\} \end{aligned} \tag{29}$$

$$\begin{aligned} A_b = \{\langle \min(t_{ij}|j = 1, 2, \ldots, n|i \in I_-\rangle, \\ \langle \max(t_{ij}|j = 1, 2, \ldots, n|i \in I_+\rangle\} = \{t_{bi}|i = 1, 2, \ldots, m\} \end{aligned} \tag{30}$$

where $I_+ = \{i = 1, 2, \ldots, m|i\}$ represents the criteria having a positive impact and $I_- = \{i = 1, 2, \ldots, m|i\}$ represents the criteria having a negative impact.

Step V: We compute the L2-distance between the target al.ternative j and the worst condition A_w.

$$d_{iw} = \sqrt{\sum_{i=1}^{m}(t_{ji} - t_{wi})^2},\ j = 1, 2, \ldots, n \tag{31}$$

The distance between the alternative j and the best condition A_b is:

$$d_{jb} = \sqrt{\sum_{i=1}^{m}(t_{ji} - t_{bi})^2},\ j = 1, 2, \ldots, n \tag{32}$$

where d_{jw} and d_{jb} are L2-norm distances from the target al.ternative i to the worst and best conditions, respectively.

Step VI: Finally, at this stage, we compute the criticality of device j (alternative j) using Eq. (33):

$$\eta_j = \frac{d_{jw}}{d_{jw} + d_{jb}}, 0 \leq \eta_j \leq 1, j = 1, 2, \ldots, n \tag{33}$$

Using the device criticality metric, we can identify and rank the critical network devices.

5.3.1 Illustration of Ranking Critical Cyber Assets Using Vulnerability Graph and TOPSIS Method

We illustrate an example of the application of TOPSIS in CPS network asset ranking using the vulnerability graph of Fig. 11. Here, we consider Fig. 11 as a vulnerability graph representation for a CPS with ten devices. We apply TOPSIS to determine the criticality and illustrate the same for the nodes (or devices) 3–8 only using Fig. 11 because of space constraints. The edge score contains two parameters: exploitability score and impact score. Table 7 shows the weights of the criteria and the parameter values of the nodes. Table 8 shows the corresponding TOPSIS computation. In Table 8, the bold italic underline value is the maximum of the criteria, and the bold only value is the minimum of the criteria. Here we find that the most critical devices are 4 and then 7 and 3, respectively, and the least critical one is 8 among the six nodes that we have considered. Again, this is a sample illustration of how we can apply the TOPSIS by choosing some criteria and corresponding weights using a vulnerability graph representation of CPS.

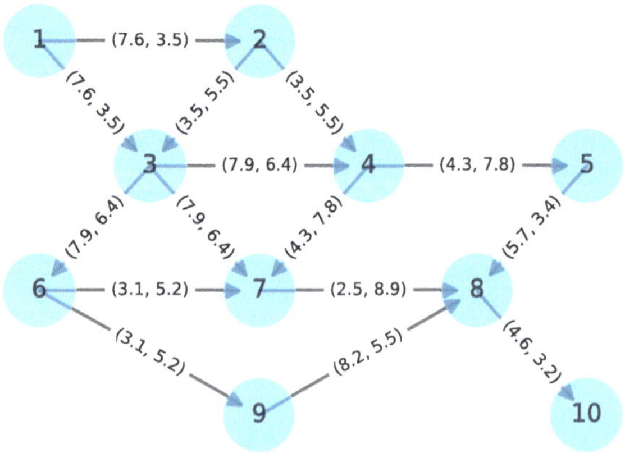

Fig. 11 A sample vulnerability graph with arbitrary edge weights. We represent the edge weights as (exploitability score, impact score). The number of nodes used in this illustration is ten. The edge weights are within the range of 0–10 to keep similar to CVSS scores

Table 7 Device criticality assessment parameters and values

Parameters	Weight (w_c)	Devices (nodes)					
		3	4	5	6	7	8
Exploitability	0.25	7.9	4.3	5.7	3.1	2.5	4.6
Impact	0.5	6.4	7.8	3.4	5.2	8.9	3.2
Betweenness centrality	0.15	0.1273	0.0671	0.0231	0.0417	0.1018	0.1111
Katz centrality	0.1	0.3243	0.3299	0.3010	0.3004	0.3635	0.3643

Table 8 TOPSIS device criticality metrics computation

Parameters/Metrics	Device (j)					
	3	4	5	6	7	8
Exploitability	***1.978***	1.075	1.425	0.775	**0.625**	1.15
Impact	3.2	3.9	1.7	2.6	***4.45***	**1.6**
Betweenness centrality	***0.019095***	0.010065	**0.003465**	0.006255	0.01527	0.016665
Katz centrality	0.03243	0.03299	0.0301	**0.03004**	0.03635	***0.03643***
d_{jw}	2.4546	2.7866	0.9708	1.4577	3.30	0.6917
d_{jb}	1.2523	1.1196	2.8202	2.2469	1.4251	2.9887
η_j	0.6622	0.7134	0.2561	0.3935	0.6984	0.1879
Criticality rank	**3**	***1***	5	4	**2**	6

The bold underline is the maximum of the criteria and bold only is the minimum of the criteria

6 Challenges in Mapping of CPS Resilience with Security Concerns and Operational Domains

The frameworks and recommended practices that we cover in this chapter provide a solid background on designing and implementing an effective cyber resilient strategy for the CPS. By correctly understanding and applying the guidance posted by the frameworks and security practices, we can transform the challenges into opportunities by using the mathematical analysis models. This section briefly discusses how to map the standards and procedures into CPS security and operational resilience.

We think that cybersecurity and cyber resilience are viewed better in a three-dimensional (3D) representation with the CPS domains (i.e., cyber, cyber-physical, and physical), as illustrated in Fig. 12. The three CPS domains, cyber, cyber-physical, and physical, have their independent security requirements. There are security concerns (i.e., threats, vulnerabilities, cyberattacks, etc.) for each domain. There are access control policies, organizational security policy, and overall security strategy in place to address the security concerns, which varies from system to system and domain to domain. The security policies and strategies evolve based on the organization and business mission and situational knowledge and awareness.

On the other hand, the resilience of the systems from cyber incidence largely depends on the organizational implementation of the policies and defense strategies according to different stages of cyber resilience (i.e., plan/prepare, absorb, recover, and adapt). Researchers utilize another set of resilience functions: identify, protect,

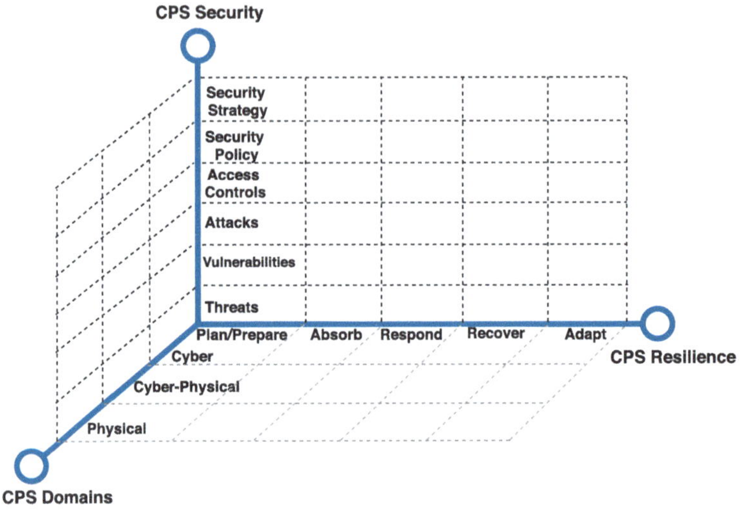

Fig. 12 Mapping of CPS resilience with the security concerns and operational domains

detect, respond, and recover for the same functionality. The frameworks presented in Sect. 3 provide concrete references for the organizations to understand the security requirements and develop cybersecurity models and strategies according to the system needs. The recommended practices and the defense-in-depth policy, as illustrated in Sect. 4 provide practical knowledge and implementation experiences required to build resilient CPS.

The overall challenge in implementing the defense-in-depth strategies into CPS is to map them to the particular system under considerations based on the functional area. For example, if the functional domain is an autonomous vehicle system, then the challenge would be to map the recommendations and strategies with the vehicle system design and specifications under consideration. Thus, with a thorough understanding of the system design specifications, devices, protocols, communications, system limitations, etc. With the help of the recommended practices and mathematical modeling, one can design and implement resilient strategies for control systems, critical infrastructures, etc. It is also imperative to utilize the formal analyses that we have presented in Sect. 5 to evaluate the system's criticality and cyber resilience to have an overall assessment of the resilience poster of the whole system.

7 Conclusions

This article discusses cyber resilience in the context of available frameworks and recommended practices proposed by the different standard bodies and cyber organizations. At first, the paper presented an in-depth analysis and review of existing cyber frameworks and recommended security guidelines for CPS systems to handle the resiliency. Then the article discusses ways to transform the challenges into opportunities by understanding and realizing the security standards and instructions. The chapter provides a three-dimensional graphical illustration among CPS security, CPS components, and CPS resilience by mapping those with the frameworks and standard practices. The article also presents formal mathematical models to assess and quantify cyber resilience analytics for CPSs to help network administrations and researchers make informed decisions. Overall, the paper would guide the researchers in the CPS domain to gain a good understanding of the relevant frameworks, CPS security measures, and modeling and simulation (M&S) constraints to overcome the challenges and utilize the opportunities within the frameworks and guidelines.

Acknowledgments This material is based upon work supported by the Department of Energy under Award Number DE-OE0000780.

References

1. Griffor, E.R., Greer, C., Wollman, D.A., Burns, M.J.: Framework for cyber-physical systems: vol. 1. Overview, Technical report (2017)
2. Shi, L., Dai, Q., Ni, Y.: Cyber-physical interactions in power systems: a review of models, methods, and applications. Electr. Power Syst. Res. **163**, 396–412 (2018)
3. Zhang, T., Wang, Y., Liang, X., Zhuang, Z., Xu, W.: Cyber attacks in cyber-physical power systems: a case study with gprs-based scada systems. In: 2017 29th Chinese Control And Decision Conference (CCDC), pp. 6847–6852. IEEE (2017)
4. Macaulay, T., Singer, B.L.: Cybersecurity for industrial control systems: SCADA, DCS. HMI, and SIS. Auerbach Publications, PLC (2016)
5. Stouffer, K., Falco, J., Scarfone, K.: Guide to Industrial Control Systems (ICS) Security, vol. 800, no. 82, p. 16. NIST Special Publication (2011)
6. Colbert, E.J.M., Kott, A.: Cyber-Security of SCADA and Other Industrial Control Systems, vol. 66. Springer (2016)
7. Johnson, A., Dempsey, K., Ross, R., Gupta, S., Bailey, D.: Guide for Security-Focused Configuration Management of Information Systems, vol. 800, no. 128, p. 16. NIST Special Publication (2011)
8. Cyware: Understanding the difference between risk, threat, and vulnerability (2019). https://cyware.com/news/understanding-the-difference-between-risk-threat-and-vulnerability-c5210e89
9. Blank, R.M.: Guide for conducting risk assessments (2011)
10. Lewis, T.G.: Network Science: Theory and Applications. Wiley (2011)
11. Haque, M.A., Gochhayat, S.P., Shetty, S., Krishnappa, B.: Simulation Foundations, Methods and Applications. SFMA) series, Cloud-Based Simulation Platform for Quantifying Cyber-Physical Systems Resilience. Springer (2020)
12. Chen, T., Abu-Nimeh, S.: Lessons from stuxnet. Computer **44**(4), 91–93 (2011)
13. Mittal, S., Tolk, A.: Complexity Challenges in Cyber Physical Systems: Using Modeling and Simulation (M&S) to Support Intelligence. Wiley, Adaptation and Autonomy (2019)
14. Haque, M.A., Shetty, S., Krishnappa, B.: Cyber-physical system resilience. In: Complexity Challenges in Cyber Physical Systems: Using Modeling and Simulation (M&S) to Support Intelligence, Adaptation and Autonomy (2019)
15. Laing, C.: Securing Critical Infrastructures and Critical Control Systems: Approaches for Threat Protection. IGI Global (2012)

16. Bruneau, M., Chang, S.E., Eguchi, R.T., Lee, G.C., O'Rourke, T.D., Reinhorn, A.M., Shinozuka, M., Tierney, K., Wallace, W.A., Winterfeldt, D.V.: A framework to quantitatively assess and enhance the seismic resilience of communities. Earthquake Spectra **19**(4), 733–752 (2003)
17. Tierney, K., Bruneau, M.: Conceptualizing and measuring resilience: a key to disaster loss reduction. TR News (250) (2007)
18. National Research Council et al.: Disaster resilience: a national imperative (2012)
19. Ross, R.S.: Recommended security controls for federal information systems and organizations [includes updates through 9/14/2009]. Technical report (2009)
20. Sedgewick, A.: Framework for improving critical infrastructure cybersecurity, version 1.0. Technical report (2014)
21. Haque, M.A., De Teyou, G.K., Shetty, S., Krishnappa, B.: Cyber resilience framework for industrial control systems: concepts, metrics, and insights. In: 2018 IEEE International Conference on Intelligence and Security Informatics (ISI), pp. 25–30. IEEE (2018)
22. Haque, M.A., Shetty, S., Krishnappa, B.: ICS-CRAT: a cyber resilience assessment tool for industrial control systems. In: 2019 IEEE 5th Intl Conference on Big Data Security on Cloud (BigDataSecurity), IEEE Intl Conference on High Performance and Smart Computing (HPSC) and IEEE Intl Conference on Intelligent Data and Security (IDS), pp. 273–281. IEEE (2019)
23. Barker, K., Lambert, J.H., Zobel, C.W., Tapia, A.H., Ramirez-Marquez, J.E., Albert, L., Nicholson, C.D., Caragea, C.: Defining resilience analytics for interdependent cyber-physical-social networks. Sustain. Resilient Infrastructu. **2**(2), 59–67 (2017)
24. DiMase, D., Collier, Z.A., Heffner, K., Linkov, I.: Systems engineering framework for cyber physical security and resilience. Environ. Syst. Decis. **35**(2), 291–300 (2015)
25. Haque, M.A., Shetty, S., Kamdem, G.: Improving bulk power system resilience by ranking critical nodes in the vulnerability graph. In: Proceedings of the Annual Simulation Symposium, p. 8. Society for Computer Simulation International (2018)
26. Haque, M.A., Shetty, S., Krishnappa, B.: Modeling cyber resilience for energy delivery systems using critical system functionality. In: IEEE Resilience Week 2019, pp. 33–41. IEEE (2019)
27. Clark, A., Zonouz, S.: Cyber-physical resilience: definition and assessment metric. IEEE Trans. Smart Grid **10**(2), 1671–1684 (2017)
28. Wei, D., Ji, K.: Resilient industrial control system (RICS): concepts, formulation, metrics, and insights. In: 2010 3rd International Symposium on Resilient Control Systems, pp. 15–22. IEEE (2010)
29. Linkov, I., Eisenberg, D.A., Bates, M.E., Chang, D., Convertino, M., Allen, J.H., Flynn, S.E., Seager, T.P.: Measurable resilience for actionable policy (2013)
30. JOINT TASK FORCE: Risk Management Framework for Information Systems and Organizations, vol. 800, p. 37. NIST Special Publication (2018)
31. Bodeau, D., Graubart., R.: Cyber Resiliency Engineering Framework. MTR110237, MITRECorporation (2011)
32. Cornelius, E., Fabro, M.: Recommended practice: Creating cyber forensics plans for control systems. Technical report, Idaho National Laboratory (INL) (2008)
33. Fabro, M., Gorski, E., Spiers, N.: Recommended practice: improving industrial control system cybersecurity with defense-in-depth strategies. In: DHS Industrial Control Systems Cyber Emergency Response Team (2016)
34. Watson, J.-P., Guttromson, R., Silva-Monroy, C., Jeffers, R., Jones, K., Ellison, J., Rath, C., Gearhart, J., Jones, D., Corbet, T., et al.: Conceptual framework for developing resilience metrics for the electricity oil and gas sectors in the united states. Technical report, Sandia National Laboratories, Albuquerque, NM, USA (2014)
35. ICS-CERT: Recommended practice: developing an industrial control systems cybersecurity incident response capability (2009)
36. Tom, S., Christiansen, D., Berrett, D.: Recommended practice for patch management of control systems. Technical report, Idaho National Laboratory (INL) (2008)
37. ICS-CERT: Recommended practice: updating antivirus in an industrial control system (2018)

38. Saaty, T.L.: Relative measurement and its generalization in decision making why pairwise comparisons are central in mathematics for the measurement of intangible factors the analytic hierarchy/network process. RACSAM-Revista de la Real Academia de Ciencias Exactas, Fisicas y Naturales. Serie A. Matematicas **102**(2), 251–318 (2008)
39. Wilamowski, G.C., Dever, J.R., Stuban, S.M.F.: Using analytical hierarchy and analytical network processes to create cyber security metrics. Def. Acquisit. Res. J.: Publicat. Def.e Acquisit. Univ. **24**(2) (2017)
40. Sun, K., Jajodia, S., Li, J., Cheng, Y., Tang, W., Singhal, A.: Automatic security analysis using security metrics. In: 2011-MILCOM 2011 Military Communications Conference, pp. 1207–1212. IEEE (2011)
41. Mell, P., Scarfone, K., Romanosky, S.: A complete guide to the common vulnerability scoring system version 2.0. In: Published by FIRST-Forum of Incident Response and Security Teams, vol. 1, p. 23 (2007)
42. Haque, M.A.: Analysis of bulk power system resilience using vulnerability graph (2018)
43. Garvey, P.R., Ariel Pinto, C.: Introduction to functional dependency network analysis. In: The MITRE Corporation and Old Dominion, Second International Symposium on Engineering Systems, vol. 5. MIT, Cambridge, MA (2009)
44. NIST: National vulnerability database. https://nvd.nist.gov/vuln/data-feeds. Accessed 14 Jan 2020
45. Arghandeh, R., Von Meier, A., Mehrmanesh, L., Mili, L.: On the definition of cyber-physical resilience in power systems. Renew. Sustain. Energy Rev. **58**, 1060–1069 (2016)
46. Bharali, A., Baruah, D.: On network criticality in robustness analysis of a network structure. Malaya J. Matematik (MJM) **7**(2), 223–229 (2019)
47. Roberson, D., Clarisse Kim, H., Chen, B., Page, C., Nuqui, R., Valdes, A., Macwan, R., Johnson, B.K.: Improving grid resilience using high-voltage dc: strengthening the security of power system stability. IEEE Power Energy Mag. **17**(3), 38–47 (2019)
48. Zobel, C.W.: Quantitatively representing nonlinear disaster recovery. Decis. Sci. **45**(6), 1053–1082 (2014)
49. Tizghadam, A., Leon-Garcia, A.: On robust traffic engineering in transport networks. In: IEEE GLOBECOM 2008—2008 IEEE Global Telecommunications Conference, pp. 1–6. IEEE (2008)
50. Chung, F.: Laplacians and the cheeger inequality for directed graphs. Ann. Combinatorics **9**(1), 1–19 (2005)
51. Bernstein, D.S.: Scalar, Vector, and Matrix Mathematics: Theory, Facts, and Formulas-Revised and, Expanded edn. Princeton University Press (2018)
52. Hwang, C.-L., Lai, Y.-J., Liu, T.-Y.: A new approach for multiple objective decision making. Comput. Oper. Res. **20**(8), 889–899 (1993)
53. Kim, A., Kang, M.H.: Determining asset criticality for cyber defense. Technical report, Naval Research Lab, Washington, DC (2011)

Key-Establishment Protocols for Constrained Cyber-Physical Systems

Carlos Andres Lara-Nino, Arturo Diaz-Perez, and Miguel Morales-Sandoval

Abstract Cryptographic keys are critical components when deploying efficient and strengthened security solutions for confidentiality, integrity, and authentication in different computer application domains. In this Chapter, we present three key-establishment protocols that are well-suited for constrained cyber-physical systems (CPSs), using wireless sensor networks (WSNs) as the particular application scope. The focus was on two-party and balanced protocols suitable for the heterogeneity and nondeterministic characteristics of WSNs. The protocols under study offer different security features that might be attractive for different applications depending on the information sensitivity and computing capabilities of the underlying devices. We studied two lightweight key-establishment protocols based on elliptic-curve cryptography (ECC), enhanced by the use of other cryptographic constructions, such as ciphers, hash functions, key derivation, and physically unclonable functions (PUFs). We also present a novel protocol for key establishment constructed on isogeny-based key-encapsulation mechanism SIKE, well-suited for operating in CPSs in the context of a post-quantum computing scenario.

1 Introduction

Cyber-physical systems (CPSs) offer multiple opportunities for deploying applications that have a direct relation with the physical world. As detailed in [1], the core idea of CPSs is the monitoring and controlling of physical objects through interconnected software systems. The idea of CPSs is tightly related to the concepts of

C. A. Lara-Nino · M. Morales-Sandoval
CINVESTAV Campus Tamaulipas, Ciudad Victoria 87130, Mexico
e-mail: carlos.lara@cinvestav.mx
e-mail: miguel.morales@cinvestav.mx

A. Diaz-Perez (✉)
CINVESTAV Campus Guadalajara, Zapopan 45017, Mexico
e-mail: adiaz@cinvestav.mx

A. I. Awad et al. (eds.), *Security in Cyber-Physical Systems*, Studies in Systems, Decision and Control 339, https://doi.org/10.1007/978-3-030-67361-1_2

ubiquitous computing, sensor networks, and the Internet of Things (IoT). The main difference is that CPSs focus on the interaction of the objects with their environment [1].

CPSs are of great relevance for applications such as the navigation and control of autonomous vehicles, the management of water resources, power systems, and smart grids, the supervision and control of oil and gas distribution systems, and remote healthcare monitoring.

Some of these applications demand the miniaturization of the devices in order to either reduce manufacturing costs, such as in management and monitoring, or to improve the user's perception of the technology, which is desirable in healthcare applications. Downsizing devices and reducing their manufacturing costs tend to impose stricter restrictions on the platform. The use of processors with lower specifications, smaller memories, and low-cost power supply translates into constraints for applications running on the device.

By *constrained devices*, we understand a network participant that must adhere to physical restrictions given by the application or environment where it is used. Such constraints might come in the form of performance, storage, bandwidth, or energy limitations. Consequently, a *constrained environment* is defined as a computational system of multiple elements that can be homogeneous or heterogeneous, and which contains devices of limited capabilities: constrained devices.

Examples of constrained devices are wireless-sensor-network (WSN) motes and radio-frequency-identification (RFID) tags. As a consequence, WSNs, RFID, and similar applications are considered constrained environments. Our work focused on WSNs, as illustrated in Fig. 1, a key enabler for CPSs [1].

A constrained CPS tasked with managing sensitive data requires at least the same security services as those of a conventional network, although these devices have less processing power [2]. In some CPS cases, constrained devices are deployed under hostile environments. This implies that an attacker can have physical access to the network. Additional security measures should be considered to patch these vulnerabilities. In big-data scenarios, the high data volume from sensors, even if it is not inherently sensitive, can be exploited for inferring knowledge about the monitored systems. Due to these reasons, all messages transmitted through sensor nodes must be provided with information security.

Resilience against the intentions of malicious actors can be obtained by providing the data with security services. *Confidentiality* can thwart eavesdropping; *integrity* and *authentication* are used to corroborate the veracity of a message; *availability* ensures that the data can be accessed on-demand. These precepts are enforced through the use of cryptographic algorithms. However, most of these cryptosystems require that the participants in the communication exchange share a common data-denominated key. As stated in [3], key management is one of the fundamental issues in CPS security.

CPS characteristics must be considered in the design of security systems: heterogeneity, real-time operation, extended threat models, interoperability, and survivability. These particularities make the design of efficient security solutions quite challenging.

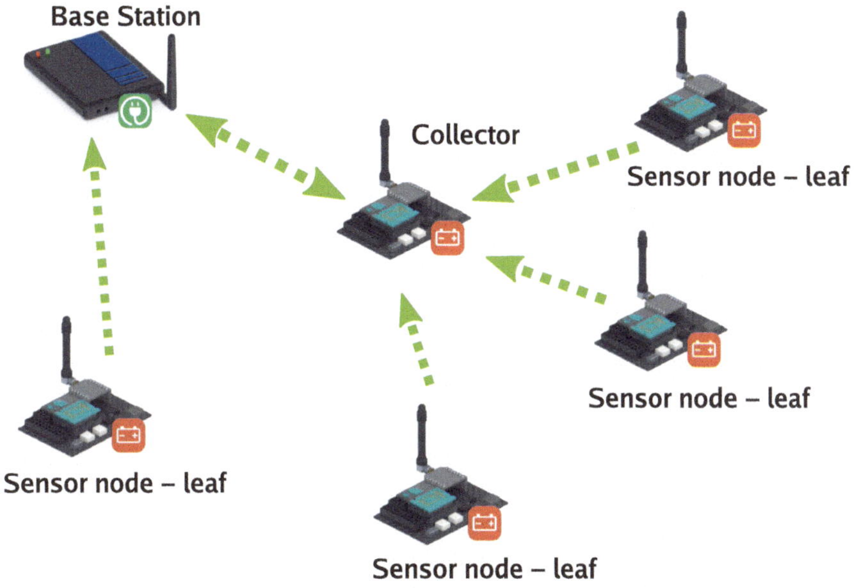

Fig. 1 WSN comprises a base station and multiple sensor nodes. Nodes that are more geographically separated from the base station must employ multi-hop links to transfer their messages to it. The network topology is nondeterministic, and sensor nodes are powered with batteries

According to [1], the great potential and envisioned benefits of CPSs stand in stark contrast to the different security threats that limit the widespread adoption of the technology by reducing the user's trust in these systems. The authors identified the divergence with the client-server model of the Internet stack as the main challenge. Thus, solutions developed for the Internet cannot be directly applied to CPSs. This poses challenges and opportunities in seeking new security solutions for these applications. Furthermore, the evolving nature of these technologies, the increment of their features, and the emergence of new ways of interaction depend on a constantly expanding threat model. The authors of [4] noted that understanding and addressing these threats is a critical challenge in order to improve a user's acceptance of the technology, which would in turn further the development of these systems.

In the past decade, the study of security solutions for constrained devices has gained popularity. Cryptographic algorithms have played an important role in providing constrained systems with the required security services for data confidentiality, integrity checks, and authenticity by means of encryption, authentication codes, and digital signatures. These cryptographic and lightweight solutions for networked environments are constructed from symmetric or asymmetric (public-key) cryptography primitives. In these scenarios, key security is critical for system safety.

Some of the challenges that must be solved by lightweight key-establishment solutions for constrained CPSs are reducing the complexity of underlying operations, decreasing storage costs, mitigating lengthy processing delays, and adapting to the relentless advance of attack threats and vulnerabilities.

The main contributions of this work are twofold:

1. We study the suitability of different solutions for providing key establishment to constrained cyber-physical systems.
2. We provide three two-party balanced key-establishment protocols that are well-suited for constrained cyber-physical systems (CPSs).

This Chapter is structured as follows. In Sect. 2, we discuss the different characteristics observed on key establishment protocols and how these make them more or less suitable for the application scope of WSNs and thus CPSs. Section 3 elaborates about notions on security services and cryptographic principles that are used in this chapter. Section 4 presents an analysis of relevant works from the literature. In Sect. 5, we describe and analyze two key establishment protocols based on elliptic curve cryptography; we explore their characteristics, assess their communications and processing overheads, and study their security properties. Section 6 presents a novel key establishment protocol created to operate with quantum-safe encapsulation mechanisms on two-party scenarios; this protocol is also evaluated and compared against the solutions described in Sect. 5. Lastly, Sect. 7 summarizes our findings and concludes this chapter.

2 The Problem of Key Establishment

Standards such as IEEE 802.15.4 [5] specify mechanisms for obtaining confidentiality and authentication on low-rate wireless-personal-area networks (LR-WPANs) by using standardized cryptographic algorithms. However, these cryptosystems require that link participants have a shared key. This can be challenging for constrained CPSs like WSNs or related technologies.

Given that the topology of a WSN is nondeterministic, it can be expected that each of its nodes is capable of creating a secure link with any other node in its proximity. A straightforward approach for key establishment consists of storing a master key in each device in the network; however, if an attacker manages to retrieve this information, the security of the whole system would crumble. On the other hand, if each node must store a session key for linking up with every possible device in the network, then the device's memory requirements would exponentially grow with the number of network participants.

The key-establishment problem (KEP) lies in the difficulty of enabling a group of two or more network participants to establish a shared piece of information in a secure fashion. As mentioned before, this key is fundamental for securing the communication channel and providing network messages with security services. Key-establishment protocols are algorithms created for solving this problem.

As a goal-driven process, key establishment can be broadly divided into key transport and key agreement. According to [6], these are defined as:

- A key-transport protocol or mechanism is a key-establishment technique where one party creates or otherwise obtains a secret value, and securely transfers it to the other(s).
- A key-agreement protocol or mechanism is a key-establishment technique in which a shared secret is derived by two (or more) parties as a function of information contributed by or associated with each of these, ideally so that no party can predetermine the resulting value.

Key-predistribution schemes are a particular class of key agreement, where shared keys are completely determined a priori by using some primordial keying materials. In this case, the key is fixed or static, and in some instances, such as for WSNs, it cannot be modified post-deployment. In contrast, dynamic key-establishment schemes are those where the key can be established by the participants on subsequent executions. Given that the key can be constructed by employing secret materials from all participants or generated by some coordinator, either key-agreement or -transport solutions can be classified as dynamic.

Another useful classification for key-establishment protocols was proposed in [7]. In this case, the discriminant characteristic was the method employed for establishing the keys. In that work, four main classes were identified:

1. Key predeployment of:
   - global key: a single key that is preloaded to all sensor nodes in the network;
   - full pairwise key: in a network of n nodes, each node has to store a key for each of the other $n-1$ nodes, thus having to store $\frac{n(n-1)}{2}$ keys; and
   - random key set: each node is loaded with a set of keys chosen randomly from a key pool.
2. Key derivation from pre-deployment information:
   - using a transitory master key that expires after some event; and
   - using a keying root that serves as provisional trust.
3. Key-management schemes based on hard mathematical problems:
   - solutions based on symmetric cryptography;
   - solutions based on asymmetric cryptography; and
   - hybrid approaches.
4. Over-the-air key-establishment protocols that:
   - extract secret keys from received signal strength; and
   - leverage channel anonymity for generating pairwise secret keys.

In most cases, dynamic key-exchange mechanisms represent the most viable solution for KEPs by enabling a node to establish shared secrets with nearby devices after deployment.

In IEEE 802.15.4, the mechanism for network participants to establish shared keys is not specified; this is also the case for other norms. Even though key establishment is an old problem, and that there are multiple standardized solutions available, most of these solutions were designed for general applications and do not consider the multiple limitations of constrained devices. Envisioned solutions for cryptographic-key establishment on constrained CPSs must be carefully designed for incurring low overheads in terms of processing, storage, and transmission costs.

Of the described KEP solution approaches, those that employ cryptographic algorithms are generally preferred. Symmetric cryptography approaches tend to be more efficient, but have shortcomings for networked environments, which can be addressed with the use of asymmetric cryptography [8]. From this class of algorithms, solutions based on elliptic curves have the main advantage of needing lower memory and processing overheads for the underlying system [9].

3 Security Notions

Providing information security greatly depends on assumptions made about attacker capabilities and system vulnerabilities. Security services ensure that certain data characteristics are protected. As introduced in previous sections, the most basic of such services are confidentiality, integrity, and authentication.

Data confidentiality implies that only authorized parties have access to the information. When an attacker gains access to the data, its confidentiality is broken, as the privacy of the information cannot be guaranteed. If the attacker's goal is to modify a message, this represents an attack on the information's integrity. Authentication is a particular case of integrity where data origin is also verified.

Most key-establishment protocols rely on these basic security services for constructing or distributing secrets in a safe manner. The strength of a protocol relies on the resilience of its building blocks against cryptographic attacks.

A data cipher is a cryptosystem formed by an encryption function $\mathcal{E}$ and a decryption function $\mathcal{D}$. The main purpose of these algorithms is to ensure privacy by means of the confidentiality service. During its operation, $\mathcal{E}$ employs a key K_E from key space $\mathcal{K}$ to map plaintext P from message space $\mathcal{M}$ into a ciphertext C in $\mathcal{C}$, the ciphertext space. Decryption function $\mathcal{D}$ and a decryption key K_D, also in $\mathcal{K}$, are necessary to retrieve P from C.

If K_E is the same as K_D, it is said that the cipher is symmetrical. On the other hand, if K_E differs from K_D, the cipher is considered asymmetric; in this case, K_E would be of the public domain, while K_D would have to be kept private. The public key of a network participant is used by third parties to encrypt messages that only the private key holder can retrieve. The public key K_E is obtained from K_D by using one-way functions that rely on hard mathematical problems so that K_D cannot be retrieved from K_E. These asymmetrical key systems conform to public-key cryptography (PKC).

Ensuring the integrity and authenticity of a message or its sender requires that the exchange participants gain an information advantage over a potential attacker. These data are either pre-distributed over a trusted channel or derived from some session data that are unknown to the adversary. Message authentication codes (MACs) are tags appended to a message so that the receiver could verify that tag and corroborate the relevant security properties. These codes can be obtained through the use of MAC functions. These cryptosystems incorporate a generation engine $\mathcal{T}$ and a verification function $\mathcal{V}$. To generate T, generator $\mathcal{T}$ employs an authentication key K_T from key space $\mathcal{K}$, and the input message, preferably a ciphertext C, so that $T = \mathcal{T}(C, K_T)$. The verification function $\mathcal{V}$ employs received tag T, received message C', and its verification key K_V for computing $T' = \mathcal{T}(C', K_V)$; if $T \equiv T'$, the verification is valid, and C can be decrypted; else, a nonalphabetic symbol is produced. When K_T is the same as K_V, the MAC function is symmetric, for example, a cipher with an authentication mode or an HMAC. If these keys are different, then the MAC function is called a signature, where K_V is derived from K_T with a one-way function. In this case, K_V is public, and K_T is private. The signer uses its private key for creating a MAC that anybody can verify by using K_V.

There are multiple ways in which the protocol itself can lead to unseen vulnerabilities, even if the underlying ciphers and MACs are secure. In [10], the authors reflected that "it is quite easy to propose protocols in which subtle security problems later emerge". Some of these problems arise from common issues:

- It is unwise to derive a shared secret from the result of a key-establishment mechanism by truncation. Even if retrieving the whole shared key could be a computationally intractable problem, an attacker might still be able to retrieve a reduced portion of it. The indirect use of the shared key also shields it from information leaking; revealing partial information about the key can lead to faulty protocols.
- In practice, an attacker can not only listen to the channel but also inject data into the line. This is the difference between passive and active attacks. The latter is closer to real-world scenarios.
- It should be assumed that a device is capable of maintaining multiple link instances with different network participants. Even if one of these keys is leaked, this should not compromise the other instances.
- The fact that a protocol is logically correct does not imply that it is secure.
- It is necessary to specify what exactly the problem being solved is. Providing a model of adversarial capabilities and a definition of security is critical for determining if a protocol is secure.

These points are relevant in the design of secure key-establishment protocols. In the particular case of constrained CPSs, such as WSNs and the IoT, additional factors must be considered:

- These networks can be deployed in nonstructured environments. In such scenarios, the availability of network infrastructure such as stable links and trusted third parties cannot be guaranteed.
- The rapid deployment of the networks and mobility make it impossible to have a defined topology. Each participant should be able to establish a secure channel with another participant at any given time.
- The evolving nature of the networks and the diversity of tasks performed implies that their composition is heterogeneous. The computational load of the protocols

needs to be even so that even the most constrained participants can consistently establish secure links.

- The wide areas where these networks are deployed and their proximity to the physical world grant attackers physical access to the devices. This is a unique characteristic of some CPS networks.
- Most network participants are under some type of processing, bandwidth, or energy constraints. Hence, protocol complexity should be kept to a minimum.

In order to outline the scope of this work, the following notions from [10] were employed:

1. The goal of key distributions considered is for the parties to simultaneously authenticate one another and come into possession of a secure, shared session key.
2. An active adversary attacks the network. The adversary controls all the communication among the players: it can deliver messages out of order and to unintended recipients, concoct messages entirely of its own choosing, and start-up entirely new instances of players. Furthermore, it can mount various attacks on a session key [...]

However, given the particular conditions of CPS networks, adversarial capabilities need to be enhanced:

2a. The adversary has physical access to the network participants.

This critical condition implies that the protocol must account for the potential capture, displacement, impersonation, and cloning of devices. These challenges are not trivial when the physical restrictions of the network participants are considered.

4 State of the Art

Solving the problem of key establishment between participants of a CPS network is regarded as the main security concern in this area [11]. As reviewed in Sect. 2, these mechanisms are classified according to four main strategies:

1. keys are preloaded;
2. challenges are employed on the basis of prior available information;
3. cryptographic algorithms are required for deriving or transmitting shared secrets; and
4. information from the channel is used to generate a key.

These solutions can also be classified depending on the general structure of the protocol. If any pair of devices can establish a shared secret, we denominate such proposals as distributed (D). Conversely, if a device requires the intervention of a central entity for joining the network, such protocol is said to be centralized (C).

Other characteristics that are relevant in the study of key-establishment protocols are:

- security fundamentals: underlying principle for assuming solution security;
- application: a particular environment for which a solution is conceived, where application constraints can guide the design process of this solution; and
- network assumptions: suppositions regarding the network composition or infrastructure that are fundamental for a particular solution and tend to restrict the solution scope.

In the following, we examined different works that proposed key-establishment solutions for CPSs.

4.1 Literature Review

In [12], the authors introduced a modified-matrix-based pairwise key-establishment scheme for wireless mesh networks. In their approach, each node was preloaded with a key seed that, together with a public matrix, was used to generate a column of a secret matrix. This matrix was created and broadcast by a network router, so any adjacent pair of nodes could obtain a key pair by selecting the respective matrix column. The main assumption of this work was that mesh routers are more powerful than the nodes; hence, offloading some matrix computations reduces storage and communication at the nodes. The computation cost for the nodes was equivalent to performing a polynomial evaluation, while the communication costs of employing a large matrix as public key were not addressed.

The authors of [13] proposed a hybrid key-distribution scheme by employing chaotic maps for key generation, and a zero-knowledge-proof protocol for authentication. The proposal claimed to provide authentication, integrity, and confidentiality. According to the authors, the protocol was less complex and required fewer message exchanges than previous schemes did, while improving security.

A key-establishment approach based on ambient wireless signals and symmetric cryptography was proposed in [14]. The authors stated that the heterogeneity of CPS manufacturers makes key-predistribution models impractical. They proposed to use a key-derivation method by [15] in order to generate a trusted root with the central authority (CA) of the network. Following this authentication step at the physical layer, the node obtained a set of credentials from the CA that were used in higher layers. One of the main concerns with this approach is the assumption from [15] that an attacker cannot obtain a trusted root unless it is in close proximity to authentic nodes. However, CPSs are often deployed in unstructured environments, so an attacker could gain physical access to the network.

Some CPSs, such as those used in automotive applications, have particularities that demand the design of ad hoc security solutions. In [16, 17], an authenticated key-establishment protocol for automotive CPSs was proposed. The described approach employed high-security asymmetric and symmetric cryptography algorithms such as ECDSA, AES, and SHA-3 for providing key establishment, confidentiality, integrity, and authentication to vehicular networks. It was assumed that only intravehicle electronic control units (ECUs) were valid network participants. These nodes were multicore processors with multithreading capabilities in charge of different systems of the vehicle. One of the main concerns of the proposal was fault tolerance, which was solved by performing redundant computations. The authenticity of the ECUs was resolved with the use of public-key certificates. However, since the network was assumed to be intravehicular, dynamic key-establishment mechanisms would not be required. The number of participants was also limited; hence, key-predistribution approaches were also be used. The authors justified their use of public-key cryptography in potential key-recovery attacks and vulnerabilities on the generation of master keys, but these could be addressed with less costly approaches, such as the use of physically unclonable functions (PUFs). While the network participants could shoulder this security overhead, more computations also convey a greater risk of operational faults.

For WSNs, the authors of [18] proposed a key-predistribution scheme based on polynomial pool-based key predistribution and random key predistribution. The approach required to preload each sensor with a set of random polynomial shares and a set of random keys. This generated better chances for nodes to establish a viable network while reducing the impact of node capture by an attacker. Nonetheless, the security and viability of the scheme still depended on node memory availability. Then, that solution was as effective as random key-predistribution models since preloaded keys were not removed from the devices after deployment. Furthermore, the proposed approach considered three mechanisms that could be used by each node upon device discovery. This not only increased the possibility of introducing unseen vulnerabilities but as the authors acknowledged, "path-key establishment is a complicated procedure. It requires more communication and computational overhead for the establishment of path keys between neighboring nodes." This contrasts with the also acknowledged "constrained memory, energy, and computational capabilities of sensor nodes".

In [19] the authors proposed a solution for authentication and key-establishment in cloud-assisted CPSs within the context of a smart grid. The protocol was designed to provide mutual authentication between user and cloud service, and between smart meters and the cloud. When the parties in any of these cases were mutually authenticated, a trusted authority was tasked with enabling these actors to establish session keys. The security of this scheme relied heavily on ECC, enhanced with biometrics on the user's end of the protocol. To prevent replay attacks, the authors considered that all the participants were synchronized with a clock. This protocol was further studied in [20], where the authors claimed to have found deadlocking errors and vulnerabilities against reply and denial-of-service attacks. The scheme was corrected in that work at the cost of increasing computation and communication costs.

For a conventional smart-grid model, the authors in [21] proposed an authenticated key-establishment protocol. This solution relied on the availability of a trusted actor for validating the authenticity of the parties. The protocol employed ECC and symmetric-cryptography algorithms for providing basic security services to the network.

Table 1 Characteristics of surveyed key-establishment proposals for cyber-physical systems (CPSs)

Year	References	Strat.	Struc.	Fundamentals	Application	Network assumptions
2013	[12]	2	D	Matrix arithmetic	Wireless mesh networks	The routers are more powerful than the clients
2017	[13]	3	D	Chaotic Chebyshev polynomials, Zero Knowledge Proof	Environmental monitoring	Availability of Machine to Machine communication
2017	[14]	4	C	Ambient wireless signals, symmetric cryptography	Generic Cyber-Physical Systems	System authority available. Attackers have restricted physical access
2018	[16, 17]	3	D	Asymmetric and symmetric cryptography	Automotive CPS	The networks are intra-vehicle. The nodes are multi-core processors with multithreading capabilities
2018	[18]	1	D	Bivariate t-degree finite field polynomials	Wireless Sensor Networks	The devices have sufficient memory resources to implement a functional configuration of the solution
2020	[19]	3	C	Biometric authentication and asymmetric cryptography	Cloud-assisted Smart Grid	A trusted authority is available. Parties are synchronized with a clock
2020	[21]	3	C	Asymmetric and symmetric cryptography	Smart Grid	A certification agency is available

Table 1 provides a summary of the characteristics of the different works from the literature.

The key-establishment protocols proposed in this work employ a hybrid approach by combining symmetric and asymmetric algorithms. Their aim is to enable any pair of devices to establish a shared secret in a secure way without extended network assumptions. The target application is WSNs; therefore, the proposed adversarial model was extended as defined in Sect. 3.

5 Lightweight Key-Establishment Protocols Based on Elliptic-Curve Cryptography

In 1976, Diffie and Hellman [22] proposed a key-establishment solution on the basis of the hardness of the discrete-logarithm problem (DLP) defined over multiplicative group Z_p^*:

Definition 1 Let a prime p and a generator $G \in Z_p^*$ be parameters of the public domain. Given $X = G^x$, compute x.

In the Diffie-Hellman (DH) protocol, let x, y, two random elements in Z_p^*, be the private keys of exchange parties A and B, respectively. The order of the group determines the complexity of computing the DLP, and consequently the security strength of key establishment.

To obtain a shared secret, A selects $x \in [1, p-1]$ at random and computes its public key $X = G^x$. This value is transferred to interlocutor B. Then, B selects $y \in [1, p-1]$ at random and computes its public key $Y = G^y$, which is transferred to A. Parties A and B then compute $K_A = Y^x$ and $K_B = X^y$, respectively. Note that

$$Y^x = (G^y)^x = G^{yx} = (G^x)^y = X^y; \tag{1}$$

thus, the exchange participants then share a common piece of information that can be used as a precursor for deriving cryptographic keys.

For a large group Z_p^*, the security of the DH protocol relies on the difficulty for an attacker of solving the Diffie-Hellman computational problem (DHCP):

Definition 2 Let a prime p and a generator $G \in Z_p^*$ be parameters of the public domain. Given G^x and G^y for x, y chosen at random from $[1, p-1]$, compute G^{xy}.

Or the Diffie-Hellman decisional problem (DHDP):

Definition 3 Let a prime p and a generator $G \in Z_p^*$ be parameters of the public domain. Given G^x, G^y, and G^z for x, y, z, chosen at random from $[1, p-1]$, decide whether $G^z = G^{xy}$.

As discussed in [10], although the DLP is considered a computationally hard problem, there is no hard proof that the DHCP can only be solved through computing discrete logarithms. Nonetheless, over the years, no such attack has been found; thus, the DHCP is also considered intractable for a computer. The corollary of this is:

> The Diffie-Hellman key exchange is secure in the sense that a computationally bounded adversary cannot compute the secret key shared by the participants.

Alternatively, gain some information advantage in distinguishing the shared key from a random string. The cost for the network participant is to perform modular exponentiation ($G^x \in Z_p^*$).

Over the years, a reduction in the computation bound for adversaries has required that the length of p be increased up to a few thousand bits. This has impacted the time complexity of the modular exponentiations that are fundamental in the exchange.

In order to improve the efficiency of this algorithm, a modification was proposed in 1986 for replacing the DH multiplicative group with abelian elliptic-curve groups [23]. This came to be known as the elliptic-curve Diffie-Hellman exchange (ECDH).

One of the main changes introduced with the use of elliptic-curve groups in the DH exchange is that the main operation, which had previously been modular exponentiation, was replaced by scalar multiplication. In the ECDH, this operation represents the consecutive addition of $k-1$ instances of the group generator or base point, where this addition is defined over the elliptic-curve group. In the following, this operation is illustrated by using the $\cdot$ operator.

Let q a large prime defining finite field $\mathbb{F}_q$. Let E an elliptic curve over $\mathbb{F}_q$, whose set of points $E(\mathbb{F}_q)$—affine coordinate pairs $(x, y) \in \mathbb{F}_q^2$ solving for $E(x, y) \in E(\mathbb{F}_q)$—together with a point at infinity $\mathcal{O}$, form an abelian group of order n. Let G a generator for this group; the public key of ECDH is $P = k \cdot G$, where $k \in [1, n-1]$ is the secret key.

Definition 4 Given adequate domain parameters (q, E, G, n), so that n is large, and the resulting value of $P = k \cdot G$, compute k.

This is known as the elliptic-curve discrete-logarithm problem (ECDLP), and it is considered intractable in polynomial time for a computationally bound adversary.

Let A and B two parties that agree on the common domain parameters: (q, E, G, n). Suppose A and B want to establish a shared key. Party A randomly chooses $a \in [1, n-1]$ and computes $P_A = a \cdot G$, while B follows the same procedure and obtains $P_B = b \cdot G$. A and B publicly exchange these intermediate results. Upon receiving P_B, A computes

$$P_K = a \cdot P_B = (a \times b) \cdot G. \tag{2}$$

Now, B obtains the same result as

$$P_K = b \cdot P_A = (b \times a) \cdot G, \tag{3}$$

so, they are both in possession of a group element P_K that can be used for creating a shared key. The interaction diagram for the basic ECDH protocol is illustrated in Fig. 2.

Due to ECDLP, a or b cannot be computed given $\{P_A, G\}$ or $\{P_B, G\}$, respectively. Due to ECDH, P_K cannot be retrieved from P_A or P_B, employing the same computational and decisional notions of DH. As a protocol, the problems that an attacker must solve are the DHCP or the DHDP [9].

The computational advantage of ECDH over DH is that it allows for selecting $q < p$. In the elliptic-curve case, field length ought to be only some hundred bits long

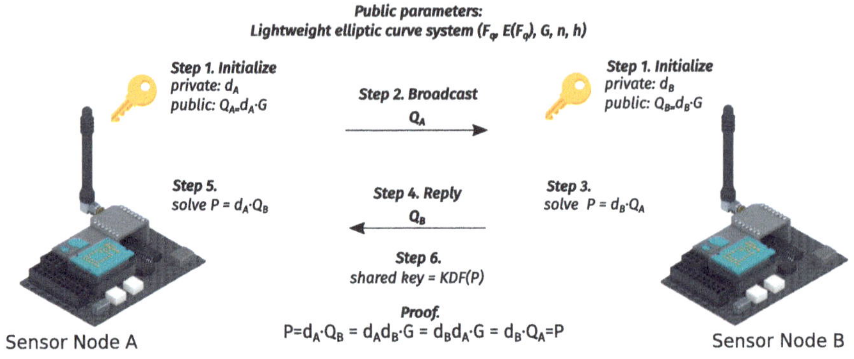

Fig. 2 Interaction diagram for the basic elliptic-curve Diffie-Hellman exchange (ECDH) protocol. In this scheme, parameters $\mathbb{F}_q$, $E(\mathbb{F}_q)$, and P are publicly known

for providing equivalent security to convectional DH instances, which would require a few thousand bits. This leads to performance improvement for scalar multiplication over modular exponentiations.

5.1 *Problem of Authenticity*

In the described key-establishment solutions, a critical concern is that users assume that the public keys they are receiving are legitimate. In the adversarial model employed in this Chapter, however, an attacker can take an active role in the channel. This can lead to man-in-the-middle-type attacks where one of the parties is impersonated. As stated in [10], "the real problem of key establishment is to exchange a key in an authenticated manner".

Network participants require some sort of information advantage to defeat active attackers. This is some data unknown to the adversary but shared by the exchange parties, a secret, or a way to verify the integrity of the message and the sender's authenticity–a tag. The first option brings us back to the main issue of key establishment in some kind of loop. The latter, as studied before, can be achieved with MAC functions, but these also employ a shared secret.

A popular approach is to offload the authentication problem to a third party that is trusted. This actor can either function as an auditor in the exchange or as a public registry of trusted parties.

In [10], the authors provided multiple examples of secure authenticated protocols. However, their scenarios supposed that a trusted actor was available for performing some of the computations or publishing an index of trusted parties. As mentioned before, our work did not make assumptions about the CPS network infrastructure.

The issue is the need to have a common piece of information agreed upon beforehand by the parties. Here, the main drawback of using a preshared secret is that, if the secret is the same for each participant in the network, a single leak would compromise the security of the whole system. Conversely, a shared secret for each possible combination of participants would require massive storage capabilities in each device.

A solution approach proposed in the literature [24] is to employ a master session key for the establishment phase and discard it before time t has elapsed. This threshold is given by the expected time for an attacker to retrieve the master session key from a compromised device. This initial trust can then be used for building simple and efficient authenticated key-establishment protocols.

5.2 Lightweight Authenticated Key Establishment

In [24], the author described a lightweight key-establishment protocol for WSNs based on ECC. Their solution combined a conventional ECDH framework with the use of symmetric algorithms and a hash chain. The author claimed that the protocol is efficient, scalable, and elastic.

The protocol used an ephemeral master key as initial trust that facilitated the authentication of the parties. This key was combined with symmetric-cryptography algorithms for enhancing ECDH with mutual authentication. The employed hash chain was part of a node rejoin scheme that addressed the network variability of WSNs. Figure 3 illustrates the interaction diagram for this protocol.

In [24], the system model was that the network was single-hop, the nodes could communicate with each other, and the link was symmetrical. When both parties perform the same scale of computations, it can be said that the protocol is balanced.

The author proposed that there is a time threshold t, defined as the time required by an adversary to retrieve K_m from a captured node according to the current technology. That is, before t is elapsed, any node in possession of K_m is considered authentic.

The protocol has three steps:

1. Initialization. A shared key K_n is preloaded to each node. This is used as initial trust and represents the last element of a hash chain $K = \{K_1, K_2, \ldots, K_n\}$, where $K_{i+1} = H(K_i)$, and H is a hash function. A node that has K_n is considered secure within a timeframe t. Time t is derived from the required time for an attacker to retrieve keying materials from a captured node. During t, every node uses ECDH to link with other devices.
2. Key establishment. Two nodes use the initial key to perform pairwise key establishment. The work proposed to utilize two *modes* of operation, *new* and *old*. These serve as tags to indicate the type of security utilized in each message. When the key establishment is complete, all nodes should operate in *old* mode. In this phase, each node broadcasts a message that contains a security tag, sender ID, and an encrypted payload containing the sender ID, its public key, and the initial encryption key K_n. The advanced encryption standard (AES) was used to encrypt the message using the starting key K_n. Both, the ID on the header and the

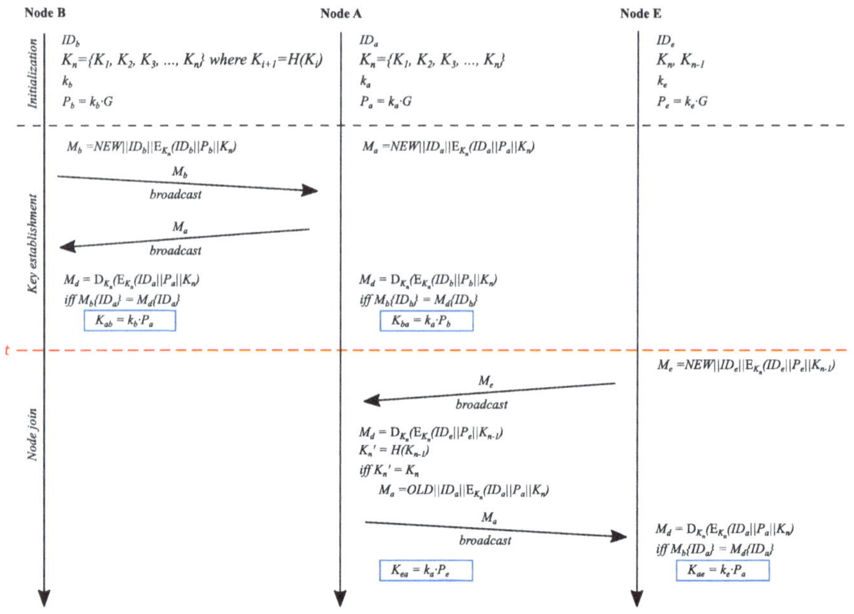

Fig. 3 Interaction diagram for the key-establishment protocol described in [24]

encrypted ID are used to authenticate the message. Once a pair of nodes exchange their public keys, they can establish a common secret using the ECDH.

3. Node join phase. When a new node tries to join the network, it broadcasts a message containing the security header, its ID, and an encrypted payload that contains its ID, its public key, and secret key K_{n-1}. The receiver verifies the new node by decrypting the message by using K_n, calculating $H(K_{n-1})$, and comparing this result with K_n. Once the identity of the joining node is verified, it is possible to establish a shared secret using ECDH.

5.3 *Revisiting Ju's Protocol*

The key issue with Ju's protocol lies in the provided authentication service. As stated in [10], the implicit authentication of encryption should not be used to replace message authentication codes. Moreover, in their work, only a small portion of the ciphertext was used for authenticating the sender. If the appropriate encryption mode is not employed, this can compromise the security of the system.

For key establishment by itself, encryption is not required when public-key algorithms are used. What is needed is a way to ensure that the received public key is authentic and that the integrity of the message is not compromised. A MAC function can be used for this end.

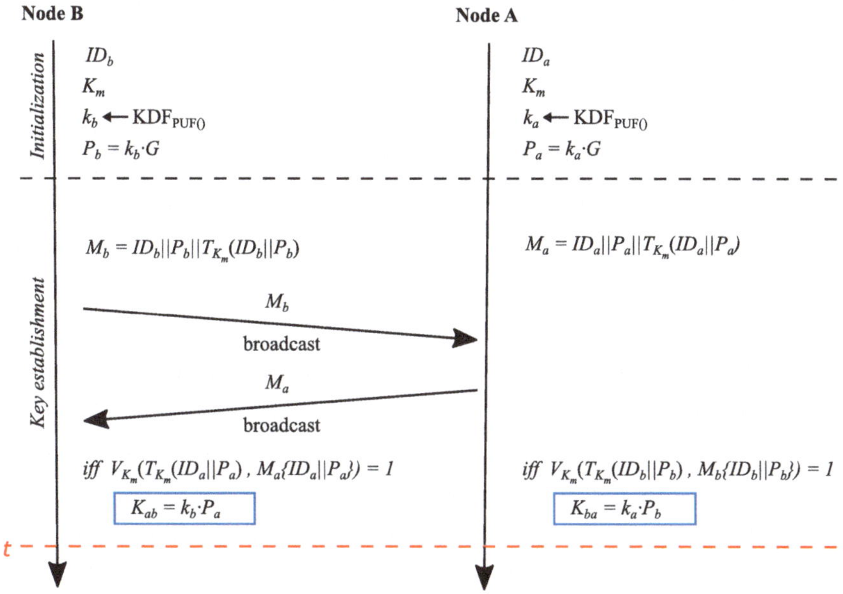

Fig. 4 Operation of the proposed key-establishment protocol based on Ju's work

Another improvement lies in enforcing the use of a key-derivation function (KDF) for obtaining the session key. In this case, a pseudorandom function (PRF) in the form of a PUF [25] was proposed. The use of a PUF prevents device cloning and impersonation attacks.

We found the broadcast system proposed in [24] is adequated for WSNs, as their topology is uncertain, and nodes should be able to collect multiple keys from any devices in their vicinity. The use of initial trust is an efficient way to ensure that MAC tags can be validated as long as t was not reached.

In our proposal, only the initialization and key-establishment steps were considered, whereas Ju's protocol has a node join phase. This phase was discarded since the proposed model does not consider that some new nodes could be introduced into the WSN.

The master key was not derived from a hash chain since it was not needed to recompute future master-key values. Once a time t had elapsed, the master key was discarded. Even if these data were eventually retrieved after t, this would not compromise the integrity of the network, as, at that point, the trust on this root would have expired.

The interaction diagram for the updated lightweight authenticated key-establishment protocol is provided in Fig. 4. This algorithm has an initialization phase when nodes compute their key pairs and a key-establishment phase when the network is formed.

5.3.1 Considerations

Each deployed node i has an identifier ID_i, a master key K_m, and a private key k_i, derived from a PRF. Each node computes a public key $P_i \in \mathrm{E}(\mathbb{F}_q)$ as $k_i \cdot G$ upon deployment. All participants possess the same information. Elliptic-curve domain parameters $\{\mathrm{E}(\mathbb{F}_q), G, n\}$ and description of MAC function $\{\mathcal{T}, \mathcal{V}\}$ are of public knowledge.

5.3.2 Steps

Based on the two first steps from Ju's protocol, the protocol consists of two steps: initialization and key establishment.

Initialization

Every sensor node i is loaded with an ID_i and an initial trust K_m. Each node derives a private key k_i from a PUF with associated public key $P_i = k_i \cdot G$.

Key Establishment

During this phase, the participants perform two key tasks. First, they construct a message containing their ID, their public key, and the MAC for these two values. There is no need to provide confidentiality for the payload since none of these data is secret. These messages are then broadcast to any device on their neighborhood. The second task consists of listening to the channel for incoming broadcasts. The receiver must verify the authenticity and integrity of these messages by means of the accompanying MAC tag. The security of this scheme relies on the secrecy of K_m up to t.

When authentication is successful, the receiver device generates a session key for the device with ID_i, and indexes this session key and their public key in an IDs directory. Logically, if the incoming-broadcast authentication was successful, then the sender device should have followed the same steps and indexed the receiver. This can be corroborated with an acknowledgement message per common network operation (ACK).

5.4 Security Analysis of Proposed Elliptic-Curve Protocol

The use of a symmetric component enhances a conventional ECDH and results in an efficient and simplified design. Here, K_m acts as a source for authentication, preventing man-in-the-middle and denial-of-service (DoS) attacks during network formation. These are two critical ECDH problems.

Key-establishment protocols that completely rely on symmetric components are vulnerable to node capture. This is addressed by using ECDH and discarding K_m after t has elapsed. Any captured node, by definition after t, does not compromise the network, as the only retrievable information by an attacker is at most a small index of session keys.

The use of a PUF as a precursor for the private key of the node provides additional security protections against physical attacks like cloning. Enhanced security can be obtained by using the PUF value and a KDF for creating the private keys.

Even though the use of a master key with a time-bound t provides some advantages for creating a lightweight protocol, this time t also prevents further exchanges to update the session key. To obtain forward security, the devices should adopt a refreshment system in order to update the session key.

6 Key Establishment in the Post-quantum World

Up to this point, the reviewed key-establishment solutions are considered secure on the difficulty of computing discrete logarithms with appropriate restrictions. This is a challenging problem for classical computers. However, that is not the case if quantum processors are involved.

In 1997, Peter Shor published quantum algorithms for computing prime factorization and discrete logarithms in polynomial time [26]. The main implication of that work is that a significant part of modern cryptography will become obsolete if a large enough quantum computer is built [27]. Banking, government, healthcare, commerce, and virtually any application deployed over the Internet would be affected. As stated in [28], cryptography has entered a race against time to adapt to this new threat. Adapting cryptography for resisting quantum attacks while maintaining low-enough overheads for constrained CPSs is a particularly difficult task.

The extent of the power of quantum computing is an open discussion. Theoretical understanding of quantum algorithms and their application to classical problems have only started receiving attention in the past decade. As a result, the reach of applications for quantum computers is still unclear. One of the few points of agreement is that quantum computing is believed to be unable to solve classical NP-complete problems. Nonetheless, quantum computers can solve problems that were believed to be unsolvable in polynomial time, such as DLP and ECDLP.

This has prompted the question of whether authenticated key-exchange protocols exist that are tailored for constrained environments that are not vulnerable to potential quantum adversaries. So far, the answer has been no. This issue is addressed in the following.

6.1 Proposed Approach

In classical cryptography, the key agreement is achieved thanks to the Diffie-Hellman key exchange (DH) and its variations. The main trait of this algorithm is to allow for two parties to establish a shared key with equal contributions in a way in which

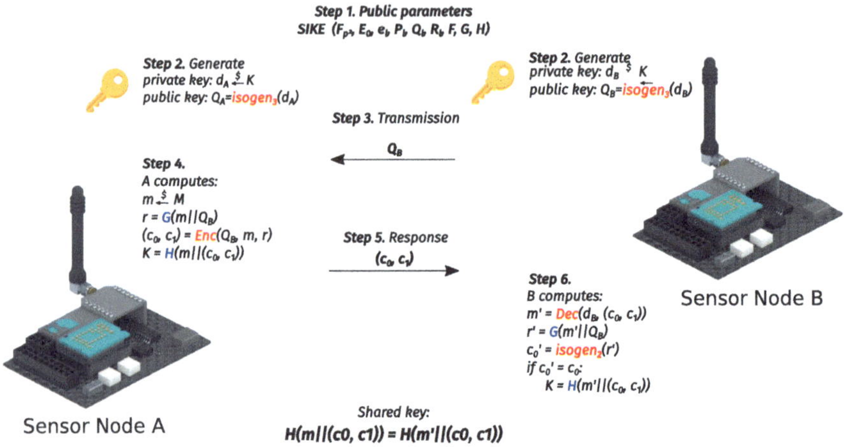

Fig. 5 Interaction diagram for the supersingular isogeny key-encapsulation (SIKE) algorithm

neither party could individually predict the resulting shared secret. However, very few quantum-resistant algorithms have the required commutability for creating DH-like constructions. Only the ding key exchange [29] and systems reliant on isogenies between supersingular elliptic curves [30, 31] offer this advantage. However, the security understanding of these constructions is still limited.

Key-encapsulation mechanisms (KEMs) are systems proposed for achieving key establishment with the use of public-key-encryption (PKE) algorithms. The Fujisaki-Okamoto (FO) [32, 33] and the Hofheinz-Hövelmanns-Kilts (HHK) [34] transforms are two constructions that were conceived for this end. The main characteristic of these systems is that they allow for converting a CPA-secure PKE into a CCA-secure KEM with tight security—see [35] for the description of these security notions. The limitation of these solutions is that, compared with DH-like exchanges, only one of the parties is responsible for creating the session key. This secret is then encapsulated and transmitted to the second party.

The supersingular isogeny key-encapsulation (SIKE) suite proposes a CPA-secure PKE system and then uses a variation of the HHK transform for obtaining a CCA-secure KEM [36]. Their modification of the HHK construction allows for reducing the complexity of the final validation step in the KEM. Figure 5 illustrates the key-establishment procedure of a SIKE KEM.

In the protocol from Fig. 5, Node A was entrusted to generate a secure session key m. Additionally, Node A assumed that the public key received from Node B was authentic, as no additional checks were performed. General applications can employ standardized authentication techniques or rely on trusted parties for corroborating the authenticity of the public key and its sender. However, constrained devices cannot afford to implement such solutions. This is a problem that has so far not been addressed in the literature.

The CCA security of SIKE KEM allows for reusing a public key in multiple exchanges without additional vulnerabilities due to reaction attacks. This can be advantageous for constrained devices since the public key can be calculated offline and then stored in the device. Furthermore, the public key does not need to be protected and can be stored in external memory.

6.2 *Protocol Design*

A modification of SIKE for obtaining an authenticated key exchange with mutual key derivation is proposed. This protocol is illustrated in Fig. 6.

The protocol is composed of two main steps:

1. Initialization. This process can be carried out offline, each device is assigned an ID, a master key computes a secret key from a physically unclonable function, and uses this secret key to obtain a public key with the public generators and base curve of SIKE.

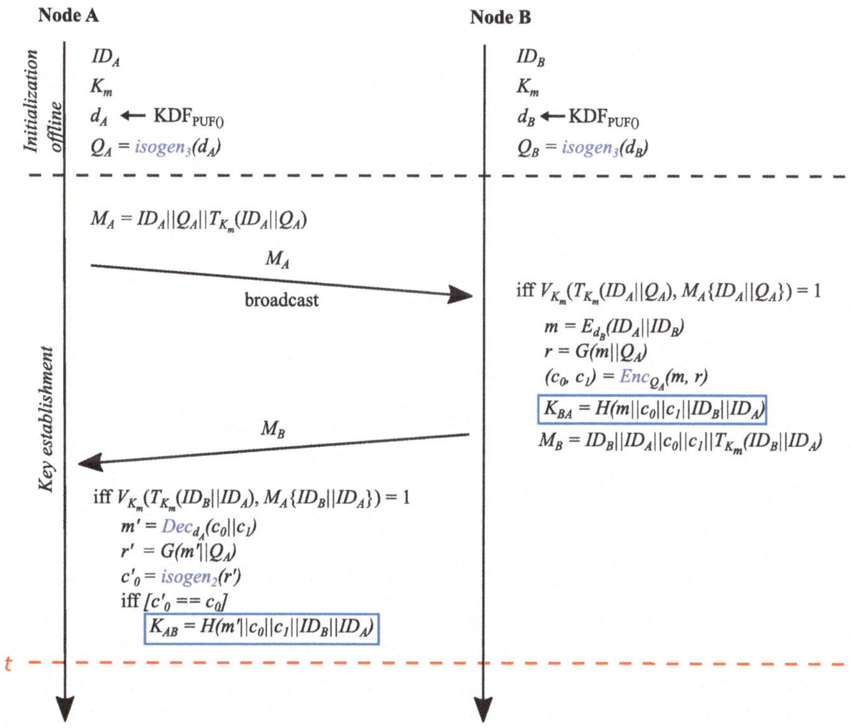

Fig. 6 Key establishment achieved with the proposed protocol. In this scheme, functions in purple are those specified in SIKE. The shared key is derived from a hash computation

2. Key establishment.
   - A device broadcasts a message containing its ID, its public key, and a MAC tag generated using the master key. This serves to authenticate the device with nearby devices. Unlike DH-like variants, SIKE follows a challenge/response approach; thus, the protocol ought to be performed once for each pair of participants.
   - The node that receives the broadcast authenticates the message with the MAC. If verification is successful, the receiver performs the encapsulation of the shared secret; this shared secret or session key is derived as the ciphertext of both devices' IDs employing the private key of the receiver. The issuer generates a new message with the ciphertext resulting from SIKE encapsulation and the respective MAC tag. The session key for the participant is generated as a side product of the encapsulation.
   - The broadcast issuer, upon receiving a reply, verifies its integrity with the corresponding MAC. It then decapsulates the secret and verifies its authenticity through partial re-encryption. If the SIKE ciphertext is valid, the device computes the session key. The new node is then authenticated and starts issuing a broadcast to allow for more nodes to join the network.

This protocol provides mutual authentication and key agreement for any pair of devices in the network. The authentication of the system relies on the difficulty of forging a MAC tag or a forensics attack for recovering the master key. Considering that the fastest of these procedures require a time t, it follows that the scheme is secure up to t. During this time, the network should be consolidated.

The proposed protocol requires encapsulation and decapsulation functions from the SIKE specification. These functions use the underlying public-key encryption scheme specified in [36]. In these procedures, the core functions perform the computation of public keys ($isogen_\ell$) and shared keys ($isoex_\ell$). These are the most expensive operations, and two of each are performed in the envisioned protocol.

6.3 Security Analysis of Proposed Post-quantum Protocol

First, systems based on supersingular isogenies can offer commutability. So, in principle, it is possible to create a Diffie-Hellman-like key exchange. Such an algorithm exists and is described in [30]. This SIDH algorithm allows for two parties to obtain shared secrets that are derived by using information from both participants. For this reason, it can be classified as a dynamic key-agreement protocol. However, the security of SIDH is limited to the CPA scenario. The main implication for a device using this algorithm is that the public key must be renewed for each new session. This involves additional storage and processing costs that are detrimental to constrained CPSs.

By employing a transformation derived from Cramer-Shoup due to [33, 34], SIDH can be transformed into the CCA-secure KEM known as SIKE. In this process, the cryptosystem acquires adaptive security under the random oracle model at the cost of becoming a key-encapsulation system. This implies that SIKE is a dynamic key-transport protocol, which might be vulnerable to key-generation faults.

The first aim of the proposed enhanced SIKE is to restore the *key-establishment* characteristic of the protocol, that is, the session key is derived with contributions from both parties. For this, taking Fig. 6 as a reference, party B derives the session key m as the result of encrypting the identities of both parties under its secret key; here, assume that B acts in good faith. In the SIKE specification, m had a length of 128 and 256 bits, which had a good relationship with the block length of most standardized ciphers.

The identity value is recommended to be at least equal to cipher block size c, so that at least two cipher blocks are processed; by doing so, the protocol is resilient against birthday attacks. The security of m relies on the strength of the selected cipher with an appropriate confidentiality mode behaving as a PRF. The length of the proposed m is then $2c$, which poses a challenge for SIKEp434, where $m = 128$ if $c = 128$. This does not affect the calculation of shared key K, since it is the result of a hash but must be considered on deriving ciphertext c_1; truncating m is not advised, so the implementer has the choice to employ an additional hash for reducing m to the appropriate length, or compute $c_1 = h \oplus m_h \oplus m_l$, where $m = m_h||m_l$. The identity values are also included in deriving the session key K by hashing; this part enforces that both parts act in good faith.

The second enhancement confers SIKE with the mutual authentication feature. In the general Internet scenario, authentication servers and trusted parties are readily available to validate the authenticity of a public key and the integrity of a message. However, in the envisioned application scope, relevant to constrained CPSs such as WSNs, assumptions regarding network infrastructure cannot be made. Hence, parties should be able to authenticate each other by themselves.

This is achieved by employing the ephemeral-master-key strategy from [24]. Every exchanged message during the key-establishment stage of the protocol carries a generated MAC using the ephemeral master key. This MAC function can be implemented by using the main block cipher of the device under an appropriate authentication mode to improve the efficiency of the system. This MAC must exhibit unforgeability and collision resistance under the Chosen Message Attack model so that the exchanged public keys and identities are trusted. In the broadcast reply, the MAC tag does not cover the SIKE ciphertext. This is done for efficiency reasons since ciphertexts alone are already authenticated by the partial re-encryption of SIKE.

Since the ephemeral master key is not used for providing confidentiality, the issue of forwarding secrecy does not need to be addressed. However, network elasticity is restricted, since no more nodes are allowed to join after t; this also implies that drastic changes in the network topology might compromise WSN operation capabilities.

Although the initial topology of a WSN is not given, it does not usually change. Other types of networks better represent problems associated with mobile targets, for example, vehicular ad hoc networks (VANETs). The main source of topology disruption can be attributed to reallocation attacks, but it can be argued that, if the attacker could access a large enough number of nodes, these would be subtracted rather than relocated. The use of PUFs can deter any attempts of sequestering or cloning the nodes.

6.4 Application Scope

The post-quantum protocol proposed in this section is aimed at filling a niche where constrained CPSs require long-term security. Even with the most optimistic forecasts for the development of real large-scale quantum computers, we are looking at a good decade-long window where modern PKCs would remain secure. However, as mentioned before, the concern lies in those applications whose data need to remain safe for longer periods.

Some of these applications include healthcare monitoring, which protects personal data, vehicular networks where exchanged messages within the network contain proprietary information critical to the product, and mobile military networks where the transmitted information by devices can be classified to protect national security interests. In these scenarios, we are looking at a good 20–50-year window where information must remain secure.

Arguably, devices used in these applications exist on the high-end profile for CPSs, but they are still bound by performance and energy constraints. The availability of solutions that can work standalone without a given topology and offer long-term security is critical for protecting sensitive data with due care and diligence. Herein lies the relevance of our work.

7 Conclusions, Final Remarks, and Future Work

The main goal of CPSs lies in connecting the cybernetic and physical worlds. These technologies offer significant advantages for applications of the management and control of public infrastructure, distribution systems, supervision of remote tasks, and healthcare. Therefore, they are intricately connected with the human world. Any data being collected, processed, and transmitted by interconnected devices must be safeguarded. This is a difficult task for constrained CPSs such as WSNs. One of the most effective approaches for ensuring information security is cryptography, which commonly relies on the use of cryptographic keys. Thus, key establishment is the main component when securing current and future CPS applications by employing cryptographic algorithms.

In this chapter, three alternatives of two-party, balanced key-establishment protocols for constrained CPSs were described. The solutions under study were analyzed under fair assumptions, and they rely on proven cryptographic principles. Two elliptic-curve-based solutions that are simple and efficient for solving the problem at hand were first reviewed. We revised the appropriateness of using these systems in the envisioned application scope of WSNs and proposed improvements for enhancing the security of an initial solution of interest in order to derive the second algorithm. We then addressed the possibility of a threat model involving quantum adversaries by proposing a novel key-establishment protocol that inherits the efficiency enhancements of ECC-based solutions but employs quantum-safe cryptographic algorithms.

The work presented here is a first in the area of security in constrained environments for modern computing scenarios and needs further study to corroborate the pertinence of the assumptions and security claims required for constrained CPSs.

Multiple challenges and opportunities can be addressed in future work. First, while it was shown that the proposed algorithms are correct, and informal security assumptions were claimed, it is necessary to demonstrate that the proposed protocols are secure through formal analysis. Second, it is necessary to quantify the operational costs for these solutions to delimit the prospective application domains where they can be used. Lastly, efficient realizations of these algorithms need to be obtained so they can be implemented in actual CPS applications.

Acknowledgements This work was supported by CONACyT under grant number 336750 and CINVESTAV. This work was also funded by "Fondo Sectorial de Investigaci6n para la Educaci6n", CONACyT México, through the project number 281565.

References

1. Henze, M., Hiller, J., Hummen, R., Matzutt, R., Wehrle, K., Ziegeldorf, J.H.: Network Security and Privacy for Cyber-Physical Systems, pp. 25–56. Wiley (2017)
2. Frahim, J., Pignataro, C., Apcar, J., Morrow, M.: Securing the Internet of Things: A Proposed Framework. Technical report, Cisco Security (2012). https://tools.cisco.com/security
3. Wang, Y., Nikolai, J.: Key Management in CPSs, pp. 117–136. Wiley (2017)
4. Ziegeldorf, J.H., Morchon, O.G., Wehrle, K.: Privacy in the Internet of Things: threats and challenges. Secur. Commun. Netw. **7**(12), 2728–2742 (2014)
5. IEEE: IEEE Standard for Low-Rate Wireless Networks. IEEE Std 802.15.4-2015 (Revision of IEEE Std 802.15.4-2011), pp. 1–709 (2016)
6. Menezes, A.J., Vanstone, S.A., Oorschot, P.C.V.: Handbook of Applied Cryptography, 1st edn. CRC Press, Inc., Boca Raton, FL, USA (1996)
7. Jilna, P., Deepthi, P.P.: Light Weight Key Establishment Scheme for Wireless Sensor Networks, pp. 124–137. Springer International Publishing, Cham (2016)
8. Yang, Y., Lu, J., Choo, K.-K.R., Liu, J.K.: On Lightweight Security Enforcement in Cyber-Physical Systems. In: Güneysu, T., Leander, G., Moradi, A. (eds.) Lightweight Cryptography for Security and Privacy, pp. 97–112, Springer International Publishing, Cham (2016)
9. Hankerson, D., Menezes, A.J., Vanstone, S.: Guide to Elliptic Curve Cryptography. Springer, New York Inc., Secaucus, NJ, USA (2003)
10. Goldwasser, S., Bellare, M.: Lecture Notes on Cryptography (July 2008). https://cseweb.ucsd.edu/~mihir/papers/gb.pdf
11. Shafi, Q.: Cyber-Physical systems security: a brief survey. In: 2012 12th International Conference on Computational Science and Its Applications, pp. 146–150 (2012)
12. Zhang, Y., Xu, L., Xiang, Y., Huang, X.: A matrix-based pairwise key establishment scheme for wireless mesh networks using pre deployment knowledge. IEEE Trans. Emerg. Top. Comput. **1**(2), 331–340 (2013)

13. Boubakri, W., Abdallah, W., Boudriga, N.: Chaotic ZKP based authentication and key distribution scheme in environmental monitoring CPS. In: Sabir, E., García Armada, A., Ghogho, M., Debbah, M. (eds.) Ubiquitous Networking, pp. 472–483, Springer International Publishing, Cham (2017)
14. Zhang, Y., Xiang, Y., Huang, X.: A cross-layer key establishment model for wireless devices in Cyber-Physical systems. In: Proceedings of the 3rd ACM Workshop on Cyber-Physical System Security, CPSS'17, pp. 43–53. Association for Computing Machinery, New York, NY, USA (2017)
15. Mathur, S., Miller, R., Varshavsky, A., Trappe, W., Mandayam, N.: ProxiMate: proximity-based secure pairing using ambient wireless signals. In: Proceedings of the 9th International Conference on Mobile Systems, Applications, and Services, MobiSys'11, pp. 211–224. Association for Computing Machinery, New York, NY, USA (2011)
16. Giri, N.K.: A Dependable and Secure Approach for Secret Key Establishment and Operation in Automotive CPS. Master's thesis, Kansas State University, Manhattan, Kansas (2018)
17. Giri, N.K., Munir, A., Kong, J.: An integrated safe and secure approach for authentication and secret key establishment in automotive Cyber-Physical systems. In: Arai, K., Kapoor, S., Bhatia, R. (eds.) Intelligent Computing, pp. 545–559. Springer International Publishing, Cham (2020)
18. Zhang, J., Li, H., Li, J.: Key establishment scheme for Wireless Sensor Networks based on polynomial and random key predistribution scheme. Ad Hoc Netw. **71**, 68–77 (2018)
19. Challa, S., Das, A.K., Gope, P., Kumar, N., Wu, F., Vasilakos, A.V.: Design and analysis of authenticated key agreement scheme in cloud-assisted Cyber-Physical Systems. Futur. Gener. Comput. Syst. **108**, 1267–1286 (2020)
20. Chaudhry, S.A., Shon, T., Al-Turjman, F., Alsharif, M.H.: Correcting design flaws: an improved and cloud-assisted key agreement scheme in Cyber-Physical Systems. Comput. Commun. **153**, 527–537 (2020)
21. Farhdi Moghadam, M., Mohajerzdeh, A., Karimipour, H., Chitsaz, H., Karimi, R., Molavi, B.: A privacy protection key agreement protocol based on ECC for smart grid, pp. 63–76. Springer International Publishing, Cham (2020)
22. Diffie, W., Hellman, M.: New directions in cryptography. IEEE Trans. Inf. Theory **22**, 644–654 (1976)
23. Miller, V.S.: Use of elliptic curves in cryptography. In: Williams, H.C. (ed.) Advances in Cryptology-CRYPTO'85 Proceedings, pp. 417–426. Springer, Berlin, Heidelberg (1986)
24. Ju, S.: A lightweight key establishment in Wireless Sensor Network based on Elliptic Curve Cryptography. In: 2012 IEEE International Conference on Intelligent Control, Automatic Detection and High-End Equipment, pp. 138–141 (July 2012)
25. Maes, R.: Physically Unclonable Functions: Constructions, Properties and Applications. Springer Publishing Company, Incorporated (2013)
26. Shor, P.W.: Polynomial-time algorithms for prime factorization and discrete logarithms on a quantum computer. SIAM Rev. **41**, 303–332 (1999)
27. Chen, L., Jordan, S., Liu, Y.-K., Moody, D., Peralta, R., Perlner, R., Smith-Tone, D.: Report on post-quantum cryptography. Technical report, National Institute of Standards and Technology (2016). http://nvlpubs.nist.gov/nistpubs/ir/2016/NIST.IR.8105.pdf
28. Bernstein, D.J., Lange, T.: Post-Quantum cryptography. Nature **549**, 188–194 (2017)
29. Ding, J., Takagi, T., Gao, X., Wang, Y.: Ding key exchange. Technical report, National Institute of Standards and Technology. NIST Post-Quantum Cryptography-Round 1 Submissions (2017)
30. Jao, D., De Feo, L.: Towards Quantum-Resistant Cryptosystems from Supersingular Elliptic Curve Isogenies, pp. 19–34. Springer, Berlin, Heidelberg (2011)
31. Castryck, W., Lange, T., Martindale, C., Panny, L., Renes, J.: CSIDH: an efficient post-quantum commutative group action. Cryptology ePrint Archive, Report 2018/383 (2018)
32. Fujisaki, E., Okamoto, T.: Secure integration of asymmetric and symmetric encryption schemes. In: Wiener, M. (ed.) Advances in cryptology-CRYPTO'99, pp. 537–554. Springer, Berlin, Heidelberg (1999)

33. Fujisaki, E., Okamoto, T.: Secure integration of asymmetric and symmetric encryption schemes. J. Cryptol. **26**, 80–101 (2013)
34. Hofheinz, D., Hövelmanns, K., Kiltz, E.: A modular analysis of the Fujisaki-Okamoto transformation. In: Kalai, Y., Reyzin, L. (eds.) Theory of cryptography, pp. 341–371. Springer International Publishing, Cham (2017)
35. Bellare, M., Desai, A., Pointcheval, D., Rogaway, P.: Relations among notions of security for Public-Key Encryption Schemes. In: Krawczyk, H. (ed.) Advances in cryptology-CRYPTO'98, pp. 26–45. Springer, Berlin, Heidelberg (1998)
36. Jao, D., Azarderakhsh, R., Campagna, M., Costello, C., Feo, L.D., Hess, B., Jalali, A., Koziel, B., LaMacchia, B., Longa, P., Naehrig, M., Renes, J., Soukharev, V., Urbanik, D.: Supersingular Isogeny Key Encapsulation. Technical report, National Institute of Standards and Technology. NIST Post-Quantum Cryptography-Round 1 Submissions (2017)

Empirical Characterization of Network Traffic for Reliable Communication in IoT Devices

Sujit Bebortta and Dilip Senapati

Abstract The massive growth in the popularity of Internet of Things (IoT) and hence expansion in the number of IoT devices has led to network control issues. The heterogeneity observed in the generated data from each device has further contributed to latency delays and network traffic concerns. An integral part of current network research encompasses the monitoring of network activities, device identification, and secure exchange of information between different devices. The recognition and administration of these persistently increasing IoT devices have posed major challenges in various fields of their application, like Cyber-Physical Systems (CPSs). Hence, the management of network traffic flow between these devices has become a concerning issue. The prolonged inconsistency in cybersecurity systems and constrained computational capabilities have further made IoT devices more vulnerable to adversarial threats. To this end, the preservation and administration of network activities become crucial to manage. In this chapter, we address the network traffic administration issue for different IoT devices. We focus on the efficient characterization of inter-arrival rates of data generated from IoT devices for packet-level and flow-level analysis. Thus, making identification and management of IoT devices exceedingly significant for securing stable functioning of network activities. We also discuss some influential works conjectured to IoT devices and network analysis. The empirical results obtained from real-world network flows have been reported to provide a precise understanding of our observations. Finally, the strengths and weaknesses of some state-of-the-art technologies are discussed along with relevant future scopes.

Keywords Internet of Things (IoT) · Network traffic analysis · Cyber-Physical System (CPS) · IoT devices · Flow control · Device identification

S. Bebortta · D. Senapati (✉)
Department of Computer Science, Ravenshaw University, Cuttack 753003, India
e-mail: senapatidillip@gmail.com

A. I. Awad et al. (eds.), *Security in Cyber-Physical Systems*, Studies in Systems, Decision and Control 339, https://doi.org/10.1007/978-3-030-67361-1_3

1 Introduction

The Internet of Things (IoT) has found tremendous applicabilities in a plethora of sectors over the last decade. However, the deployment of smart sensory devices has witnessed phenomenal existence in many real-life applications, whether it may be the medical informatics sector or the industrial revolution [2]. The increased usage of these services along with some communication protocols like Zigbee, Z-wave, or Bluetooth, has made them popular for smart homes as well as public spaces. These interconnected smart devices usually lack in their computational capabilities and provide a secure software platform to the devices, making them more vulnerable to security and network adversaries [3, 10]. Towards this end, software defined networks (SDNs) have proven to be the right choice in achieving a centralized control over the network activities. This technology has found large-scale applications in data centers for monitoring and redirecting any suspicious traffic flow [4, 6, 17]. However, these models have also faced a substantial amount of limitations over their use inaccurately and securely characterizing network infrastructures.

Network analysts and administrators are using several network traffic analysis tools for capturing the network traffic flow between different IoT devices [29, 30]. Wireshark, OpenWRT, NetFlow are some popular network analysis tools that assist in capturing the traffic flow data between different devices through wireless access points. In [1], techniques for the isolation and regulation of IoT devices was proposed to preserve these networks against attacks. However, these techniques entail some challenges like computational complexities and latencies associated with evaluating massive traffic flows. In [32, 33], some modern strategies based on port identification protocols were employed to detect any malicious activities in the network. Cisco was a frontier in the field of network traffic analysis, which first came up with the network flow characterization technique for IP based applications [5]. Cisco's NetFlow tool provided a diverse set of traffic flow analysis and monitoring services, including network planning, network usage accounting, network security analysis, monitoring Denial of Service (DoS) attacks, and many more. Till the present day, the NetFlow tool continues to be a popular choice for analyzing and measuring network traffic flows.

Considering the diverse number of devices and the disparate addresses pertaining to each device, the network traffic data are usually vulnerable to provide poor classification while working with traditional classification algorithms. In this view, we discuss strategies for identifying IoT devices using different state-of-the-art approaches. We provide a descriptive framework for network traffic analysis in IoT based environments by exploring the dynamics of IoT networks for analyzing traffic flow between each connected device. It gives the network administrators and network engineers more control over network activities and traffic flow between the networked devices for maintaining an account of increasingly growing IoT devices. The recent day dense IoT networks are vulnerable to several challenges like network congestion, device authorization issues, data theft, etc. This requires proactive characterization for identifying the streaming data originating from IoT devices and detecting any security breaches in the network for efficient and secure functioning of the network. Thus, towards this end, we exploit the capabilities of different network analysis approaches for characterizing information flow between different IoT devices. The primary motive is to obtain a high-level knowledge of the amount of data streaming between different devices and to identify the authenticity of the communicating devices and their device type. This would provide a robust framework for many real-life scenarios like defense agencies and government organizations for the efficient administration of IoT devices. The compliance of the framework discussed in this article is shown for real network traffic data originating from different IoT devices. The empirical results corresponding to real-world sober network traffic flow environment have been presented. An insight on some more recent and evolving technologies which can be used in convergence with the conventional models for further escalating the performance of IoT systems is provided. Some important real-world application areas of the network traffic characterization are addressed. We also focus on some of the grueling challenges encountered with existing frameworks and suggest appropriate countermeasures for overcoming them.

1.1 Motivations

The large-scale growth in the networking devices and IoT applications has induced the exchange of heavy network traffic between the devices. This has imposed an enormous load on the computing systems and network management resources [2]. Thus, prompt strategic considerations are required for monitoring and characterizing the flow type and network devices involved in these wide-spread environments. Towards this end, we discuss the applications of different statistical tools and learning methodologies that can provide proactive measures for efficiently characterizing the network traffic flow. This can also assist in administering the associated network applications, thereby providing secure and delay-free communication. The network flow entails a high-level characterization of different networking devices, mostly communicating through internet connections. However, do not carry any information regarding the actual data being transmitted. Hence, capturing and evaluating the network flow is more vital for understanding the logical dynamics of the network, which may lead to a better interpretation of any violations or misconfigurations in the system for network controllers. This would be beneficial for most real-life applications like the military, business, healthcare, and government sectors where the reliability and security of communication networks are highly inevitable. The approaches discussed in this study can significantly mitigate the latencies associated with these traffic intense communication networks. Apart from this, statistical methods can also curb most of the limiting factors associated with existing techniques commonly caused due to low sample size and intricate features.

The organization of the chapter is as follows: Sect. 2 provides a detailed discussion of some vital studies made in the direction of network traffic analysis and monitoring. In Sect. 3, the IoT based network flow monitoring framework is discussed. Section 4 deals with the application areas of network traffic characterization. The empirical results obtained for a real-world network traffic scenario is provided in Sect. 5. In Sect. 6, the strengths and limitations of the discussed network traffic monitoring strategies are provided. Section 7 provides the future research scope and possible improvements that can be used for addressing some of the limitations of existing approaches. Finally, in Sect. 8, the conclusive remarks are provided.

2 Background Study

In this section, different network analysis tools, and techniques for characterizing the network traffic are presented. A comprehensive account of some relevant studies using machine learning approaches, statistical methods and software defined network (SDN) based approaches for capturing the network traffic generated from IoT based devices and applications are discussed. Figure 1 provides a generalized high-level architecture for the network traffic flow mechanism between different devices.

2.1 Tools for Network Traffic Monitoring

IoT has found enormous applications in diverse fields including healthcare, industrial automation, smart homes, smart cities, agricultural industries, as well as government organizations [23–28]. The IoT devices used in these diverse environments are usually low powered and IP-connected. These devices often suffer from security and design challenges making them vulnerable to privacy and security breaches. In [10], a novel technique for managing and controlling IP-enabled IoT devices was proposed. The data from a total of 27 different IoT devices were collected over a Kali Linux system using OpenWRT and Open VSwitch (OVS) for redirecting the traffic flow between the client devices and fingerprinting the connected devices respectively. A K-fold cross validation scheme was used for evaluating the data captured from all the devices using the random forest (RF) classifier. Further, the proposed device fingerprinting scheme was validated against different performance parameters like network latency, CPU utilization, and so on. In [29], a distributed network traffic analysis scheme was proposed. The study used two network traffic acquisition and analysis tools namely, TOPAS and Wireshark. Here, a real-time network analysis framework was proposed which was used for monitoring and configuring the devices responsible for transmitting data packets in a network. A network centric strategy for privacy preservation of IoT devices against different adversarial and vulnerability attacks was provided [30]. A device monitoring framework acronymed as "Honeyscope" was proposed which provides a fine grained control in administering communication of IoT devices in a network. This architecture uses a virtual deception scheme which is also referred to as honeypot to identify potential attackers in the network.

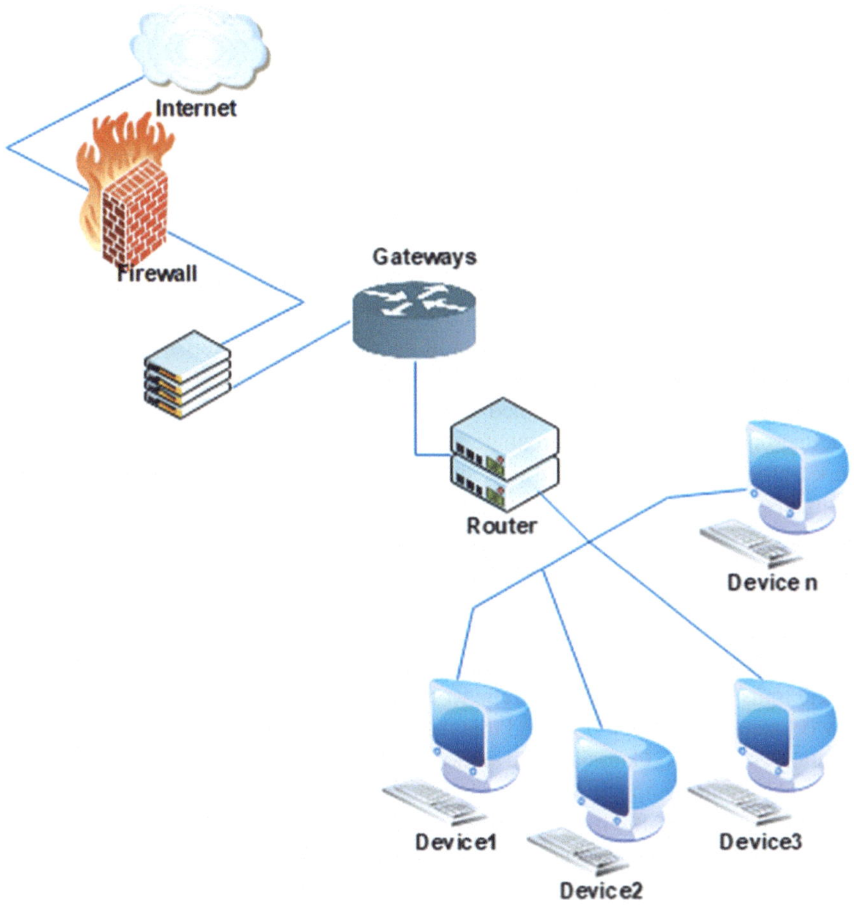

Fig. 1 An overview of the network traffic flow between various devices

Network traffic monitoring tools can also benefit network administrators in controlling and identifying devices that tend to misbehave. In [31], a strategy for identifying suspicious traffic flows between IoT devices was proposed which was acronymed as "IoTGuard". This framework uses a semi-supervised fuzzy C-mean learning algorithm for identifying malicious and benign devices. The model was validated by considering 39 features captured from IoT network logs using OpenWRT tool. The algorithm used in this study was further used to obtain clusters from the supplied featureset for benign and malicious behavior of devices. A collaborative, seamless and adaptive sentinel for IoT acronymed as COSMOS was proposed in [34]. Considering the challenges faced by existing network infrastructure, the work provides a sentinel shield based approach for protesting the IoT devices against attacks. The framework has been implemented over Raspberry Pi, and has been experimentally proven to achieve better performance even under heavy network traffic conditions.

2.2 Statistical Models for Network Traffic Characterization

In order to administer the massively growing IoT networks, several statistical as well as predictive models have been employed. Most of these models are concerned with the identification of the devices connected to these networks. An integral part of network traffic classification is to identify the network services and manage communication flow between devices. However, network traffic characteristics are still crucial to many applications as they may largely influence the QoS parameters of communication channels. In [35], the issues for automatic network anomaly detection were addressed. This work modeled the network traffic flow parameters as a finite Gaussian mixture model. The variations in the network characteristics in context to the presence of anomalies was studied for local area networks (LANs). Considering the network parameters obtained using the proposed mixture model, the normal baseline operations were differentiated from the malicious activities from the network traffic data. Further, a real-time online algorithm was employed for detecting anomalies. This work provided low false alarm rates and timely detection of anomalies. An entropy based detection of network information was proposed in [36], this study considered network features corresponding to short-term network statistics. The chances of compromises in the network induced due to anomalies are localized using adaptive Wiener filtering model and auto-regressive moving average (ARMA) approach. This allows network administrators to easily capture and analyze the statistical traits of network anomaly. The work has been validated against the real-time network data. It has been observed that the network features follow probability distributions similar to Gaussian distribution. In [37], a theoretical hybrid framework using telescoping graph approach with noncentral t-distribution was used for modeling the network traffic data. It was observed that the noncentral t-distribution can capture heavy tails and skewness in the traffic flow data much accurately. This leads to reduction of false alarm rates, as the probability of network anomalies can be better accommodated by considering heavy tails. Numerous studies have focused on the role of family of probability distributions with heavy tail behavior for IoT as well as wireless communication channels [38–40]. The prime goal is to entail any chances of compromise in the network or identifying network intrusions. A statistical framework for anomaly detection in cellular networks was proposed in [41]. The feature distributions corresponding to data collected for 3G cellular networks was derived. This work used a change detection algorithm for analyzing each distribution for the presence of anomalies determined by observing the deviations from the empirical distributions. The relative entropy measure was used for determining the deviations between the distributions for identifying anomalies.

2.3 Machine Learning Models for Network Traffic Classification

The real-time characterization of network traffic for large-scale communication networks is a vital research area which can significantly enhance the network security management strategies, the network size, and QoS parameters. Several conventional traffic classification tools have considered port numbers of the connected devices and packet payloads for addressing some of the above issues. However, with the increase in IoT devices and complex encryption schemes, these techniques have suffered some disadvantages. In [9], a hybrid architecture using machine learning algorithms, network classification tools, and heuristic based co-clustering was proposed. It involves the identification of port numbers and inspection of the packet payloads transmitted between the devices. This technique violates the users' privacy constraints and increases the computational workload. A machine learning based mobile application fingerprinting strategy for IoT environments was suggested in [11]. The study provided strategies for identifying device type and ambiguous traffic flow across different devices. The framework achieved an accuracy of approximately 96% in precisely classifying a collection of 110 applications. In [12], a supervised machine learning framework for classification of network traffic was proposed. Here, Wire Shark tool was used for capturing the network flow traffic originating from different remote devices. Four supervised learning algorithms viz., Naïve Bayes (NB), BayesNet, Support Vector Machine (SVM), and C4.5 based decision tree were used. The C4.5 algorithm was experimentally observed to provide highest classification accuracy of 78.91%.

IoT based devices mostly rely on application programs at the back-end for furnishing a control mechanism to the IoT devices. These applications are typically used for obtaining sensory data and other notifications regarding the sensed events. With the popularity of IoT, software applications associated with them have also considerably increased. This imposes certain challenges for the identification of applications and traces of traffic flow from these applications. In [13], a machine learning approach was suggested for identifying malicious network activities and applications. It was experimentally observed that most of the traffic generated from malicious applications was benign whereas only few were categorized as malicious. This leads to the class imbalance issue, which was addressed by using the synthetic minority oversampling technique (SMOTE). The balanced dataset was tested using SVM classification technique, cost sensitive SVM and cost sensitive C4.5 algorithm. In [14], a high precision network traffic classification framework was developed. The network traffic generated from mobile applications was considered and was mirrored over the access point to server where the analysis was performed. The C4.5 supervised machine learning technique was used which provided a detection accuracy of 97.89% in identifying malicious network activities. In [15], an optimal strategy for the selection of feature sets from network flow traces was provided. The model uses the information gain ratio to estimate imbalance in the feature set. This scheme can increase the robustness of network traffic classification methods by handling unstable features. A real-time classification scheme for multimedia traffic was proposed in [16]. The work addressed some of the issues associated with manually capturing the network traffic data which often results in noisy and mislabeled datasets. An unsupervised feature selection and instance purification scheme was introduced for classification of the flow fragment corresponding to different features. The framework was validated using six well known UCI machine learning repository datasets.

2.4 SDN Based Network Traffic Classification

With the increase in demand for advanced IoT based information acquisition and dissemination techniques, the networked IoT devices have witnessed challenges in efficiently integrating these services and monitoring information flow across large networks. To this end, the software defined networks (SDNs) have been proven to provide a centralized control over the networked devices and for dynamically monitoring the network traffic flow between these devices [17]. In [4], the authors provided the convergence of deep learning models viz., recurrent neural networks (RNN) and convolutional neural networks (CNN), for classifying network traffic flow between different Internet devices. Their proposed framework was further compared with different deep learning models to identify frequently occurring features in the network traffic. This framework achieved an overall accuracy of 96.32%. To overcome some of the limitations in wireless sensor network (WSN) frameworks, the SDNs were introduced as a paradigm for achieving more tractability and reliability in managing the network traffic and flow statistics. However, due to some of the intrinsic challenges associated with SDNs, the WSNs suffer several major drawbacks [6]. In this perspective, a close scrutiny into some of the existing network traffic classification techniques was provided in [7]. Further, the limitations posed by machine learning techniques for network traffic classification was provided and some possible emerging techniques for efficient classification of these traffic was presented. In [8], some exploratory techniques for identification of IoT devices were discussed. Their proposed model used knowledge inferred from the servers regarding IP addresses and DNS names corresponding to different IoT devices. These techniques can preserve the safety of the network against different adversaries.

In [18], a comprehensive study of SDNs for handling traffic based anomalies and unauthorized activities on-demand at the data centers was provided. In order to provide a better utilization of network resources and to achieve high quality of service (QoS) a network traffic engineering framework was proposed in [19]. Initially a reference framework based on SDNs was proposed which focused on traffic intensity monitoring and measurement. This facilitates the real-time acquisition of network traffic, granular traffic scheduling and controlling. The framework is highly influenced by the packet forwarding mechanism in communication networks for improving capabilities of traditional network applications. In [20], a machine learning model was used for detection of distributed denial of service (DDoS) attacks. Here, SDNs served as the central controlling mechanism for the network to monitor the threat vectors. A multiclass SVM classifier was used for classifying attacks in the network traffic flow. Six different learning classifiers viz., random forest (RF), SVM, radial basis function (RBF) classifier, NB, C4.5, and Bagging were used, out of which SVM provided an accuracy of 95.11% in predicting the attacks. A controller clustering based framework for multi-controller SDN networks was provided in [21], which was used for handling DDoS attacks in these networks. This framework is organized into three phases where the first phase is responsible for identifying overloaded controllers, the second phase elects the best controller for initiating the DDoS control process, and finally the third phase is responsible for minimizing the effects of attack by using operational controllers. This scheme was experimentally claimed to reduce 52.39% CPU usage as compared to some benchmark traffic flow based DDoS attack mitigation strategies. In [22], the authors provided a strategy to estimate network traffic matrix from traffic traces for end-to-end traffic flow. The authors used fractal interpolation scheme for reconstructing granular network traffic for SDN applications. Further, the weighted geometric average method was applied to the model for improving the reconstruction accuracy.

Considering the above studies, a descriptive framework for the characterization of network traffic flows for IoT based environments has been provided in this article. This framework can be useful for attaining an extensive analysis of the network traffic generated from large-scale IoT based systems like industries and government agencies. It is observed that the use of statistical models have largely benefitted in characterizing the network adversaries as well as any possibility of compromises arising in the network due to irregular flows. Apart from this, several machine learning approaches as well as SDN based network monitoring approaches have also been employed to derive a precise knowledge of the chances of compromise, or for localizing the anomalies in the communication networks. Although most of these models are effective till date and usually rely on extensive network statistics for their analysis, however some of these models entail some limitations due to the lack of appropriate real-time network traffic datasets. Further, it is also evident that most conventional models fail to capture the large number of variations that occur in densely populated networks due to the large correlation between the perceived data, thus the use of statistical models would prove to be a robust approach for such situations. In this study, we provide strategies for capturing and monitoring heavy network flow data. We also provide some empirical results corresponding to real-world scenarios for illustrating the effects of network anomalies leading to compromises in the communication networks.

3 Network Flow Monitoring and Analysis Framework

The growth in the popularity of networked devices providing information exchange over large-scale communication channels have induced the requirements for network monitoring and device profiling techniques. IoT has served as a central paradigm for facilitating communication over long-haul links extending information flow of large network boundaries [3, 10, 32]. These communication may consist of traffic flows originating from different remotely located sensory devices. Hence, the characterization of these flows is crucial to understand the intrinsic properties and composition of data packets. An extensive analysis of the network flows can also provide network engineers knowledge regarding the devices connected to the network and individual devices responsible for forwarding the packets. The QoS of the network infrastructure is highly dependent on the connected devices and applications associated with them [26, 39]. Network measurement strategies are important to identify any anomalous behavior giving rise to network compromise. These behavior causing compromises in communication networks may include unusual, or voluminous traffic flow between devices, packet routing issues, DDoS attacks, and so on. Apart from this some other dominant issues for which network measurement strategies may be inevitable are providing on-demand services to meet the users' requirements, for deciding pricing constraints, for endorsing and configuring new network models, and many more. This also allows the service providers to keep a track of their customers' behavior considering their network activities. In context to these factors, we discuss two important active network flow analysis and monitoring strategies which are important for capturing traffic in real-time IoT based environments.

3.1 Packet Level Analysis

IoT based transmission networks may constitute of several data packets accumulated from multiple IoT devices. These data may entail significant amount of users' information like health status, location, biometrics, and other sensitive information, hence making them vulnerable to malicious activities and privacy breaches [17, 24]. Therefore, the individual packets being transmitted through these networks require some thorough measurement strategies to identify the points, or devices across which the packets travel. This requires fine grained knowledge of the underlying network infrastructure as well as the demographical characteristics of the connected devices like their IP addresses, duration of transmission, the ISPs connected to them, port numbers, and so on. Several popular tools like Ethereal, WireShark, TCPdump, Network Packet Monitor (NPM), WinDump, etc., have been widely used as packet sniffing tools to capture the network traffic [54, 55]. These tools are eminent for network analysts and network engineers to assess the networks' performance against anomalies and identify the underlying issues. These tools mostly depend on APIs for capturing the network traffic corresponding to different operating system platforms. The Unix based tools leverage the `pcap` libraries, whereas the Windows based tools provide `libpcap` libraries. Figure 2 provides a generalized network traffic monitoring architecture for IoT environments. The architecture constitutes of IoT devices from which the data packets are generated across different platforms. The traffic is channelized through the IoT gateways to measurement routers, from where the data packets are captured by the packet sniffing tools and are stored over a database. These

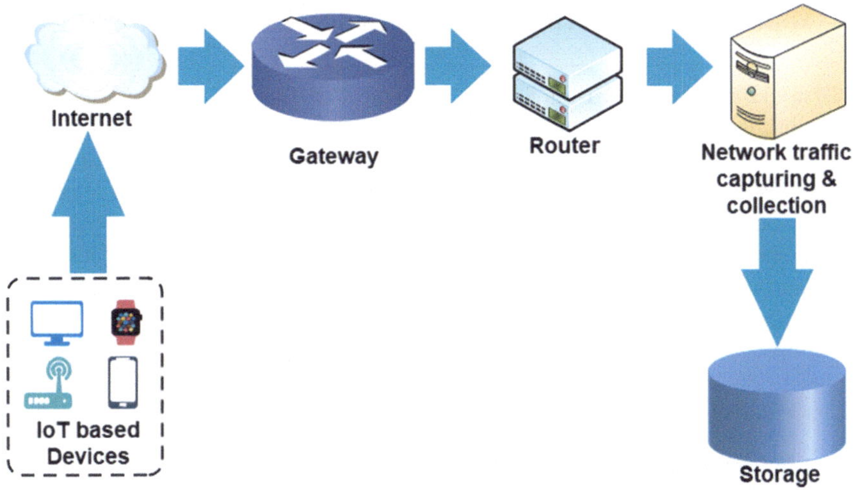

Fig. 2 The network traffic flow capturing and collection mechanism for IoT based devices

databases may either be located on a local device, or may be remotely located which can be used by network administrators for traffic analysis.

3.2 Flow Level Analysis

The analysis of the traffic flow between the IoT devices provides a high-level view of the network dynamics [50, 55]. In this approach, the individual data packets generated from different sources are concentrated into composite traffic flows. Here, it would be vital to consider aggregation schemes at the network traffic monitoring end to capture the flows across different network models. The network traffic flow data entails flow based attributes like the duration of flow, inter-arrival rates of individual flows, bitrates, and so on. Some popular tools used to capture network traffic flow include Cisco's NetFlow, J-Flow, sFlow, and so on [54]. The traffic flow capture is similar to that illustrated in Fig. 2, except that the network monitoring tool is only concerned with the analysis of the traffic flow rather than the data packets. At this point, it would be vital to mention that device identification is highly influenced by traffic monitoring strategies, as in order to profile the devices connected in a network one needs to characterize the traffic generated by these devices after which an efficient classification of the devices can be made. In the flow level analysis, the flow information captured using flow analysis tools are stored into the database for future reference. Unlike in packet level analysis, the flow analysis tools need not always require to be connected to the routers as it is independent of traffic measurements.

4 Applications of Network Traffic Characterization

IoT based traffic characterization has become quite challenging due to the advancements in communication protocols like Zigbee, Z-wave, LoRaWAN technology, and so forth, which have provided increased network capabilities. It provides an extensive knowledge regarding the composition and dynamics of network traffic. Network traffic characterization for IoT environments extends the capabilities of bandwidth management, regulate network capacity, ensure the secure delivery of data packets, improve network security, and ensure QoS to the clients, or end-users. Below we discuss some applications of network traffic characterization in context to IoT networks.

4.1 Information Flow Monitoring

The devices involved in an IoT network are heterogeneously scattered over different remote locations. Hence, there is little knowledge of the devices from which the traffic is generated and the remotely located ISPs. Therefore, it is essential to know the information flow between the devices for estimating network flow capacity and identifying malicious traffic flow. Considering a specific network, network traffic data statistics can be used to identify IoT devices involved in the network. The inter-arrival time between the incoming traffic flow, transmission rate, flow size, etc., can facilitate the network administrators to identify malicious activities within the network. Furthermore, network analysis tools provide the administrators an account of IP addresses corresponding to the recipient and the source device from which the data was originally generated.

IoT enabled environments require real-time network analysis as these devices are mostly used for performing active sensing activities. This requires the traffic monitoring tools to continuously capture dynamic changes in the network traffic flow. Mostly these real-time network monitoring tools are used for assessing the fault tolerance of the network, identifying vulnerabilities, and for conducting network performance tests. At this end different statistical and machine learning approaches can be used for characterizing the traffic flow dynamics. For improving the performance of the network both real-time as well as historical data can be analyzed. However, most IoT networks require a real-time analysis to quickly probe into a network issue.

4.2 Efficient Bandwidth Utilization

It is essential for the network administrators to track the usage of network bandwidth, so that the users have optimal experience. Attackers may illicitly consume your network bandwidth, resulting in outages while performing crucial operations. For instance in IoT based business organizations high bandwidth services are required to facilitate the transmission of massive business data. In this case, bandwidth hogs may lead to the failure of crucial processes. Hence, the network administrator needs to have an in-depth knowledge of the users utilizing the bandwidth, their authorization details, and the time for which they use the bandwidth. Packet sniffing tools can prove to be a good choice for this scenario as they may trace into wireless access points and client side ports to accurately footprint the utilization of bandwidth. These tools also resolve some of the routing problems to facilitate secure delivery of data packets to their respective destination.

The efficient management and planning of network traffic can also lead to the better utilization of bandwidth. Route planning can also be implemented by considering some QoS aware routing protocols [43], to make optimal usage of the bandwidth. This prevents the network from being overwhelmed with traffic flows and facilitates quick transmission of data packets for serving most real-time applications.

4.3 Device and Application Identification

Most IoT applications encompass low-cost sensory devices for carrying out the sensing operations. These devices are mostly IP based devices and are usually overlooked in terms of the security and implementation aspects. The weak infrastructure and lack of firmware updates may make such devices prone to vulnerabilities such as firmware attacks, or DDoS attacks. Further, the presence of such devices in the network may also induce potential vulnerabilities in the normal functioning of the network. Hence, such devices need to be identified and isolated to prevent the network and it's connected devices from threats. In large-scale IoT networks these issues are crucial for enforcing regulatory measures over the vulnerable devices and mitigating the chances of network compromise.

Several incidence of hacking smart devices and applications have been witnessed in recent times. As an exploratory strategy Avast hacked a smart coffee maker and also turned it into a ransomeware for using it as a gateway to reach all the Wi-Fi connected smart home devices resulting in hacking the entire home network [42]. This proves how a single flaw could result in affecting an entire range of devices connected to the network.

4.4 Monitoring Network Performance

The present day IoT based data networks are drastically distinct from telephone dial-up networks, as these networks posses high network capacity. As far as the network performance is concerned, these networks require high network traffic control capabilities to manage the continuous flow of traffic across heterogeneously located devices. Different revolutionary networking approaches have been proposed for enhancing the performance of these networks and for minimizing the network loads [35, 38, 39]. The ubiquity of IoT devices further adds to the complexity of traffic features which are essential to be analyzed in order to achieve better understanding of multifaceted features pertaining to the traffic's dynamic behavior. By capturing the actual behavior of the network traffic, a precise prediction of the network's performance can be achieved. Network monitoring tools like NetFlow, J-Flow provide the users some essential network statistics which can be utilized for analyzing the performance of the network.

In [44, 45], network performance techniques using synthetic traffic injection were suggested. All the flows constituting the traffic were identified prior to the experiment. The traffic traces were then verified for possible vulnerabilities using different inference rules. The network performance assessment is crucial to many real-world applications including assessing the network traffic for identifying usage of applications, detecting network anomalies, identify the network's performance against realistic workloads, and so on. These applications are critical to both the network operator as well as the users since the security and integrity of the data packets is involved. Therefore, in order to assess the quality of a network, different parameters are to be considered like choice of a proper reference baseline network, trustable network monitoring tools, and so forth.

4.5 Addressing Security Aspects

In IoT environments the ubiquity of IoT devices leads to the generation of a voluminous amount of outbound network traffic, which could make them susceptible to attackers. Considering the flaws in the firmware of IoT devices and lack of implementation benchmarks, the attackers may dig into the network and may use other IoT devices over the network for generating malicious traffic to the network. These issues may last to the severity of causing failures in the network due to overload if not handled timely. Hence, an accurate and complete monitoring framework is required which can segregate and identify such flaws to prevent complete outages.

The most obvious solution to this is using complete protocol parsing techniques. In [46], a Markovian process based network protocol parsing technique was proposed. However, these techniques have several disadvantages like these protocols may prevent packet sniffing tools from working, additional computational complexities may be introduced for parsing the network protocols for individual users connected to a network, and so on. These techniques may provide promising results when used with less computationally intensive tools.

4.6 *Role of Network Traffic Analysis for Cyber-Physical Systems*

In order to facilitate uninterrupted connectivity among physical world devices and other IoT based computational objects, it is essential to provide a secure communication infrastructure between these devices and the networking processes. The cyber-physical system (CPS), is one such collaborative framework which combines interconnected computing devices as well as the physical world entities. Some popular examples of CPS include robot assisted surgery, autonomous vehicles and so on. A CPS leverages pervasiveness in handling autonomous processes, like those in IoT based industries, to facilitate an intelligent control mechanism for smart environments.

The CPSs are mostly dependent on Internet technology for facilitating exchange of information and control signals between the connected objects. This may make the system susceptible to malicious attacks arising from the propagation of malwares. The severity of such attacks may span from delay in transmission to major system failures. The transmission delays are mostly caused due to flooding of network bandwidths caused as a consequence of flooding attacks in such networks. Further, some attacks may also be imposed from the physical environment. Hence, this makes CPS more prone to attacks which in turn may intercept the appropriate functioning of CPSs. Several attack control mechanisms have been implemented for CPSs, which mitigate the chances of attacks over such systems. This prevents complete or partial failure of the components associated with these systems. Towards this end, network traffic analysis plays a crucial role in identifying the propagation of malware and malicious traffic flows in the network.

5 Empirical Results and Discussions

In this section, the empirical results corresponding to a real-world network traffic data are provided. The data has been acquired from publicly available data bases. The data was collected over a duration of three months for 10 local workstations. Almost half of the data was subject to some sort of network compromise arising due to botnet attacks. Figure 3 provides the count for remote ISPs connected to local IP based workstations. They are allotted device IDs ranging from 0 to 9 accounting to a total of 10 IP based workstations with heavy traffic flow data. These workstations further have different computing devices connected to them, resulting in heavy traffic flows, which perform computations and exchanging data packets. It is essential for most large-scale IoT based organizations to keep an account of the number of ISP connections to ensure efficient connectivity throughout the transmission session. In this view ISP monitoring tools can prove to be vital for achieving an elementary idea regarding network traffic. This is also essential for managing the router configurations remotely for implementing customized network policies. The more number of

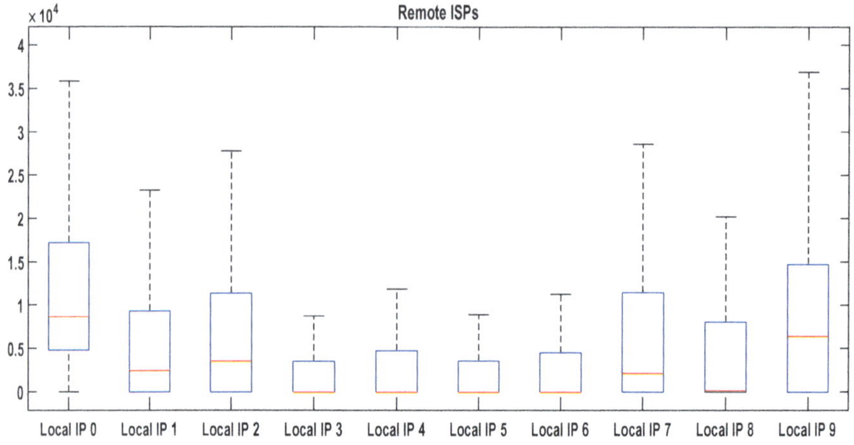

Fig. 3 Different remote ISP counts corresponding to 10 local IP based devices

ISPs indicate greater transmission speed and volume of data being transmitted. With the transformation of traditional traffic infrastructure, to highly composite network model the requirement for more number of ISP connections is becoming inevitable to facilitate transmission of greater traffic payloads across heterogeneously scattered clients [47]. Many network performance monitoring applications require to account for ISP traffic links and the ISP connections across the network to asses network performance metrics like latency, secure delivery of data packets, service outages, and so on. Table 1 provides a summary of 10 IP based workstations along with the connected devices and the minimum and maximum number of connections associated with each workstation represented as "Min. Connections" and "Max. Connections". In Table 2,

Table 1 An example of connection counts along with the minimum and maximum connections between different devices

Device IDs	No. of connections	Min. connections	fMax. connections
0	105177	1	5059
1	195691	1	4718
2	195713	1	5214
3	4904	1	313
4	1175417	1	784234
5	5780	1	322
6	6674	1	530
7	35362	1	1057
8	120203	1	7902
9	108777	1	1027

Table 2 An example of remote ISP counts along with the minimum and maximum remote ISPs between different devices

Device IDs	Remote ISP count	Min. ISP count	Max. ISP count
0	47323625	4	39834
1	24244232	3	40028
2	31618621	3	39484
3	13016730	3	40028
4	19041250	3	39484
5	12954638	4	40028
6	14748130	3	40028
7	28033353	3	36856
8	24001599	3	40092
9	36472764	3	36856

the total number of remote ISPs associated with each workstation are provided along with the minimum remote ISP count represented as "Min. ISP count" and maximum ISP count represented as "Max. ISP count". In Fig. 4, the device connection counts for different IP flows (between 0 and 9) are provided.

The traffic flow analysis is crucial for identifying attacks in a network with heavy traffic flow. This provides more convincing insights towards identification of devices in the network, which may further assist in identifying vulnerable devices. Figure 5 summarizes the traffic flow analysis performed corresponding to the number of devices considered in the dataset. The flows for each device represent the total number of data packet flows analyzed for botnet attack environment. The peak traffic flow values in the analyzed data indicate network compromises arising due to unusual traffic flow in the network. It can be observed that local IP devices 1, 7, and 9 realized much frequent peak values for the traffic flow. These abnormal flows are induced due to botnet attacks which is common to most IoT environments. It is observed that botnet attacks usually include DDoS attacks where the attacker may implement some malicious instructions for targeting the devices' IP address [48]. Further, the vulnerable device may generate a high amount of traffic causing network overload. The traffic flow behavior observed by our analysis could be efficiently used in convergence with many statistical and machine learning models for constraining the attack instances.

6 Strengths and Challenges

The recent state-of-the-art studies have considerably addressed some of the rising issues for large-scale network traffic analysis and device detection. This has assisted in influencing the network traffic flow and packet monitoring mechanisms

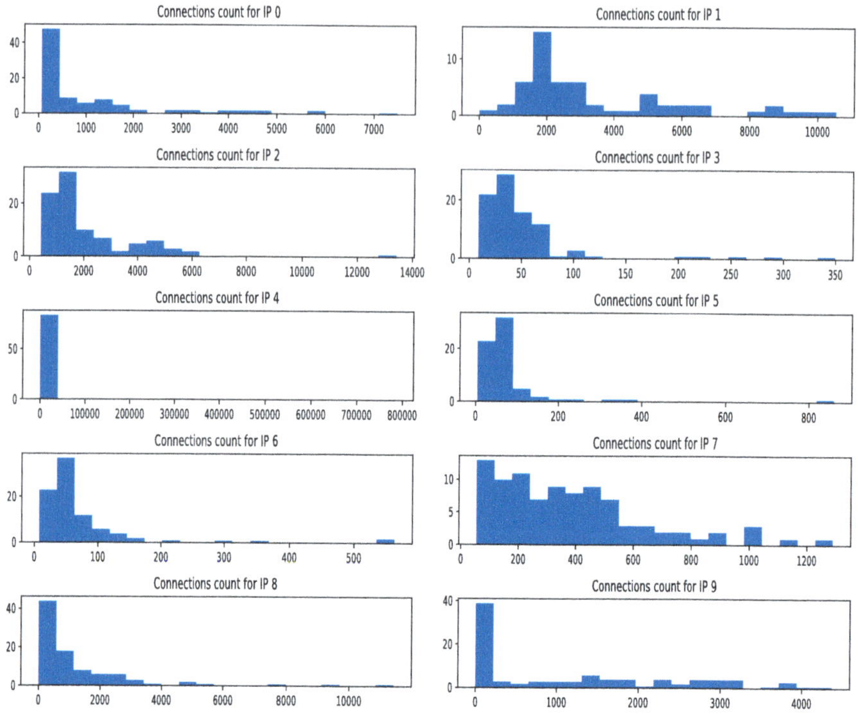

Fig. 4 Representation of different IP flows corresponding to different connection counts

immensely. Such mechanisms involve the capture, storage and analysis of network traffic flow and data packets for deducing inferential knowledge regarding the underlying network infrastructure. IoT based applications are highly heterogeneous and require real time traffic analysis schemes. Hence, traditional centralized traffic monitoring systems cannot efficiently capture malicious activities across these networks. Therefore, a highly dynamic, scalable, decisive and real-time monitoring technology is required. Strategies for network level identification of vulnerabilities in IoT environments using network traffic characteristics were addressed in [8, 49]. These traffic capturing mechanisms enhance the accountability, scalability, and performance of network management scheme. Further, in [50] a botnet detection framework for preventing DDoS attacks was presented.

As the IoT devices continue to grow and generate voluminous data, it becomes difficult for network administrators to keep track of the connected devices. Such instances pose threats for IoT devices, network assets and users leading to data theft, violation of network privacy policies, generation of malicious network traffic, and so forth. In [9, 31, 35, 49], different IoT device identification schemes were proposed considering the network flow statistics. These models mostly concentrated on statistical and machine learning based predictive models for categorizing and identifying the devices for a specific network. It was observed that these frameworks

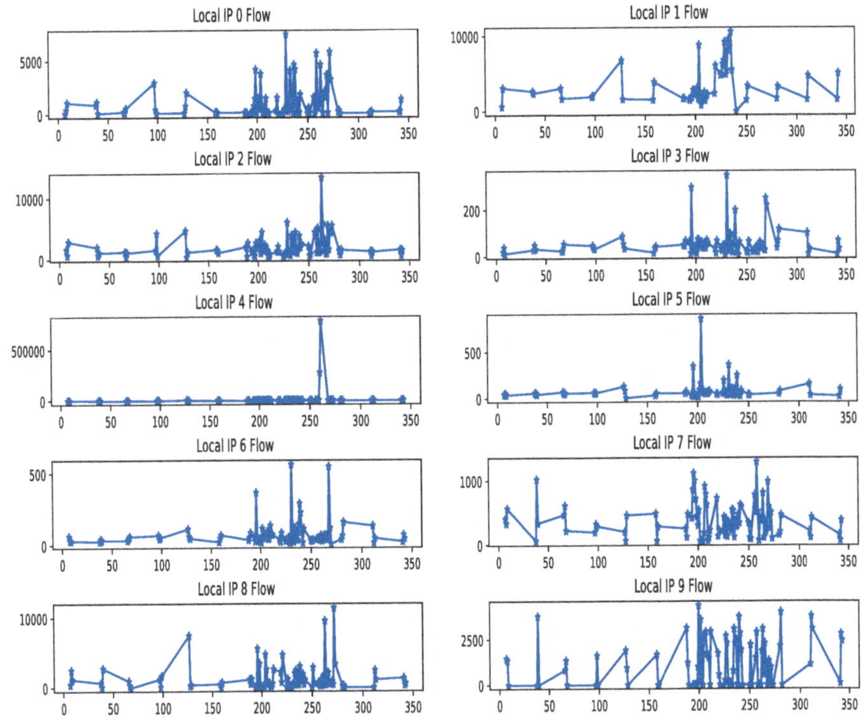

Fig. 5 Traffic flow for different local IP based devices

performed consistently for capturing the traffic demographics and precisely analyzing the underlying statistics. Further, different inferential rules were developed using these strategies for identification and isolation of vulnerable devices in heterogenous IoT environments.

Although these models can efficiently handle the current network monitoring and device identification requirements; however some more holistic approaches are required for meeting the future network perspectives. The constantly changing network infrastructures, and introduction of protocols and devices have posed several challenges on the present network monitoring frameworks. Some of these challenges include increased network workload, high computational demands, limited storage capabilities, scalable resources and analysis tools, contemporary application softwares, increased security demands, and so on. Furthermore, effective aggregation and filtration techniques are required to be implemented at network traffic acquisition level for specifying more accessible traffic statistics. In this scenario, machine learning models have proven to be a good solution, however with the lack of appropriate training featuresets the predictions provided by these models is imprecise. Also the selection and processing of individual traffic features may be time consuming and computationally intensive.

In dynamically growing IoT networks, the volume of data generated is usually associated with high storage costs. Hence, some more adaptive futuristic mechanisms are required for scaling down these costs by only considering the most relevant traffic features for analysis. This can also improve the analysis outcomes, since it is beyond the scope of traditional database systems to process such voluminous data. Most studies rely on packet level analysis as it provides a more logical insight for device identification. This technique however imposes additional CPU workloads and resource outages in highly constrained computing environments hence resulting in potential integrity deterioration of the captured packets. Further, user privacy issues also intercept in achieving a real-time and complete idea of the network traffic characteristics. Hence, most researchers confine their experimentations only to simulated, or synthesized traffic data which does not provide consistent and complete results in understanding actual characteristics of the underlying network.

7 Future Research Scope

Several researchers and standards developing organizations (SDOs) have suggested various solutions towards characterization and mitigation of network attacks. However, attackers unceasingly find new techniques for distorting the network infrastructure. A significant advancement in network traffic analysis involves deep packet inspection (DPI) technique for implementing extreme packet capture capabilities and network analytics. DPI tools usually employ sensors deployed at the Internet gateways or Wi-Fi access points for efficiently monitoring the network activities. Traditional network monitoring tools provide only a network level analysis considering the data packets and traffic flow between devices. However, DPI tools go much beyond them, providing application and port level analysis of the connected network devices. This is most suitable for IoT based environments like large business organizations, industrial application, smart cities, smart energy grid lines, etc., to facilitate network performance and security. NetFort (https://www.netfort.com/), is a popular DPI tool for large-scale commercial applications, which provides organizations an integrated platform to monitor user activities, analyze network performance and accountability, report network intrusions, and so on.

Some recent studies have suggested the use of signature based network traffic monitoring for characterizing botnet induced DDoS attacks in IoT environments [48, 51–53]. This technique provides more flexibility for traffic analysis by triggering customized actions for identification of network adversaries. Further, the use of machine learning approaches in convergence with statistical sampling techniques has proven to be vital in most scenarios for monitoring network traffic as well as for identifying IoT devices. Emergence of network monitoring and analysis tools like NetFlow, Wireshark, TCPdump, etc., have assisted significantly in deriving statistical inferences regarding the port addresses, device type (i.e., IP based, or non-IP based), traffic flow characteristics, network capacity, and so on. Considering these outcomes several futuristic models can be developed which can satisfy the real-time requirements of IoT environments.

8 Conclusion

The substantial growth observed in IoT has largely revolutionized the inception of several smart societies globally giving rise to new IoT enabled perspectives like smart industries, smart cities, smart governance and enterprises. However, due to the lack of traceability and visibility of IoT devices and underlying network infrastructure, network administrators face difficulty in analyzing the network characteristics and chances of network compromise arising due to vulnerable devices. Towards this end, a descriptive framework encompassing the different network monitoring tools and strategic measures for identifying vulnerable devices in heterogeneously scattered IoT networks was suggested. An account of some relevant studies focused towards network traffic monitoring and IoT device identification was provided. A comprehensive discussion on the rising applications for network traffic characterization, device identification, along with some of their substantial benefits was presented. The empirical results corresponding to a real-world network traffic dataset were provided. Some fundamental flow characteristics of these complex network traffic subject to botnet attacks have been reported. The traffic traces were collected over a span of three months for different IP based devices operating over remote ISPs. It was observed that these attacks resulted in network performance compromises by targeting vulnerable devices in the network. Later, the strengths and challenges of some state-of-the-art technologies were discussed. Finally, some relevant future scopes for improving the conventional network traffic data capture, analysis and storage capabilities were discussed. From the above observations and empirical results, an insight towards improving performance of current network monitoring and traffic analysis strategies can be conjectured. To this end, some proactive measures can be implemented for identifying and isolating the network vulnerabilities. This can also assist in satisfying some crucial network QoS constraints like scalability, reliability, timeliness, and traceability of resources.

References

1. Roux, J., et al.: Toward an intrusion detection approach for IoT based on radio communications profiling. In: 2017 13th European Dependable Computing Conference (EDCC). IEEE (2017)
2. Pammi, A.A.: Threats, countermeasures, and research trends for BLE-based IoT devices. Dissertation, Arizona State University (2017)
3. Miettinen, M., et al.: IoT sentinel demo: automated device-type identification for security enforcement in IoT. In: 2017 IEEE 37th International Conference on Distributed Computing Systems (ICDCS). IEEE (2017)

4. Lopez-Martin, M., et al.: Network traffic classifier with convolutional and recurrent neural networks for Internet of Things. IEEE Access 5, 18042-18050
5. Cisco, I.O.S.: NetFlow, Introduction to Cisco IOS NetFlow-a technical overview, May 2012 (2007), http://www.cisco.com/c/en/us/products/collateral/ios-nx-os-software/iosnetflow, 30 Apr 2014
6. Nguyen, T.M.C., Hoang, D.B., Chaczko, Z.: Can SDN technology be transported to software-defined WSN/IoT? In: 2016 IEEE International Conference on Internet of Things (iThings) and IEEE Green Computing and Communications (GreenCom) and IEEE Cyber, Physical and Social Computing (CPSCom) and IEEE Smart Data (SmartData), pp. 234-239. IEEE (2016)
7. Thupae, R., Isong, B., Gasela, N., Abu-Mahfouz, A.M.: Machine learning techniques for traffic identification and classifiacation in SDWSN: A survey. In: IECON 2018-44th Annual Conference of the IEEE Industrial Electronics Society, pp. 4645–4650. IEEE (2018)
8. Guo, H., Heidemann, J.: Detecting IoT devices in the Internet (extended). USC/ISI Technical Report ISI-TR-726, July 2018
9. Lu, W., Xue, L.: A heuristic-based co-clustering algorithm for the internet traffic classification. In: 2014 28th International Conference on Advanced Information Networking and Applications Workshops, pp. 49–54. IEEE, May 2014
10. Miettinen, M., Marchal, S., Hafeez, I., Asokan, N., Sadeghi, A.R., Tarkoma, S.: IoT SENTINEL: automated device-type identification for security enforcement in IoT. In: 2017 IEEE 37th International Conference on Distributed Computing Systems (ICDCS), pp. 2177–2184. IEEE, June 2017
11. Taylor, V.F., Spolaor, R., Conti, M., Martinovic, I.: Robust smartphone app identification via encrypted network traffic analysis. IEEE Trans. Inform. Forensics Secur **13**(1), 63–78 (2017)
12. Shafiq, M., Yu, X., Laghari, A.A., Yao, L., Karn, N.K., Abdessamia, F.: Network traffic classification techniques and comparative analysis using machine learning algorithms. In: 2016 2nd IEEE International Conference on Computer and Communications (ICCC), pp. 2451–2455. IEEE, October 2016
13. Chen, Z., Yan, Q., Han, H., Wang, S., Peng, L., Wang, L., Yang, B.: Machine learning based mobile malware detection using highly imbalanced network traffic. Inform. Sci. **433**, 346–364 (2018)
14. Wang, S., Chen, Z., Yan, Q., Yang, B., Peng, L., Jia, Z.: A mobile malware detection method using behavior features in network traffic. J. Netw. Comput. Appl. **133**, 15–25 (2019)
15. Liu, Z., Wang, R., Japkowicz, N., Cai, Y., Tang, D., Cai, X.: Mobile app traffic flow feature extraction and selection for improving classification robustness. J. Netw. Comput. Appl. **125**, 190–208 (2019)
16. Wu, Z., Dong, Y.N., Wei, H.L., Tian, W.: Consistency measure based simultaneous feature selection and instance purification for multimedia traffic classification. Comput. Netw. **107190**, (2020)
17. Bull, P., Austin, R., Popov, E., Sharma, M., Watson, R.: Flow based security for IoT devices using an SDN gateway. In: 2016 IEEE 4th International Conference on Future Internet of Things and Cloud (FiCloud), pp. 157–163. IEEE, August 2016
18. Satasiya, D.: Analysis of software defined network firewall (SDF). In: 2016 International Conference on Wireless Communications, Signal Processing and Networking (WiSPNET), pp. 228–231. IEEE, March 2016
19. Shu, Z., Wan, J., Lin, J., Wang, S., Li, D., Rho, S., Yang, C.: Traffic engineering in software-defined networking: measurement and management. IEEE Access **4**, 3246–3256 (2016)
20. Kokila, R.T., Selvi, S.T., Govindarajan, K.: DDoS detection and analysis in SDN-based environment using support vector machine classifier. In: 2014 Sixth International Conference on Advanced Computing (ICoAC), pp. 205–210. IEEE, December 2014
21. Macedo, R., de Castro, R., Santos, A., Ghamri-Doudane, Y., Nogueira, M.: Self-organized SDN controller cluster conformations against DDoS attacks effects. In: 2016 IEEE Global Communications Conference (GLOBECOM), pp. 1–6. IEEE, December 2016
22. Jiang, D., Huo, L., Li, Y.: Fine-granularity inference and estimations to network traffic for SDN. PloS One **13**(5) (2018)

23. Catarinucci, L., De Donno, D., Mainetti, L., Palano, L., Patrono, L., Stefanizzi, M.L., Tarricone, L.: An IoT-aware architecture for smart healthcare systems. IEEE Internet of Things J. **2**(6), 515–526 (2015)
24. Kaur, N., Sood, S.K.: Cognitive decision making in smart industry. Comput. Indus. **74**, 151–161 (2015)
25. Qi, R., Feng, C., Liu, Z., Mrad, N.: Blockchain-powered internet of things, e-governance and e-democracy. In: E-Democracy for Smart Cities, pp. 509–520. Springer, Singapore (2017)
26. Bebortta, S., Singh, A.K., Mohanty, S., Senapati, D.: Characterization of range for smart home sensors using Tsallis entropy framework. In: Advanced Computing and Intelligent Engineering, pp. 265–276. Springer, Singapore (2020)
27. Bebortta, S., Panda, M., Panda, S.: Classification of pathological disorders in children using random forest algorithm. In: 2020 International Conference on Emerging Trends in Information Technology and Engineering (ic-ETITE), pp. 1–6. IEEE, February 2020
28. Bebortta, S., Rajput, N.K., Pati, B., Senapati, D.: A real-time smart waste management based on cognitive IoT framework. In: Advances in Electrical and Computer Technologies, pp. 407–414. Springer, Singapore (2020)
29. Munz, G., Carle, G.: Distributed network analysis using TOPAS and wireshark. In: NOMS Workshops 2008-IEEE Network Operations and Management Symposium Workshops, pp. 161–164. IEEE, April 2008
30. Al-Shaer, E., Wei, J., Hamlen, K. W., Wang, C.: HONEYSCOPE: IoT device protection with deceptive network views. In: Autonomous Cyber Deception, pp. 167–181. Springer, Cham (2019)
31. Hafeez, I., Ding, A.Y., Antikainen, M., Tarkoma, S.: Real-Time IoT device activity detection in edge networks. In International Conference on Network and System Security, pp. 221–236. Springer, Cham (2018)
32. Meidan, Y., Bohadana, M., Shabtai, A., Guarnizo, J. D., Ochoa, M., Tippenhauer, N.O., Elovici, Y.: ProfilIoT: a machine learning approach for IoT device identification based on network traffic analysis. In: Proceedings of the symposium on applied computing, pp. 506–509, April 2017
33. Kim, M.S., Won, Y.J., Hong, J.W.: Characteristic analysis of internet traffic from the perspective of flows. Comput. Commun. **29**(10), 1639–1652 (2006)
34. Nespoli, P., Useche Pelaez, D., Daz Lpez, D., Gmez Mrmol, F.: COSMOS: collaborative, seamless and adaptive sentinel for the Internet of Things. Sensors **19**(7), 1492 (2019)
35. Hajji, H.: Statistical analysis of network traffic for adaptive faults detection. IEEE Trans Neural Netw. **16**(5), 1053–1063 (2005)
36. Celenk, M., Conley, T., Willis, J., Graham, J.: Predictive network anomaly detection and visualization. IEEE Trans. Inform. Forensics Secur. **5**(2), 288–299 (2010)
37. Djidjev, H., Sandine, G., Storlie, C., Vander Wiel, S.: Graph based statistical analysis of network traffic. In: Proceedings of the Ninth Workshop on Mining and Learning with Graphs, August 2011
38. Senapati, D.: Generation of cubic power-law for high frequency intra-day returns: maximum Tsallis entropy framework. Digital Signal Process. **48**, 276–284 (2016)
39. Bebortta, S., Senapati, D., Rajput, N.K., Singh, A.K., Rathi, V.K., Pandey, H.M., ... Tiwari, P.: Evidence of power-law behavior in cognitive IoT applications. Neural Comput. Appl. 1–13 (2020)
40. Mukherjee, T., Singh, A.K., Senapati, D.: Performance evaluation of wireless communication systems over Weibull/q-Lognormal shadowed fading using Tsallis entropy framework. Wirel. Person. Commun. **106**(2), 789–803 (2019)
41. D'Alconzo, A., Coluccia, A., Ricciato, F., Romirer-Maierhofer, P.: A distribution-based approach to anomaly detection and application to 3G mobile traffic. In: GLOBECOM 2009—2009 IEEE Global Telecommunications Conference, pp. 1–8. IEEE, November 2009
42. Hron, M.: The Internet of Thing: How a single coffee makers vulnerabilities symbolize a world of IoT risks, June 2019. https://blog.avast.com/avast-hacked-a-smart-coffee-maker
43. Chen, L., Heinzelman, W.B.: QoS-aware routing based on bandwidth estimation for mobile ad hoc networks. IEEE J. Sel. Areas Commun. **23**(3), 561–572 (2005)

44. Badr, M., Jerger, N.E.: SynFull: Synthetic traffic models capturing cache coherent behaviour. ACM SIGARCH Comput. Architect. News **42**(3), 109–120 (2014)
45. Yoshigoe, K., Dai, W., Abramson, M., Jacobs, A.: Overcoming invasion of privacy in smart home environment with synthetic packet injection. In: 2015 TRON Symposium (TRON-SHOW), pp. 1–7. IEEE, December 2015
46. Estevez-Tapiador, J.M., Garca-Teodoro, P., Daz-Verdejo, J.E:. Detection of web-based attacks through Markovian protocol parsing. In: 10th IEEE Symposium on Computers and Communications (ISCC'05), pp. 457–462. IEEE, June 2005
47. Wang, J.H., Chiu, D.M., Lui, J.C.: A gametheoretic analysis of the implications of overlay network traffic on ISP peering. Comput. Netw. **52**(15), 2961–2974 (2008)
48. Ceron, J.M., Steding-Jessen, K., Hoepers, C., Granville, L.Z., Margi, C.B.: Improving IoT Botnet investigation using an adaptive network layer. Sensors **19**(3), 727 (2019)
49. Sivanathan, A., Gharakheili, H.H., Loi, F., Radford, A., Wijenayake, C., Vishwanath, A., Sivaraman, V.: Classifying IoT devices in smart environments using network traffic characteristics. IEEE Trans. Mob. Comput. **18**(8), 1745–1759 (2018)
50. Franois, J., Wang, S., Engel, T. BotTrack: tracking botnets using NetFlow and PageRank. In: International Conference on Research in Networking, pp. 1–14. Springer, Berlin, Heidelberg, May 2011
51. Pour, M.S., Mangino, A., Friday, K., Rathbun, M., Bou-Harb, E., Iqbal, F., Ghani, N.: On data-driven curation, learning, and analysis for inferring evolving internet-of-Things (IoT) botnets in the wild. Comput. Secur. **91**, 101707 (2020)
52. Yousefi, S., Derakhshan, F., Karimipour, H.: Applications of big data analytics and machine learning in the Internet of Things. In: Handbook of Big Data Privacy, pp. 77–108. Springer, Cham (2020)
53. Tuan, T.A., Long, H.V., Kumar, R., Priyadarshini, I., Son, N.T.K.: Performance evaluation of Botnet DDoS attack detection using machine learning. Evol. Intell. **1–12** (2019)
54. DAlconzo, A., Drago, I., Morichetta, A., Mellia, M., Casas, P.: A survey on big data for network traffic monitoring and analysis. IEEE Trans. Netw. Serv. Manag. **16**(3), 800–813 (2019)
55. David, J., Thomas, C.: Efficient DDoS flood attack detection using dynamic thresholding on flow-based network traffic. Comput. Secur. **82**, 284–295 (2019)

Machine Learning for Fostering Security in Cyber-Physical Systems

Akash Dhiman, Kanishk Gupta, and Deepak Kumar Sharma

Abstract Cyber-Physical Systems (CPSs) are developed by the amalgamation that comprises computing elements in physical systems and establishing an intricate communication network alongside them. With the rapid advancements in the field, CPS is being employed in healthcare, grid control systems, autonomous vehicles, and much more. The fairly new technologies used and areas of its application make it a favorable target to be exploited. The outcome of leaving its vulnerabilities unchecked can be detrimental and henceforth demands efficient ways to establish security. A vast and ever-changing system like CPS demands a modern solution to tackle the problem of establishing security and that is what brings machine learning (ML) based methods into the picture. This chapter begins with an introduction towards ML and explains the motivation towards using ML-based methods to establish security. The chapter having made the reader familiar with the key terms of ML will then go on to talk about their applications in fostering security for CPS. This will be done in two phases of classification, where firstly we will discuss ML methods from the perspective of the security domain they aim to tackle, i.e. methods of thread monitoring, mitigation techniques, etc. Then we will talk about ML security based on the architecture of CPS and explain where various methods sit inside the whole architecture, consisting of the application, communication, and the physical layers. The chapter will follow these methods with real-world examples and promising research of them being used, while also discussing their effectiveness in doing the same. We then discuss parameters that help decide if ML is applicable for a specific use-case while also making recommendations regarding the model selection and

A. Dhiman · K. Gupta
Department of Computer Engineering, Netaji Subhas University of Technology (Formerly Netaji Subhas Institute of Technology), New Delhi, India
e-mail: akash.d0407@gmail.com

K. Gupta
e-mail: kanishkgupta2000@gmail.com

D. K. Sharma (✉)
Department of Information Technology, Netaji Subhas University of Technology (Formerly Netaji Subhas Institute of Technology), New Delhi, India
e-mail: dk.sharma1982@yahoo.com

A. I. Awad et al. (eds.), *Security in Cyber-Physical Systems*, Studies in Systems, Decision and Control 339, https://doi.org/10.1007/978-3-030-67361-1_4

training process. These sections have been designed to provide readers with a start to explore this domain and hopefully make their contributions in the future.

Keywords Cyber-Physical systems · Cyber security · Network security · Machine learning · Threat mitigation

1 Introduction

Cyber-Physical Systems (CPSs) are autonomous systems that control real-world physical processes using computers. This is done using a network of embedded computers, microcontrollers, and mainframe computers to establish a feedback loop in which physical processes influence computer systems, which in turn causes them to influence their surroundings [1]. Done properly they have a wide variety of applications in the current world, being utilized in fields of automated manufacturing industries, health care facilities, energy generation fields like grid control systems, etc. The key drivers for CPS include sensors that take measurements in various dimensions of the system's surrounding, a network of interconnected components which can communicate reliably amongst one another quickly as well as in real-time, various processing units that are programmed to achieve the goal of the system and the actuators that finally deliver appropriate responses to the surrounding. Ensuring the proper functioning of all these components and workflow becomes critical when designing a CPS.

While working with such a vast amalgamation of different technologies and the real-time nature of the system, it becomes difficult to simply employ conventional methods to security and often results in security compromises that were overlooked by humans or were novel for conventional security methods to easily bypass them.

This chapter will take a look at how Machine Learning (ML) plays an important role in ensuring the stable functioning and security of a CPS that may augment or even be better than the conventional methods. ML enables the system to have scalability and quick real-time response towards anomalies in the CPS and hence provides a better way to target the issue of establishing security. ML techniques are especially useful in the fields where the task is too general and vast to simply be done by a human or even by automated programs. Moreover, ML enables us to take advantage of continuously generated data by the CPS to provide us with real-time analysis for any kind of anomalous behavior, while also providing a rapid course of action in case of a discrepancy. Therefore, it becomes important for us to first understand some key terminologies and demands of ML, which in itself is a growing field used in many areas besides security like modern agriculture [2], medical industries [3], etc. This chapter will first discuss basics in regards to ML and then it will dive deeper into its application for security in CPS. The chapter's main aim and contribution are to stand as guidance for Cyber-Physical System design teams looking to integrate defenses to their systems, which are enabled via the power of Machine learning, by carefully written case studies, such as that of Vehicular ad-hoc networks, which is a real-time critical environment prone to disasters if it is left vulnerable to external intrusions, we aim to demonstrate the power of Machine learning and how it is currently being employed in such production environments and being heavily relied upon for safety protocols. No two CPS can be identical and hence the defenses employed need to be different, hence the chapter aims to tackle ML application in the field of CPS in the most generalized way as possible, helping Machine learning engineers understand how they could employ their skills into designing such a system. The chapter aims to categorize this domain in a two-pronged approach, the first being based on security type, which takes a blend of the cybersecurity domain and helps cybersecurity experts understand what are the different contexts of using machine learning, such as its application in direct security intrusion, or anomalous behavior tracking. The second approach is to study the generalized three-layer approach followed by most CPS, which helps readers build upon a strong example for them to realize how Machine learning could not only theoretically be used but is being practically implemented.

The chapter starts with an overview of the basics of Machine learning, understanding the metrics used to quantify the precision and accuracy of our predictions, which is followed by a brief introduction of standard Machine learning Algorithms, which are more importantly relevant to the applications of ML being discussed in the subsequent sections, hence providing with a basic foundation to understand the further discussion carefully. In the following section, we discuss the application of Machine learning Algorithms in the context of Security Types, such as direct security threats such as malware and intrusion, Predictive Analysis, and anomalous behavior, as well as Risk and Damage Assessment. This section is followed by understanding the application of Machine learning from a system design approach, where we discuss categorically how Machine learning defenses employed in Autonomous Vehicle at namely three layers which also form the basis for most CPS designs today, those being the application layer, network Layer, and the physical layer. While the chapter enthusiastically looks forward to the future of Machine learning in CPS defense, in our last sections, we focus on informing the users a clearer picture of the current challenges faced in this domain, to make sure that readers take an informed and realistic decision, this is followed by our last section which talks about certain guidelines we recommend to follow for CPS designers to keep in mind, when aiming to integrate ML-powered defenses into their CPS.

2 Machine Learning

Machine learning comes under the widely known umbrella term that is Artificial intelligence. It is a data analysis technique to perform predictive analysis on a given dataset. Essentially it is a technique that tries to model a mathematical correlation function between two correlated quantities, namely, features ($\mathbf{X}$) that are used as input to the model and the labels ($\mathbf{y}$) that are the correlated values one expects to find in the presence of those particular features. Notice here that $\mathbf{X}$ does not have to be single-valued, it can very well be and is most of the time a collection of carefully chosen n-dimensional features that allows us to produce a reasonably good model.

2.1 Overview

Modeling can be created by a variety of different methods, all under the domain of ML. In general, these include starting with a random approximation of the function to be modeled and iteratively improving upon its variables by comparing the output that it generates with the actual label associated with the data into consideration. This comparison is adequately captured by what is termed as a *loss function*, the value of which determines how the parameters of our models are adjusted, there are many different loss functions at our disposal allowing us to fit our model most accurately based on the dataset at hand. This type of ML technique is called *supervised learning*; examples of this include regression models, Support vector machines, Naive Bayes classifier, etc. Contrary to this we also have *unsupervised learning* algorithms where we model a mathematical function that tries to find natural correlations or classes or patterns within our dataset, examples of this include association and clustering algorithms. Another class of ML that is worth noting is of the *reinforcement learning algorithms* which works on the principles of game theory where we have an artificial agent trying to achieve a goal that yields the best reward out of a reward function, this function can be programmer-defined or defined using another ML model (Table 1).

Quantifying a model performance is just as critical as creating a model. It allows us to measure and compare different models rigorously and allows us to in turn get relevant information about the best correlation and features to look for in our dataset for predictive analysis. Hence the discipline of ML provides us with many ways

Table 1 Important Metrics

Metric	Formula	Use case
Accuracy	$\frac{TP+TN}{TP+TN+FP+FN}$	Net Performance
Precision (Positive Predictive Value, PPV)	$\frac{TP}{TP+FP}$	Measures accuracy of true positive prediction over classified positives
Recall (True Positive Rate, TPR)	$\frac{TP}{TP+FN}$	Measures accuracy of true positive prediction over real positives
Specificity (True Negative Rate, TNR)	$\frac{TN}{TN+FP}$	Like recall but for negatives
False Positive Rate (FPR)	$1 - TNR$	Measures amount of false prediction over real positives
F1 score	The harmonic mean of PPV, TPR	Measures false prediction accuracy

Table 2 Confusion matrix

	Actual Classes	
Predicted Classes	True Positive	False Positive (Error - I)
	False Negative (Error - II)	True Negative

of doing so. To understand deeper concepts, we need to grasp the concept of the *confusion matrix*. Simply put, it is a tabular representation of the relations between true values and the predicted values. It is illustrated in Table 2.

An ML model can generate predictions which when compared with actual cases give us a confusion matrix. The Quantities listed inside Table 2 can then be used to generate meaningful evaluation metrics shown in Table 1.

2.2 *Important Techniques*

At this point, it would be beneficial for us to talk about a few ML algorithms that are used throughout this chapter and which might help us in the design of CPS security.

2.2.1 Linear Regression

It is perhaps the simplest model we can discuss. As the name suggests this model is used to formulate a simple linear relation between output values and input values as can be seen in Fig. 1. Mathematically the dependency is shown in Eq. 1.

$$W \cdot X + b = Y \tag{1}$$

Here X is the feature matrix, Y is the label vector, and weights (W) and bias (b) are trainable factors. At the start of each training iteration, the trainable factors are initiated randomly, based on these random factors a hypothetical label prediction ($\hat{Y}$)

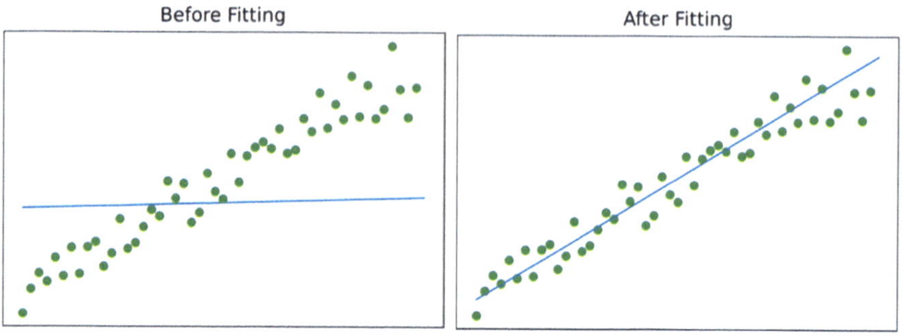

Fig. 1 Demonstrating how Linear Regression works

is calculated, this is compared with the original Y and a loss function is determined, an example of such a loss function is given in Eq. 2.

$$C(W, b) = \left(Y - \widehat{Y}\right)^2 \tag{2}$$

This cost function is then minimized in the process of training the model, following a specific technique of optimization. One method of optimization is called gradient descent in which weights are updated as given in Eq. 3.

$$W' = W - \eta \cdot \frac{\partial C}{\partial W} \tag{3}$$

2.2.2 Support Vector Machines

A Support vector machine (SVM) is primarily used when a programmer aims to deal with classification-based problems or for outlier detection. It works by finding a linear equation that represents a hyperplane able to separate different classes of labeled data, see Fig. 2. You may notice its resemblance with the previously discussed Linear Regression model, both of them try to model a linear equation albeit for different purposes. The linear regression model tries to fit the linear equation along with the dataset, while SVM tries to use this linear equation to create a separation between two different classes present in our dataset. The difference is conveyed using a different loss function to be minimized. It is given in Eq. 4.

$$\max(0, 1 - Y \cdot (W \cdot X - b)) + C \cdot ||W||^2 \tag{4}$$

Here $Y = \{1, -1\}$, the correct label, C is a special parameter that adjusts the margin of separation between classes, the larger its value the smaller the margin. SVM works

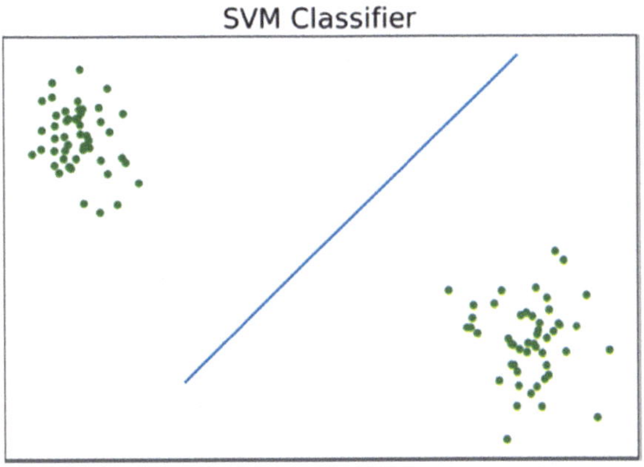

Fig. 2 SVM classifier over two separate clusters

great in higher dimensions and hence oftentimes lower-dimensional data is mapped into higher dimensions via preprocessing to increase the effectiveness of SVM.

2.2.3 Neural Networks

Neural Networks can be thought of as a connection of multiple linear models with nonlinear activation at every node, this allows us to theoretically model any complex nonlinear relation between different quantities. It is optimized based on the back-propagation of error and gradient descent. Simple Artificial Neural Network (ANN) contains 3 layers, the input layer which intakes the dataset for training testing and predictions, the hidden layer which acts like a black box, and the output layer which provides the outcome of predictions. If a neural network contains more than 3 hidden layers, it is considered a Deep Neural Network (DNN). They can be further specialized by the use of convolution layers to create CNNs [4] or by using feedback layers to create Recurrent Neural Networks [5] which are fairly good for time series analysis. Figure 3 denotes a simple Neural Network.

2.2.4 Decision Trees

A decision tree assumes that data can be differentiated one way or the other, based on rigid rules. The model then tries to apply these rules to identify special properties of data, manifesting these decisions in a tree-based data structure. For simplicity, it can be thought of as automatically creating an excessive amount of if-else conditions to classify data. A decision tree can further be augmented by using multiple decision trees to cover up for the shortcomings of any single decision tree or by applying regression methods. These approaches give rise to random forest classifiers [6] and boosted decision trees [7].

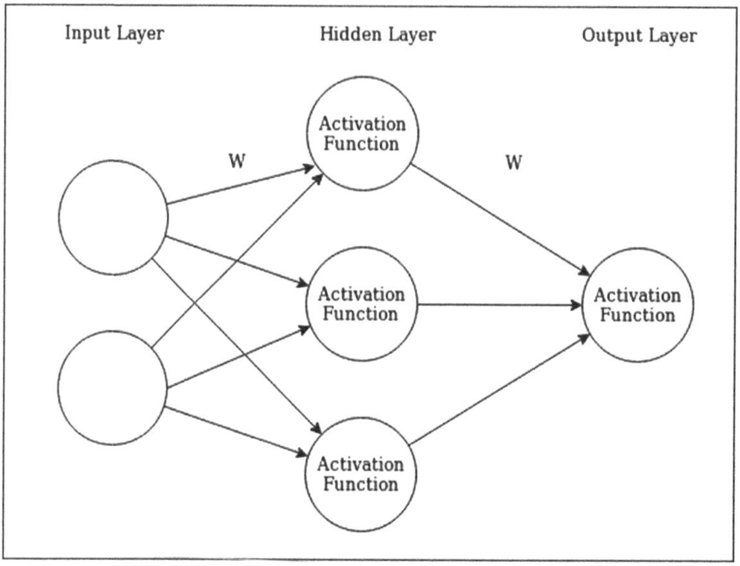

Fig. 3 Demonstration of a typical neural network with 3 layers and 2-3-1 node sequence

2.2.5 Naive Bayes Classifier

This supervised learning model has been developed on the principle of Bayes theorem [8] as well as the "naive" assumption that there does not exist any correlation between the input features. It gives us the probability of occurrence of each label given a unique set of observations. Mathematically its prediction can be written as a decision rule given in Eq. 5.

$$\hat{y} = argmax_{\forall Y}(P(Y) \cdot \prod P(x_j|Y)) \tag{5}$$

This equation associates a given output with the features that result in the highest probability of occurrence of that output.

2.2.6 KNN

K Nearest Neighbor or KNN is primarily used as a classification method or an outlier detector. It is more of an estimation technique in which a data point of the unknown class is compared with k closest neighbor in its surroundings to decide which class this unknown datapoint may best fit in.

3 Security in the Domain of Cyber-Physical System

In the introduction section, we discussed the basic components of a Cyber-Physical System and sought out to explore the applications of Machine learning in securing Cyber-Physical Systems, before moving forward into the applications, it is important to understand why this domain is essential to talk about, the reader should understand the criticality of mass-scale Cyber-Physical Systems being setup and operated around the world by Governments and major industries, and how exploitable vulnerabilities have had an impact or can have an impact on human capital, more importantly, human life. It is important to note that there is no one way when it comes to exploiting a system, hackers tend to choose to inflict change to a system that could provide them with the desired impact, for instance, falsifying data fed to a sensor to get the desired result from a CPS, or in the case of time-critical CPS, inducing a delay of seconds could help achieve the attacker's objective of causing harm to the system or the environment the CPS's actuators and sensors interact with. This section discusses such cases that show the impact of exploitative vulnerabilities in the past as well as experiments that show how external hackers can critically disrupt a CPS environment.

Jeep Cherokee. This is probably one of the most talked-about examples in recent years when it comes to exploitable system vulnerabilities in a smart car. In 2015, two researchers Charlie Miller and Chris Valasek were able to remotely gain control of essential controls of a Jeep Cherokee via its connection to the Internet. They were able to control the acceleration, braking, and steering of the vehicle. What is interesting to note is how these hackers were able to indirectly take command of these functionalities via exploiting the self-driving features of the vehicle. For instance, in the case of Jeep Cherokee, they exploited Jeep's cruise control features to perform acceleration, and use its automated parking tools to turn the steering wheel of the vehicle, where they essentially illusioned the system to be in the parking mode, while it was cruising at highway speeds. The security researchers used the same approach and found the ability to manipulate Electronic control units (ECU) of other internet-connected vehicles such as the Toyota Prius and Ford Escape. This experiment is important as it gives us insight into the kind of approaches hackers can employ to get their way through a system. The first thing to understand is that as technology in the automotive industry is working towards self-driving and internet-connected cars, the threats will eventually increase on the roads, if attackers can find the right vulnerabilities, we can imagine how one car malfunctioning on a highway can cause a chain of reaction at such high speeds and suddenly turn into a catastrophe causing harm to infrastructure and human life. The second important **point** to understand is that attackers tend to find innovative ways to indirectly impact the system, here we can see how cruise control and automated parking tools, and others such as braking systems are being exploited and can be the cause of fatal accidents on the road. Hence, we get to see the intricacy of the Cyber-Physical System, and how important it becomes to safeguard every point of attack that can be exploited by an attacker since any such vulnerability could lead to fatal consequences. This is just one domain we have talked about since our world is moving towards the age of Cyber-Physical

Systems as it is possibly the way to go in the coming years, such opportunities will keep on increasing and hence automated security defenses like Machine Learning will gain more prominence in the coming years.

Baku–Tbilisi–Ceyhan pipeline (BTC). This is an example of just how failing to secure a CPS in the past has been used to the advantage of Cyber-Terrorism. The 1100 miles BTC pipeline, once considered the most secure pipeline in the world, with robust infrastructure and camera surveillance covering it. The hackers infiltrated the system, disabled alarms, and communication systems, and then subjected the oil in the pipeline to extremely high pressures. They were able to get past the camera networks, after the attack, they were able to erase any proofs that could help the security officials trace it back to the attackers. This example clearly shows how important it is to safeguard a CPS not only from one end, but both, the owners of this massive undertaking tried to safeguard the pipeline from a physical security point of view, but they did not give due importance to the Cybersecurity end of it. When it comes to CPS, a comprehensive analysis of the security measures become important in the right context, there have been many cases where a non-holistic evaluation has led to the inability to safeguard these systems. It also helps the reader observe the consequences of such attacks on a political level that could destabilize a region and indirectly on a global level.

Stuxnet. The Stuxnet Cyber-attack on Iran nuclear facility is another example of how highly critical infrastructures like a nuclear facility, with layers of military security and Cybersecurity to safeguard it, can be broken down, as the attacker just needs one point of attack to gain access to the system. The Stuxnet Worm is said to be developed in 2010, while the origins of the virus have never been identified neither any government nor other organizations have taken responsibility for its development, it is said that Stuxnet was developed for the sole purpose of disrupting Iran's nuclear program. As many as 15 Iranian facilities were attacked by Stuxnet, the Stuxnet infiltrated the systems by the USB of a worker which was plugged into the facility's systems, through that one point of entry, it was able to spread through the internal network of these facilities, which did not have any external network connectivity in the first place, to avoid compromising the systems, hence a physical medium was chosen to conduct the attack, via a USB plug-in.

4 Application of ML-Based on Security Type

For a CPS or any system per se, security can mean multiple things, more specifically it can exist in multiple domains, namely:

- Direct security threats like intrusions and malware, which are concerned with malicious users trying to exploit a CPS.

- Anomalous behavior of the CPS itself might suggest a direct security threat or a complication within the elements of the system which may increase the risk of critical failure.
- Risk Assessment and Damage Control is concerned with quantifying the likelihood of a system getting compromised and providing the best course of action in case it does eventually fail.

In this section, we will focus on the application of ML for these topics in detail.

4.1 Direct Security Threat Detection

When we consider the security of any system, we mainly think about direct security threats. Exploitable bugs in our system that a mal-intent user might take advantage of. Such a user would typically try to find entry points unintentionally introduced due to bugs in the system and try to get into the system using various intrusion techniques. If they succeed, the next course of action would be to exploit from within the system to gain full root access, this can typically be done by using malicious software commonly called malware, which is further developed over the vulnerabilities of the CPS. The pressing priority for an ML system employed to ensure security becomes to monitor, classify, and prevent intrusion attempts, and to classify different software operating within the system as benign or malicious. Here we only discuss the principle concepts, Sect. 5 will have a much more in-depth analysis of these along with their associated examples.

Before considering the ML approach for any problem it is a good idea to assess the shortcoming of the heuristic approach. A heuristic method [9] of intrusion detection would focus on finding rigid rule-based abnormal patterns of interaction with the System. These include allowing only certain protocols to be operational within different communication channels, limiting communication between specific nodes in the system, etc. All these heuristic methods should be updated regularly because systems are updated from time to time. This task is performed by humans making it a slow and tedious process while also leaving room for mistakes; on top of it all, we always get rule sets that are not flexible. The heuristic approach towards malware detection also works in a similar sense where we typically keep records of every known malware and anomalous behavior and try to look for their presence in our system. This focuses on finding specific malware more so than the behavior a malware typically exhibits and thus is not much effective towards detecting zero-day exploits. ML excels in such tasks because of its characteristic to generalize well.

The methodology to perform malware analysis can be done in two primary ways. First is static analysis of the structure of any software, looking into its metadata data, or disassembled executable and tracing out unusual patterns in the structure itself. These would include critical permission requirements or a dangerous obfuscated sequence of instructions. The second method for malware analysis is dynamic malware analysis looking at ways of how the software affects the overall system or what potential it possesses to affect the system. Authors in [10] used a semi-supervised ML algorithm using K-means clustering and deep learning for performing static file analysis. With K means clustering we identify clusters of benign and malign software and extract the distances of observation from the cluster center, using these as inputs to a deep learning algorithm we create a generalization of patterns to identify malware with 97.15% Accuracy, 96.17% recall score, and 96.27% F1 score. More practical methods along with examples of their application are discussed in detail in Sect. 5.

4.2 Predictive Analysis and Anomalous Behavior

Security is not just about protecting a system from the most imminent threat but rather also involves the prediction of a possible attack before it happens. This comes under the field of Anomaly detection. Anomaly detection is however slightly different than regular pattern recognition. ML for pattern recognition requires a dataset of fairly equal bias for making an appropriate classification of say a threat-based behavior and a benign behavior, but the complication arises when the system has to deal with novel vulnerabilities, even with a large amount of malicious training behavior we cannot guarantee our ML model will be able to classify malicious activity, hence we reframe our problem statement to not train the model for simple binary classification but rather train it to predict the future state of the system based on current data, now if the future prediction keeps on following the trend that the system is following than most likely the system is stable and no malicious or abnormal behavior has taken place, however, if the trend seems to diverge we might want to have a closer examination on the system. This is called Time series analysis or Forecasting. These are done using specialized deep neural networks which are known as Long Short-Term Memory (LSTM) [11]. LSTMs have a feedback loop based deep learning architecture where parameters in the previous iteration influence the current parameters, thereby creating a persistent memory of the past within the model.

Looking into the features to do time series analysis, malware tends to influence basic system-level processes the most and hence some features to look out for while training the ML model include daemons (background processes), Active nodes in the entire system, Total network connections, bandwidth and communication channel, CPU/Memory consumption of each process, etc.

LSTM is however not the only method of doing future prediction, observing the recent research papers we can see many other methods as well. Authors at [12] used a clustering method called DBSCAN as well as Bidirectional Recurrent Neural Network (BRNN) [13] for fault detection. In DBSCAN we consider all the features as dimensions of an n-dimensional vector space, now we plot each observation inside of this vector space, DBSCAN allows us to group various data points as belonging to a single class based on how far apart points are from all the other points, and how many points are surrounding any given point, see Fig. 4. These two qualities can be adjusted according to the designer's satisfaction, the goal is to realize that

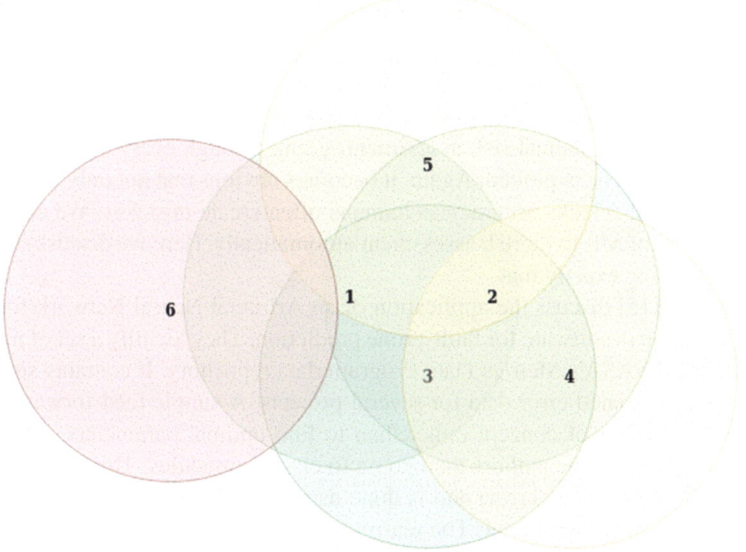

Fig. 4 DBSCAN to detect outliers, green represents core points of the cluster, yellow represents non-core part of the cluster, and red represent outlier. The radius of the circle and the number of neighbors

an abnormal data point would deviate from clusters of standard data points and will be isolated without a unique cluster. Then in the same paper BRNN, a special kind of neural network with regularization to prevent overfitting is used to create a generalization of various anomaly patterns based on the abnormal data points. One hot encoded representations of faulty data points are fed into a 3 layered BRNN with *tan(h)* activation at a hidden layer. This approach results in an accuracy of 99.7% in clustering and 93.13% accuracy for generalization, which according to the paper is state of the art. Authors at [14] tried to improve the reach of this approach as it becomes difficult to cluster data in higher dimensions. They accomplished this by using Principal Component Analysis (PCA) which is an ML-based dimensional reduction scheme. PCA enables us to find the most essential features of the given data. PCA takes normalized data as input and tries to find an arbitrary axis by the process of linear regression that maximizes the spread of the range of the data, it does this as many times as there are dimensions in the original dataset but for each axis that it finds they are automatically sorted from the best to the worst, i.e. axis 1 obtained from PCA would be the most crucial, then axis 2 and so on. Hence after obtaining all the datasets we can decide which ones to consider for our classification and which one to discard with minimal loss.

4.3 Risk Assessment Using Machine Learning

Another way to foster security comes from the risk analysis of a given system. Here by risk analysis, we mean how likely is the system to be compromised. Typically, humans would do a manual risk assessment, going through every component that has the potential to be exploited. Again, it becomes obvious that not only is this time consuming but also lacks accuracy as humans often create mistakes. We can again take advantage of ML to do risk assessment automatically. Here we discuss some of the ways of doing exactly that.

Authors at [15] discuss the application of an Artificial Neural Network for best feature selection in software for fault-prone prediction. They identify a set of metrics on modules of NASA's Metrics Data Program data repository. It contains software metrics and associated error data for several projects. A simple feed-forward ANN was used as a proof of concept rather than to find optimal parameters. Using the said optimal features the authors then go on to classify modules. This classification is binary (error-prone, no error) and is done using SVM to achieve an accuracy of 87.4% on the validation dataset. The reason to use SVM stems from the fact which is also acknowledged by the authors, include the ability to model non-linearity, able to need fewer data points to converge, able to work in higher dimensions easily, and its natural architecture to be able to perform well in a classification task.

Methods involving Adversarial Neural Network are fairly novel yet effective ways of tackling the task at hand. In such methods, we create 2 agents operating over a network of nodes. The task of one is to identify and exploit vulnerabilities in these nodes while the task of the other agent is to identify and fix these vulnerabilities. These two agents compete with each other and try to optimize their functionality to outshine the other, the condition of optimization is simple—to succeed in their task as much as possible. The end result of training a security system based on this model is a robust and self-regulating way to deal with vulnerabilities. This method is discussed in [16] where they associated the cost of exploiting nodes, attack strategies, and defensive measures. The goal then becomes to optimize these numerical values based on different optimization algorithms. The authors discuss and measure the effectiveness of various optimization techniques involving Monte Carlo, Q-Learning, and Neural Network.

It often happens that the elements creating the most risk in a system are not related to the underline system design as much as the users that interact with it. It is often noted in security that the weakest link to exploit is the human in the system and hence it becomes vital for us to explore some methods that would be useful to identify risk associated with the same. Authors at [17] proposed a way to identify risk based on the user of the system. The research quotes using 20 different models with the best performance obtained from an ANN having 16 input layers, 14 hidden layers, and 2 output layers, the output layer having exponential activation function, and the model compiled on Broyden Fletcher Goldfarb Shanno algorithm [18]. Features as input included categorical features like age, education level, computer expertise, etc. These parameters can be used to assess humans in the loop components.

Some other noteworthy mentions include—The Naive Bayes Model [19] for Scrum-based Software development and Deep Neural Network discussed here [20]. All of these methods are primarily discussed for software development processes but with appropriate modification can be extended in the design of CPS.

5 Application of ML-Based on System Design

The general categorization of a CPS is usually done into three layers, namely the application layer, network layer, and the physical layer. Figure 5 shows the diagram of a classic generalization of a CPS. A general CPS is bound to have all its components easily classifiable into these 3 components. Autonomous vehicle systems are no exception to this train of thought. The level of automation has been developing at a rapid pace and recent exploration has been encouraged in the direction of Connected Autonomous Vehicle (CAV) [21], which has focused on an additional information collection source by forming Vehicular ad-hoc Networks (VANET) [22]. A VANET enables information sharing between the connected vehicles and enables what is popularly known as Vehicle to Vehicle communication (V2V). As expected, a new arena of networking capabilities brings with it more opportunities for vulnerabilities for attackers to exploit.

5.1 Application Layer

In CPS, the application layer comprises the computing core of the whole infrastructure; it is instrumental in gathering data from sensors via networking layers. It comprises various software that possesses the required algorithms for processing the data collected and generating control commands that help in the execution of functionality by actuators for various applications of the CPS in question. We shall read about sensors and actuators later when we discuss the physical layer. In the context of a CAV, the application layer of a CAV consists of a plethora of software, which is dependent on the functionalities required to be performed via the Electronic Control Units (ECU) [23], such as Anti-braking systems, Cruise Control, Lane management, Parking Assistance and so on.

When the application layer is taken as an attack vector via malicious parties, they can indulge in false code injection attacks, or introduce Viruses such as the Trojan horse [24], introduce Malware [25] that could compromise the system security or other ways depending on the category of vulnerability exploited. Figure 6. Illustrates various security threats to the application layer. In the context of a CAV, we take the instances of possible application-layer attacks and state of the art ML methods that are currently being employed in autonomous driving systems or from a different domain but have a scope of implementation given the similarity in use cases.

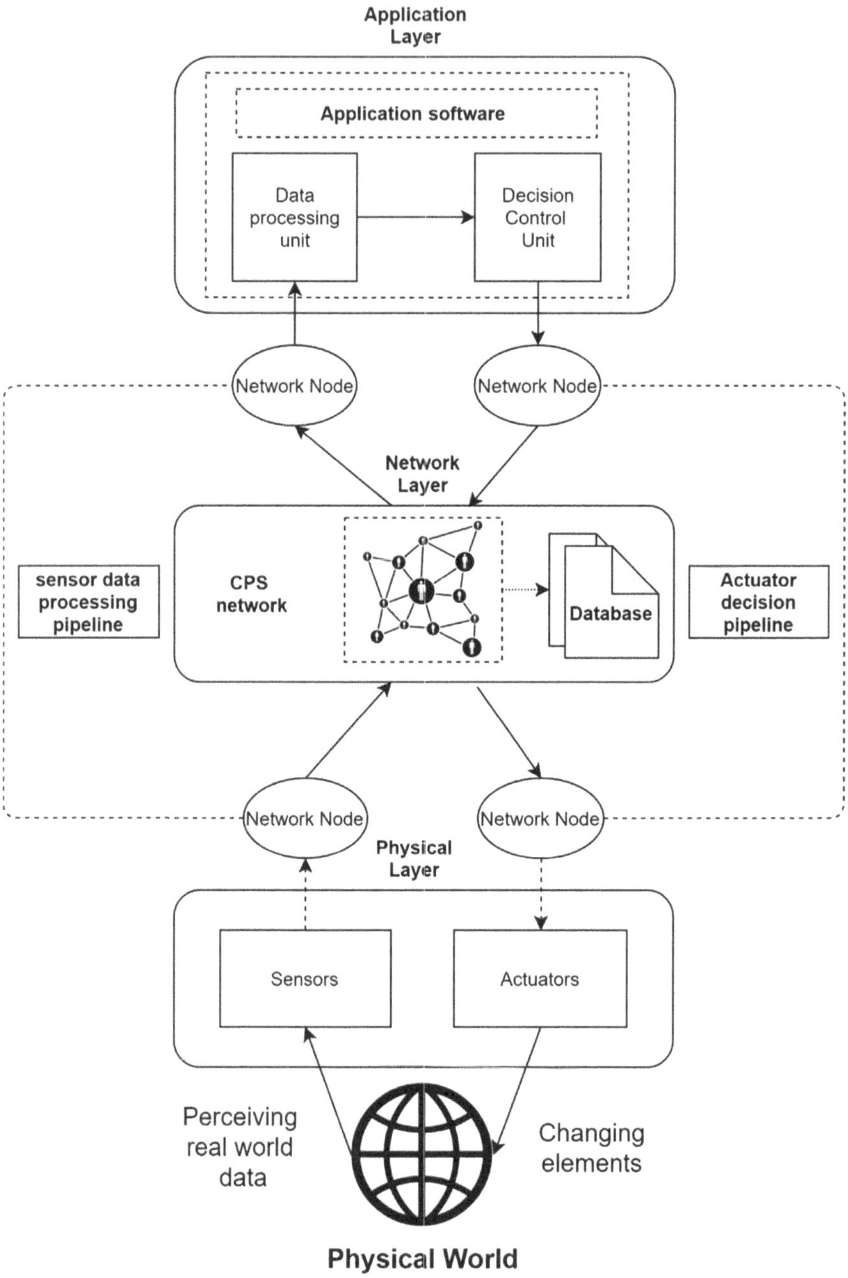

Fig. 5 Layer-wise design of a CPS

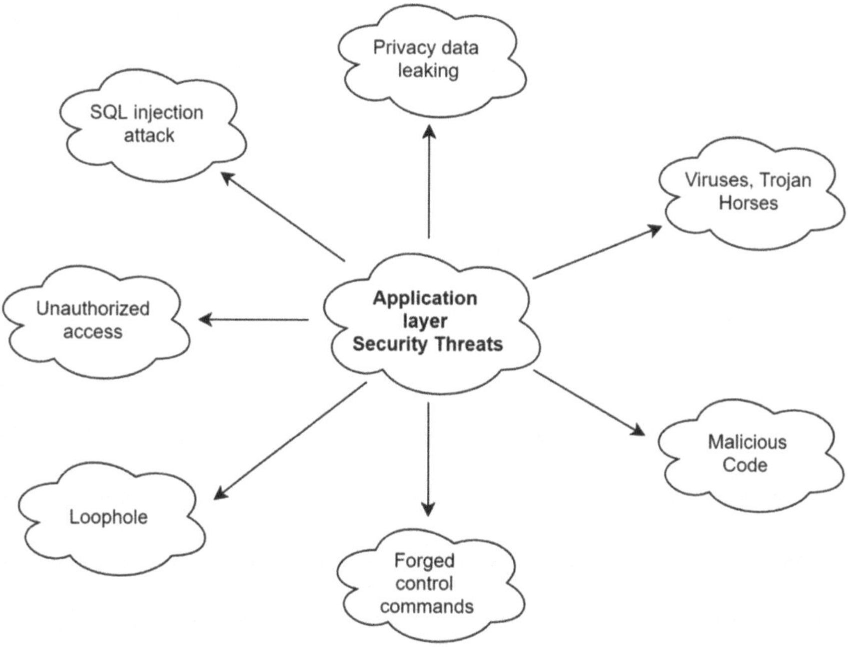

Fig. 6 Application Layer based security threats

Cooperative Adaptive Cruise Control. By taking VANET into the scope of discussion in this chapter, we can discuss an upgrade to the existing Adaptive Cruise Control Systems that took input from the inbuilt sensors supported by the vehicle to make velocity and acceleration variation decisions, VANET allows the CACC to take into account the vehicles in their vicinity as they communicate the physical state of the vehicles among each other to further optimize the decisions. It is highly apparent how a falsification attack on CACC is not only harmful to the vehicle in question but the whole stream of traffic in the network. Fuzzy decision-making systems [26] have a huge upside in this case as braking mechanisms cannot be initiated on a singular pivot point but require a buffer for activation, which is it's the basic advantage over Boolean logic and hence is popular in the automation driving domain. The authors of [27] proposed a neural network-based fault detection scheme followed by a decision support system implemented via fuzzy logic, the scheme of which has been illustrated in Fig. 7.

It is important to understand that the course of action taken after a successful False Data Injection (FDI) [28] attack by the intruder is fault detection followed by attack response or mitigation which readjusts the vehicle speed and ensures maintaining the appropriate gap required. The neural network's core algorithm was also modeled after a fuzzy decision-making system.

Malware detection. In-Vehicle infotainment systems running on embedded Android or other OS can be vulnerable to malware attacks via malicious code introduced via

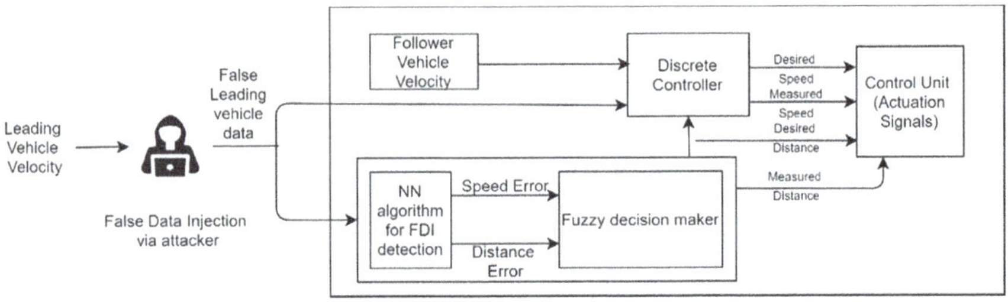

Fig. 7 Scheme for the proposed FDI attack detection and attack recovery scheme [27] for Cooperative Adaptive Cruise Control

hacked mobile devices plugged in by the unaware user or driver of the vehicle. The authors of [29] introduced an ML method aimed at anomaly detection on the data flowing from the user mobile into the in-vehicle infotainment system, as shown in Fig. 8. It aims to prevent malicious attacks aiming to send false commands to the ECU accessible to the infotainment system, the attack flow of which has been illustrated in Fig. 9. The proposed model is based on the efficient preprocessing of data; by initially employing a 10 cross-fold validation method based on correlation measuring methods. This validated dataset is used in Improved Feature Selection (IFS). A train-test split of 3:1 is applied to the dataset and used to train 6 different classification models using scikit-learn [30]. Table 3 shows the F1 scores of various algorithms employed on the given dataset. The F1 score was obtained via a newly proposed score function that could evaluate while the model trains to make the detection algorithm feasible in a real-time scenario where the speed requirements are often in milliseconds(ms). Two test cases were observed, case 1 consisted of two classifications of malicious content as well as benign. Case 2 ignored the malicious

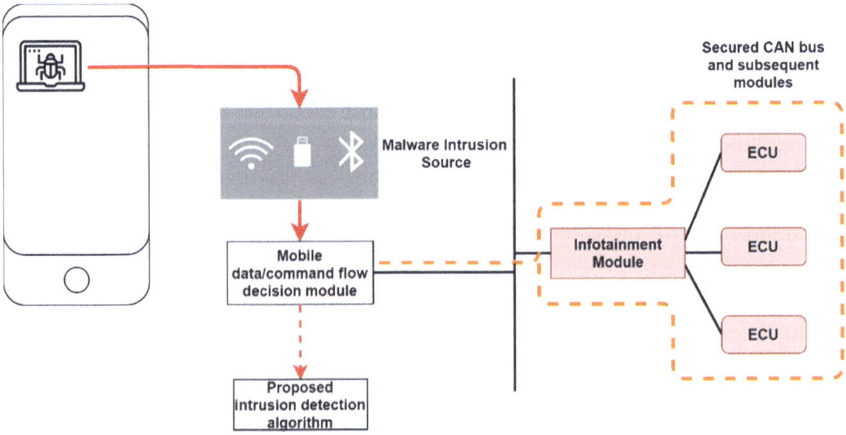

Fig. 8 Figure depicting the application point for the malware detection scheme safeguarding the infotainment module

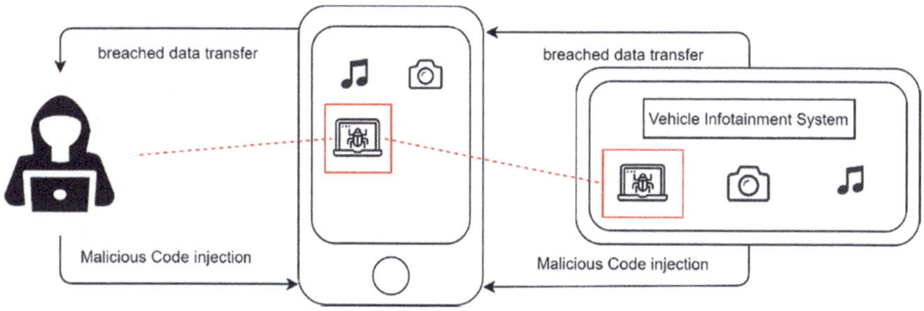

Fig. 9 Attack flow diagram for malware injection into the vehicle infotainment system

Table 3 Performance evaluation of various algorithms employed on the improved feature set (IFS) obtained in the given malware detection scheme

Machine learning algorithm	Multi-class classification	Estimated time (s)	Binary classification	
	F1 score		F1 score	Estimated Time (s)
Random forest	0.938	1.94	0.920	1.82
Gradient boosting	0.83	256	0.883	3.98
Bagging classifier	0.817	1.93	0.919	1.82
K-Nearest neighbors	0.813	6.10	0.914	12.91
Decision tree	0.810	0.32	0.912	0.29
Extra tree classifier	0.782	010	0.906	0.10

content classification with a lesser ratio of the total dataset (0.8%). The Random Forest algorithm turned out to be the most efficient in terms of its F1 score (93.8%) as well as a low elapsed time (1.94 s) in the Multi-Class classification scenario (Case 1). In case 2, we observe that Extra Tree Classifier is the preferred choice of ML algorithm with an optimum accuracy (90.6%) which is in the vicinity of the highest precision accuracy (92%), it is preferable because of the significant advantage of estimated time over the little decreases in accuracy.

5.2 *Network Layer*

This layer can be easily considered as one of the most essential components of CPS since it can be credited for the integration of the computational capabilities and decision logic of the application layer with the sensors and actuators of the physical layer, and ML plays an important role in not only its security but also architectural design and logic as well [31–33]. They enable real-time communication between the various nodes of the system. The range of networks can depend on the use case, be it a Body Area Network (BAN) that is seen in wearable devices to connectivity spread across the globe via Wide Area Network (WAN), a popular example of WAN being the internet. Figure 10 illustrates the various security threats to the network layer. In our example of a CAV, interestingly there are two networks at play essential to the setup. The first being the VANET as introduced previously, which is responsible for the inter-vehicle connectivity which facilitates information sharing. The second being the intra-vehicle communication facilitated by the Controller Area Network (CAN) bus, which is responsible for the communication between the ECUs, sensors, and actuators. The security threats to a network layer have far more catastrophic consequences in comparison to the application layer, the reason being that it can have a distributed impact on various components of the system. This stands as a purely cyber-security breach domain, and most documented famous network attacks like DoS, DDoS, man-in-the-middle attacks, or signal jamming are applicable here. In our example case of a CAV, extensive research has been conducted on securing VANETs by the application of ML in Network Intrusion Detection Systems (NIDS) [34]. We shall discuss the core attack vectors and some applied ML methodologies in these fields. The attack vectors are as follows.

Intrusion Detection System. The authors of [35] deployed an immunity algorithm for efficient data preprocessing followed by the SVM approach to detect any network intrusion. The immunity algorithm provides a better result after combining with

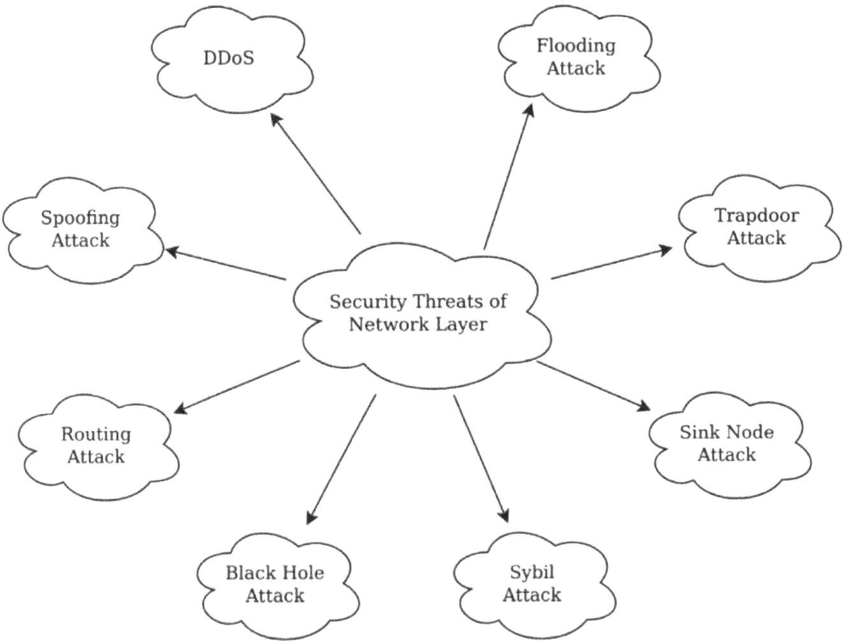

Fig. 10 Security threats to network layer

SVM in comparison to plain SVM implementation, the reason being that redundancy in datasets can result in reduced performance. The immunity algorithm facilitates the preprocessing and extraction of important characteristics from the data. This state-of-the-art approach provides a tremendous improvement in comparison to the conventional use of SVM as the former gave a recognition rate of 95.8% while the latter managed a recognition rate of 81.4%.

Denial of Service Attack. The authors of [36] proposed a DoS attack detection methodology comprising two components as illustrated in Fig. 11. The first is a triangle-area-based multivariate correlation analysis (MCA) [37] method which extracts the hidden correlations in the features of the training dataset geometrically, which helps in improved feature selections and better characterization of the data. This is followed by an anomaly detection method that could detect DoS in cases where the attack pattern is already known to the algorithm as well as the unknown attack patterns. It was observed that when normalized data were used, the accuracy of detection was as high as 99.95% and more for the various DoS attacks. These observations were compared with two other approaches, the first being the nearest neighbor approach with an accuracy of 92.15% and a Euclidean map-based approach which had an accuracy of 99.87%. The data normalization step is important considering that without it, the detection accuracy in the proposed method falls to 95.2%. DoS is very common with CPSs involving opportunistic networks, and some methods, involving fuzzy logic and game theory, to prevent them are also worth looking into [38–40].

Jamming. The authors of [41] proposed a model that was implemented via multi-agent reinforcement learning (MARL), which has been derived from Q-learning [42], a popular reinforcement learning technique. The model was instrumental in countering the attacks carried out by a sweeping jammer as well as optimum utilization of the spectrum ensuring that any unintentional inferences from other nodes are avoided. It has been employed in Wideband Autonomous Cognitive Radios (WACR) [43]. To understand the area of application, we should know that the purpose of cognitive radios is the maximum utilization of the spectrum via Dynamic Spectrum Sharing (DSS) and WACR are upgraded classes of Software Defined Radios that build upon

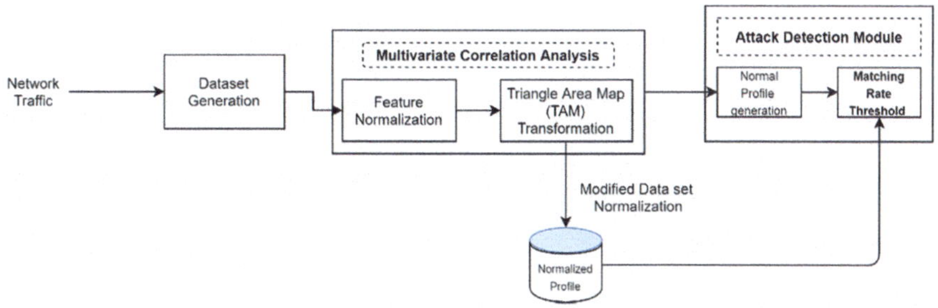

Fig. 11 Multivariate Correlation Analysis based Feature Generation Scheme for detecting DoS attacks

CRs. The transmission by a node performs spectrum sub-band switching to ensure the transmission is uninterrupted via any interference or sub-band jamming for the longest time possible. The success of the model can be understood properly by the test case which is as follows. Two nodes are taken in this test case which communicates over a bandwidth of 200 MHz, divided into 5 sub-bands of 40 MHz A sweeping jammer is taken as the adversary that takes 0.25 ms to detect transmission in each sub-band which sweeps the total bandwidth from lower to higher frequencies. For instance, as shown in Fig. 12, if the jammer starts sensing and jamming the 4th sub-band, the transmission from a node should occur on the 3rd node, since the jammer will sweep up to the 5th band and then start back from the lower end of the bandwidth up to the 3rd band. In this process taking 1 ms which is the maximum time, a node

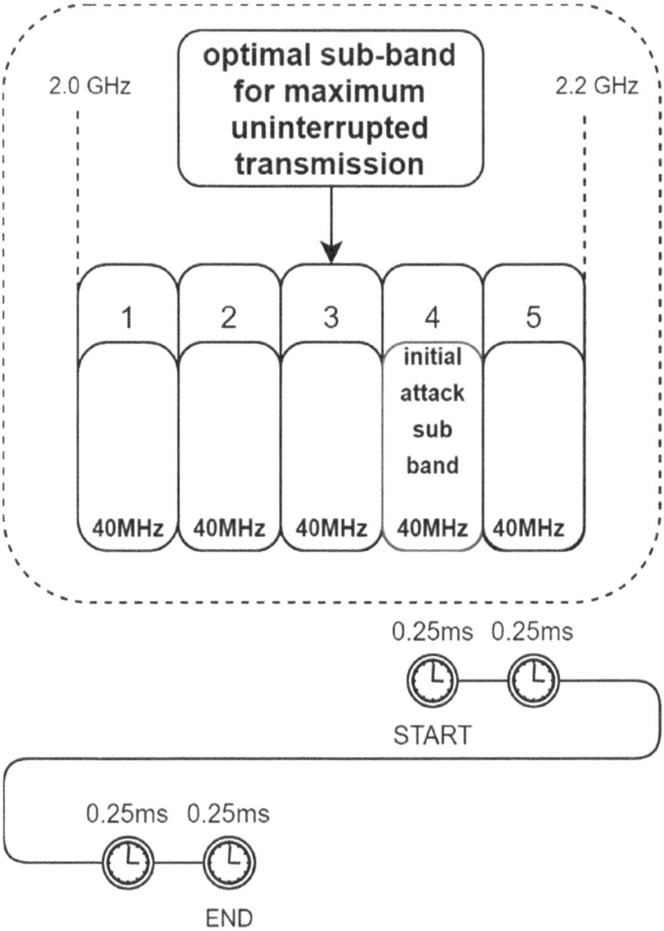

Fig. 12 Optimum Spectrum Utilization Strategy is employed by the proposed MARL ant-spamming method, when unintentional interference is not considered

can perform uninterrupted transmission. By MARL implementation, it was found that 75–90% of the maximum time was able to be utilized as uninterrupted transmission. In contrast, algorithms that randomly selected sub-bands for transmission of data, only about 60% of the maximum time was obtained, which corresponds to an average time of 0.6 ms as calculated via tests.

Spoofing attacks. The authors of [44] proposed an Intrusion detection system that introduced a new parameter of input, that being Position Verification using Relative Speed that was analyzed by extracting information from both the application layer as well as the physical layer of an Electric Vehicle (EV). We account for data such as Received Signal Strength Indicator (RSSI) and Signal-to-interference-plus-noise ratio (SINR) which is obtained from the physical layer of the vehicle. We also incorporate Geo-positioning data and relative speed that is consolidated via the application layer. The models were trained on a 7:3 train to test split. K-Nearest Neighbors and Random Forest Algorithms were employed against a dataset accounting for PVRS and the results showed a significant over a test scenario where it was not implemented. While the accuracy achieved for both models was found to be the same (91.3%), the AUC scores displayed a higher performance via the RF algorithm which was 0.986, as compared to the KNN algorithm that achieved a score of 0.935. This technique is significant as it is an excellent example of how the application layer and physical layer abilities are being synchronously used together to defend the network layer. The maximum accuracy achieved among the two **algorithms** when trained on a conventional unmodified dataset is 85.6%, achieved by the RF algorithm. Note that both the models used lie under the category of Supervised learning.

5.3 Physical Layer

Now we talk about the last layer of a CPS, the physical layer. This layer facilitates the interaction of the CPS with its surroundings. Earlier in the chapter, we mentioned sensors and actuators that belong to this layer. These are the two classifications of physical layer components. Sensors are the components that take input from the real world followed by feeding it to the application layer for processing via communication enabled by the network layer, while actuators are the physical components that enable physical functionalities of the CPS by executing the component control commands received from the application layer via the network layer. In the context of a Continuous Autonomous Vehicle, these sensors and actuators are of utmost importance since essential applications and software heavily depend on the data sent by the sensors for estimating the required dynamic state of the vehicle and actuators are essential for meeting these requirements. The sensors in the CAV include Cameras, SONAR, RADAR, and LIDAR sensors. An Autonomous vehicle has Electronic Control Units (ECU) which are responsible for receiving commands from the Controller Area Network (CAN) Bus and further facilitates actuators such as motors to perform required functions. Figure 13 illustrates the various threats to

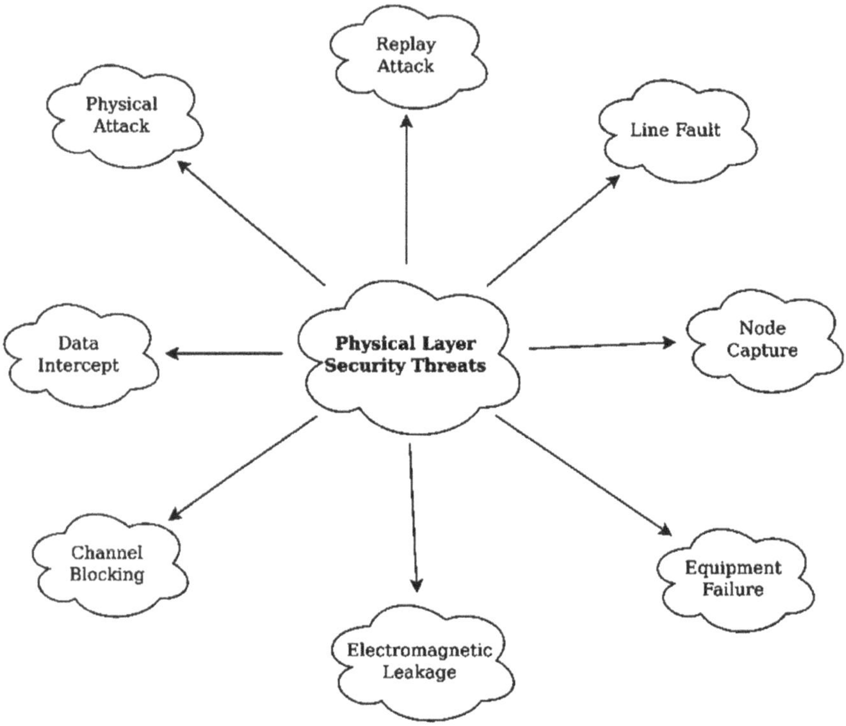

Fig. 13 Security threats to physical layer

the physical layer. We will discuss ML methods for physical layer security in two important domains, the first being authentication control and the second being Fault Detection Isolation (FDI).

Authentication. We focus on Wireless Sensor Networks (WSN) [45], in research work done in this area, we find the main objective of the sensors and the actuators to be authenticating received messages from the sender node. Classic cryptographic methods such as RSA [46], DSA [47] which are key-based digital signatures as well as methods which depend on time-efficient stream loss-tolerant Authentication (TESLA) [48] can pose a challenge in terms of implementation in the physical layer components since these nodes are not equipped for such computationally intensive algorithms. Hence the approach taken is the utilization of the spatial and temporal properties of the channel between two nodes which is unique and can act as a key, better known as the Channel State Information (CSI). It has been observed that previous research done in physical layer authentication (PHY-AUC) is suitable for industries with static components, but not in the case of sensors and actuators that are mobile in nature. The authors of [49] proposed a threshold-free method which performs binary classification by training models in mobile scenarios using CSI as the training data in the form of channel matrices derived from CSI to maximize data obtained from CSI. Ensemble learning was employed, which is essentially a combination of other ML-based classification algorithms. Bagged trees (BT) [50] is an ensemble learning method that uses a cluster of base simple trees. Voting is used to make the final decision from the consolidated base tree decisions. The conclusion of the proposed model shows that Bagged Tree (BT) reached 100% accuracy in some cases with an average accuracy of 0.77, while the prediction time is 5x multiple of the standard approach, it is well within the vicinity of 10 ms which is considered suitable in many cases. It is important to note that the distance between the moving components is inversely proportional to the authentication accuracy of the proposed model.

Fault Detection Isolation (FDI). The authors of [51] proposed a CNN based LeNet-5 [52] for fault diagnosis, this ML model is built upon a proposed method which performs data pre-processing by converting the raw time-domain signals into 2-D gray images which are used as input for the CNN model proposed. Three datasets were taken, the first case of the motor bearing dataset which gave a prediction accuracy of 99.79%, the second being a self-priming centrifugal pump dataset with an accuracy of 99.481% and the third is the axial piston hydraulic pump dataset which gave ideal prediction accuracy (100%). The results of the proposed model were compared against other traditional deep learning methods and ML methods like SVM, sparse filters.

6 Limitations of Machine Learning Based Security in CPS

In previous sections, we have stressed deeply about how ML can be instrumental in CPS security. We have learned in-depth about the wonders an ML algorithm can do, but this is where many newcomers to the field of ML misinterpret ML as a one-stop-shop for all solutions in terms of algorithmic requirements. This chapter is not about promoting ML into every field but is aimed to provide a fair analysis and communicate the status quo of the current capabilities of ML, which shows that while significant progress has been made, there is still room for improvement. It is important to understand that tasks that could be performed easily without an ML algorithm should not be performed with an ML algorithm. Conventional methods for computer security including forensics [53] hold their important place and should not be overlooked in favor of ML.

The extent of Security. We should also observe that multi-front security is not possible via employing a single ML algorithm, as observed in the case of Continuous Autonomous Vehicle (CAV) in the previous sections. Although we have discussed various novel approaches that can act to safeguard various components of a CPS, it is important to note that there is no CPS that relies solely on ML approaches for security. As a matter of fact, there can be many industrial CPS that don't employ ML and rely on conventional techniques instead. For instances, the BTC pipeline blast mentioned in the previous sections, while a cyber domain defense was critical to avoid that security failure, one cannot simply say that an ML employed solution could defend it on its own, there still will have been a need of surveillance and security forces patrolling the stretch of pipeline, as the chapter has mentioned before how safeguarding every component of a sophisticated CPS is important, the designer should not focus on only one component of security, but each component possible, the very advantage of the cyber and physical domain interacting together to form a CPS is also a potential vulnerability where the interaction between these two domains needs to be bonded with robust security, leaving no vulnerability behind.

Transparency. While ML algorithms might promise precision and high accuracy results, they are not only the requirement for an algorithm's employment in the domain of security. For instance, while an ML algorithm might provide an accurate classification for a test case to be faulty or not, it might not help the system in explaining the classification if required. Transparency of a security system is important since, without it, there is no accountability of the actions taken by the system, hence no one will be held responsible if the security systems are flagging activities that are not illegal due to reasons unknown, this is not new, there have been instances in the past where machine learning models were found to have racial biases when it came to credit ratings and face recognition systems. Since there is little transparency in how an ML algorithm works, hence it is difficult to find if such biases are introduced by the Engineers or not, hence security mechanisms need to be made accountable and should strive to avoid the introduction of any such inherent biases into the building of a system.

Computational Costs. We have to also remember that ML algorithms can be computationally intensive, and it might not be possible to employ these algorithms in the case of physical layer components such as sensors and actuators since they might not have the computational required for the same. A cost-benefit analysis becomes of the essence here where the designer will have to compare if a physical measure or a Machine learning-powered computationally expensive security defense will be cheaper. From an economical point of view, the question does not only pertain to maximizing the security of a sensor by any means possible, the designer needs to tradeoff between mass scalability of the system and the security if computational costs become too high, but with the current advancements in embedded systems, there might be an increasing scope of ML defenses in the coming years.

The uniqueness of Different Attack Scenarios. As iterated upon before, ML is not a one-stop-shop which can be a solution to all things related to security, which can be seen by the highly unique models discussed in this chapter, each attack type, each layer, and each security classification has its requirements and correspondingly appropriate ML models that are generally employed in the particular use case. As discussed in the previous section when we took up the example of the hacking of the Jeep Cherokee, we can see how attackers could approach a target with unconventional methods, use functionality not exactly designed to drive a vehicle but as autonomous driving assistance tools, and finding these methods can cause potential damage to human life. We need to realize the importance of good datasets required for their training and how they might act as a barrier in their implementation at large-scale across various industries. While many ML models proposed in a different field can find its way here in the context of CPS security, for that matter even cloning models from one CPS structure to another, this transition would require a system-specific dataset.

Context-Based limitations. While ML might perform tasks not possible for humans to do like finding hidden features and correlation in a dataset, there are certain domains where competing with human intelligence is difficult. For instance, natural language processing techniques can struggle to account for the context of a given statement which might lead to misclassification.

Cost of Error. As observed, the error rates always exist in ML models, because the basic idea is to predict an outcome based on previous outcomes, which is not something that can always be possible in a real-world scenario, we should consider how an ML security protocol employed in highly critical infrastructure might lead to disasters of unbounded proportions if an error manages to seep in the model's prediction, hence it is highly important to see the trade-off between the benefits of ML and the criticality of a CPS.

7 Guidelines for Application of Machine Learning in Cyber-Physical Security Systems

Throughout this chapter, we have sought to build a guiding map for the reader to help them find a starting point to introduce and integrate ML's capability to provide security into a CPS. In the previous subsection, we talked about how ML is not suitable for all environments. It is solely the responsibility of the reader to take note of the instances where ML can or cannot be applied in the security perspective, which preferably should be decided via a well thought cost-benefit analysis and a thorough risk assessment of installing such a safeguard. This section is meant to provide considerations for readers to keep before designing their models.

7.1 Use-Case Analysis

A designer should always keep in mind the viability of the model they propose, it can be dependent on what kind of a setup is it being proposed to be deployed in. As we can understand from the previous section, there is a difference between a model proposed for academic research and the ones that are applicable for a roll out in industrial systems. The kind of CPS in question is also a deciding factor. For instance, nuclear facilities are one of the most critical CPS infrastructures on the planet, be it nuclear missile silos or nuclear power plants. Their security is a top priority for governments around the world and their mere existence is a huge threat in itself. Understanding the magnitude of the issue at hand, proposing complete automation for security via a one-layer safeguard provided by ML for such infrastructure seems dangerous, and additional hardware or software safeguards might be required along with human monitoring, for instance, the requirement of synchronous key insertion and unlocking by two individuals for launching a missile. We need to understand why we prefer these additional safeguards in this example. The first reason is the Cost of Error discussed in the previous section, the cost of error is catastrophic, secondly, since this infrastructure is of utmost importance, it also attracts huge crime organizations or terrorist groups with huge resources and top of the line cyber-hacking personnel and computing power, which makes it necessary to question if the ML models proposed to be applied in this case could compete with these situations. Furthermore, in this example, the incentive of the attackers is extremely high and they will extensively research for vulnerabilities and try to approach the security with multiple combinations and attack vectors.

7.2 *Scope of Implementation*

It is essential that before the programmer begins with the task of designing the defense algorithms, they take a clear picture of what problem is their algorithm aiming to tackle. The designer should know the classification of the attack they are dealing with. The more the targeted analysis, the higher the capabilities of the ML system as such analysis helps designers to pursue a new novel approach to make their systems robust. This assessment should be followed by an unbiased judgment of optimal security methods that could be implemented to satisfy the safeguard requirements. The priority of the designer is not implementing ML but to safeguard the CPS with the optimal approach.

If the optimal method is applying ML, the next step is deciding which specific ML algorithm is to be employed. The programmer should have a clear reason and valid explanation for the implemented algorithm in the given use-case. The algorithm doesn't need to decide the final algorithm just via theoretical analysis. They can compare their top contending algorithms and analyses their results side by side, as done in the case of Malware Detection in Sect. 4.1. The researchers took 6 algorithms and employed them on their prepared dataset, analyzed the algorithms based on their F1 score and estimated time, and calculated the optimal algorithms for two different classification scenarios. Here the time estimation parameter heavily influenced their conclusions.

7.3 *Balancing Error Rates*

We have discussed the cost of error in the previous section and we understand how the criticality of the system influences it. This calls for detailed analysis for managing thresholds and which is more acceptable to the system, more False positives while compromising on operational speeds or higher operational efficiency while lowering down False positives. While the question posed seems to be an easy one where the reader might think that the latter option is the obvious way to go about, but it is important to know that the number of False Positives that seep into the predictions is threshold-based which is set by the programmer. The idea is that considering the priority of security over efficiency for critical infrastructure, it is preferable to let more False Positives cases to be detected while compromising operational speed, which in turn lowers the number of False Negatives that could have damaged the CPS significantly.

7.4 *Dataset Selection*

This step is one of the factors that will determine the integrity and performance of the finally trained ML model. Hence the programmer should extensively research for good datasets, higher quality dataset implies better outcomes.

Feature selection can consist of outlining important factors that might affect the outcome you may require from the system, asking for surveys amongst the stakeholders, and doing exploratory analysis over the data gathered, similar to what is done here [54]. Once the dataset is taken, it needs to be carefully analyzed for possible faults or noise in the dataset, this is followed by undergoing operations suitable to clean the dataset such as filtering. The size of the dataset is also of importance as it will determine if the programmer can produce a successful model using the dataset taken, in the case of lack of datasets the programmer's bound to use algorithms that could be trained in even when the dataset lacks enough data points. Hence it is needed to also account for the total dataset being divided into an appropriate train-test split.

8 Conclusion

CPSs are becoming a part of life more than ever in today's day and age. Whether we realize it or not, CPSs influence our day to day activities, be it indirectly through being used in factories and power delivery services, or by directly being closer to us as Internet of Things (IoT) devices or home device networks. This raises security concerns and prompts us to find newer ways of establishing security requirements. Application of machine learning within the field of cyber-physical system security, although promising is a novel idea that still requires a lot more exploration. This chapter aims at providing an adequate level of knowledge to the reader so that they become self-reliant towards understanding new and innovative research in the same field. The chapter quotes examples and explains a multitude of sub-domains within the field to facilitate the same. It goes on to also highlight some crucial limitations that we currently face from such defense mechanisms which can themselves become a topic of research. Finally, it provides guidelines for new researchers trying to get into the field themselves.

References

1. Sanislav, T., Miclea, L.: Cyber-physical systems—concept, challenges and research areas. Control Eng. Appl. Inform. **14**, 28–33 (2012)
2. Kumar, S., Yadav, A., Sharma, D.K.: Deep learning and computer vision in smart agriculture. In: Modern Techniques for Agricultural Disease Management and Crop Yield Prediction, IGI Global, pp. 66–88 (2020)
3. Sinha, U., Singh, A., Sharma, D.K.: Machine learning in the medical industry. In: Handbook of Research on Emerging Trends and Applications of Machine Learning, ed. Arun Solanki, Sandeep Kumar and Anand Nayyar, IGI Global, pp. 403–424 (2020)
4. Albawi, S., Mohammed, A., Tareq & ALZAWI, Saad: Understanding of a convolutional neural network (2017). https://doi.org/10.1109/icengtechnol.2017.8308186
5. Bianchi, F.M., Maiorino, E., Kampffmeyer, M., Rizzi, A., Jenssen, R.: Recurrent neural network architectures (2017). https://doi.org/10.1007/978-3-319-70338-1_3
6. Cutler, A., Cutler, D., Stevens, J.: Random forests (2011). https://doi.org/10.1007/978-1-4419-9326-7_5
7. Drucker, H., Cortes, C.: Boosting decision trees. Adv. Neural. Inf. Process. Syst. **8**, 479–485 (1995)
8. Nyberg, S.: Bayes' theorem (2018). https://doi.org/10.1002/9781119246909.ch6
9. Mukhopadhyay, I.: Heuristic intrusion detection and prevention system (2015). https://doi.org/10.1109/iemcon.2015.7344479
10. Sharmeen, S., Huda, S., Abawajy, J.: Identifying malware on cyber-physical systems by incorporating semi-supervised approach and deep learning. IOP Confer. Ser.: Earth Environ. Sci. **322**, 012012 (2019). https://doi.org/10.1088/1755-1315/322/1/012012
11. Hochreiter, S., Schmidhuber, J.: Long short-term memory. Neural Comput. **9**, 1735–1780 (1997). https://doi.org/10.1162/neco.1997.9.8.1735
12. Liang, Z., Fei, H.E., Yifei, T., Dongbo, L.: Fault detection and diagnosis of belt weigher using improved DBSCAN and Bayesian regularized neural network. Mechanics, 21 (2015). https://doi.org/10.5755/j01.mech.21.1.8560
13. Schuster, M., Paliwal, K.: Bidirectional recurrent neural networks. Signal Process. IEEE Trans. **45**, 2673–2681 (1997). https://doi.org/10.1109/78.650093
14. Maier, A., Schriegel, S., Niggemann, O.: Big data and machine learning for the smart factory—solutions for condition monitoring. Diagnosis Optim (2017). https://doi.org/10.1007/978-3-319-42559-7_18
15. Gondra, I.: Applying machine learning to software fault-proneness prediction. J. Syst. Softw. **81**, 186–195 (2008). https://doi.org/10.1016/j.jss.2007.05.035
16. Elderman, R., Pater, L.J., Thie, A.S., Drugan, M.M., Wiering, M.A.: Adversarial reinforcement learning in a cyber security simulation. ICAART (2017)
17. Levesque, L., Fernandez, F., Somayaji, J., Anil.: Risk prediction of malware victimization based on user behavior. In: Proceedings of the 9th IEEE International Conference on Malicious and Unwanted Software, MALCON, 128–134 (2014). https://doi.org/10.1109/MALWARE.2014.6999412

18. Fletcher, R.: Practical Methods of Optimization (2nd ed.), Wiley, New York (1987). ISBN 978-0-471-91547-8
19. Perkusich, M., Soares, G., Almeida, H., Perkusich, A.: A procedure to detect problems of processes in software development projects using Bayesian networks. Expert Syst. Appl. **42**, 437–450 (2015). https://doi.org/10.1016/j.eswa.2014.08.015
20. Paltrinieri, N., Comfort, L., Reniers, G.: Learning about risk: machine learning for risk assessment. Safety Sci, 118 (2019). https://doi.org/10.1016/j.ssci.2019.06.001
21. Elliott, D., Keen, W., Miao, L.: Recent advances in connected and automated vehicles. J. Traffic Trans. Eng. (English Edition) 6 (2019). https://doi.org/10.1016/j.jtte.2018.09.005
22. Bitam, S., Mellouk, A.: Vehicular Ad Hoc Networks (2014). https://doi.org/10.1002/9781119004967.ch1
23. Dipl.-Ing, Martin & nat, Ulrich & (FH, Gerhard.: Electronic control unit (2015). https://doi.org/10.1007/978-3-658-03975-2_3
24. Tyler, TRJ: Trojan Horses (2017). https://doi.org/10.1007/978-3-319-73380-7_5
25. Ozkaya, Erdal & Islam, Md Rafiqul: Malware (2019). https://doi.org/10.1201/9780367260453-5
26. Poliakov, A.: An example of fuzzy decision-making system. Catalysis Commun. CATAL COMMUN **2**, 382—384 (2003). https://doi.org/10.1109/korus.2003.1222641
27. Sargolzaei, A., Crane, C., Abbaspour, A., Noei, S.: A Machine Learning Approach for Fault Detection in Vehicular Cyber-Physical Systems, 636–640 (2016). https://doi.org/10.1109/icmla.2016.0112
28. Wolf, M., Serpanos, D.: False Data Injection Attacks (2020). https://doi.org/10.1007/978-3-030-25808-5_6
29. Park, S., Choi, J.-Y.: Malware detection in self-driving vehicles using machine learning algorithms. J. Adv. Transp. **2020**, 1–9 (2020). https://doi.org/10.1155/2020/3035741
30. sci-kit learn. (n.d.). Retrieved April 14, 2020, from https://scikit-learn.org/stable/
31. Vashishth, V., Chhabra, A., Sharma, D.K.: A machine learning approach using classifier cascades for optimal routing in opportunistic internet of things networks. In: 16th IEEE International Conference on Sensing, Communication, and Networking (SECON), 10–13 June 2019, Boston, MA, USA
32. Sharma, D.K., Dhurandher, S.K., Woungang, I., Srivastava, R.K., Mohananey, A., Rodrigues, J.J.P.C.: A machine learning-based protocol for efficient routing in opportunistic networks. IEEE SYSTEMS JOURNAL, December 2016, ISSN (Print): 1932–8184, ISSN (Online): 1937–9234, pp. 1–7. https://doi.org/10.1109/jsyst.2016.2630923
33. Vashishth, V., Chhabra, A., Sharma, D.K.: GMMR: a Gaussian mixture model based unsupervised machine learning approach for optimal routing in opportunistic IoT networks. Comput. Commun. Elsevier **134**(15), 138–148 (2019). https://doi.org/10.1016/j.comcom.2018.12.001
34. Sharma, A.: Intrusion Detection System (2019). https://doi.org/10.13140/rg.2.2.14638.87360
35. Chen, Y., Qin, Y., Xiang, Y., Zhong, J., Jiao, X.: Intrusion detection system based on immune algorithm and support vector machine in wireless sensor network, 372–376 (2010). https://doi.org/10.1007/978-3-642-19853-3_54
36. Tan, Z., Jamdagni, A., He, X., Nanda, P., Liu, R.: A system for denial-of-service attack detection based on multivariate correlation analysis. IEEE Trans. Parallel Distrib. Syst. **25**, 447–456 (2014). https://doi.org/10.1109/TPDS.2013.146
37. Downton, F., DuBois, P., Anderson, T., Roy, S.: Multivariate correlational analysis. Mathematical Gazette **44**, 154 (1960). https://doi.org/10.2307/3612602
38. Chhabra, A., Vashishth, V., Sharma, D.K.: A game theory based secure model against Black hole attacks in opportunistic networks. In: Proceedings of 51st Annual Conference on Information Sciences and Systems (CISS), 2017, 22–24 March 2017, Baltimore, MD, USA, pp. 1–6
39. Chhabra, A., Vashishth, V., Sharma, D.K.: A fuzzy logic and game theory based adaptive approach for securing opportunistic networks against black hole attacks. Int. J. Commun. Syst. Wiley **31**(4), 10 (2018). https://doi.org/10.1002/dac.3487
40. Sharma, D.K., Agarwal, S., Pasrija, S., Kumar, S.: ETSP: Enhanced trust-based security protocol to Handle Blackhole attacks in opportunistic networks. In: Jain V., Chaudhary G.,

Taplamacioglu, M., Agarwal, M. (Eds.) Advances in Data Sciences, Security and Applications. Lecture Notes in Electrical Engineering, vol. 612. Springer, Singapore (2020)
41. Aref, M., Jayaweera, S., Machuzak, S.: Multi-Agent Reinforcement Learning Based Cognitive Anti-Jamming, 1–6 (2017). https://doi.org/10.1109/wcnc.2017.7925694
42. Clifton, J., Laber, E.: Q-learning: theory and applications. Ann. Rev. Statist. Appl. **7**, 279–301 (2020). https://doi.org/10.1146/annurev-statistics-031219-041220
43. Li, Y.: Wideband Autonomous Cognitive Radios: Spectrum Awareness and PHY/MAC Decision Making (2013). https://doi.org/10.13140/rg.2.2.14883.71202
44. Kosmanos, D., Pappas, A., Maglaras, L., Moschoyiannis, S., Aparicio-Navarro, F., Argyriou, A., Janicke, H.: A novel intrusion detection system against spoofing attacks in connected electric vehicles. Array (2019). https://doi.org/10.1016/j.array.2019.100013
45. Dahane, A., Nasr-eddine, B.: Wireless Sensor Networks: A Survey (2019). https://doi.org/10.1201/9781351190756-1
46. Effinger, G., Mullen, G.: RSA Cryptographic System (2019). https://doi.org/10.1201/9780429324819-20
47. Alajbegović, H., Zečić, D., Jamak, H.: Digital Signature Algorithm (DSA) (2006)
48. Jakimoski, G.: Some Notes on the Security of the Timed Efficient Stream Loss-Tolerant Authentication Scheme, 342–357 (2006). https://doi.org/10.1007/978-3-540-74462-7_24
49. Pan, F., Pang, Z., Wen, H., Luvisotto, M., Xiao, M., Liao, R.-F., Chen, J.: Threshold-free physical layer authentication based on machine learning for industrial wireless CPS. IEEE Trans. Industr. Inform., p. 1 (2019). https://doi.org/10.1109/tii.2019.2925418
50. Hothorn, T., Lausen, B.: Bundling classifiers by bagging trees. Comput. Statist. Data Anal., 1068–1078 (2005). https://doi.org/10.1016/j.csda.2004.06.019
51. Wen, L., Li, X., Gao, L., Zhang, Y.: A new convolutional neural network based data-driven fault diagnosis method. IEEE Trans. Industr. Electron., p. 1 (2017). https://doi.org/10.1109/tie.2017.2774777
52. Feng, S., Wu, J., Zhou, S., Li, R.: The Implementation of LeNet-5 with NVDLA on RISC-V SoC (2019). 39–42. https://doi.org/10.1109/icsess47205.2019.9040769
53. Sharma, D.K., Kwatra, K., Manwani, M.: Smartphone security and forensic analysis. In: Forensic Investigations and Risk Management in Mobile and Wireless Communications, IGI Global, pp. 26–50 (2020)
54. Khera, A., Singh, D., Sharma, D.K.: Information security and privacy in healthcare records: threat analysis, classification, and solutions. Security and Privacy of Electronic Healthcare Records: Concepts, Paradigms and Solutions, IET, pp. 223–247 (2019)

A Model for Auditing Smart Intrusion Detection Systems (IDSs) and Log Analyzers in Cyber-Physical Systems (CPSs)

Joshua Ojo Nehinbe

Abstract Suitable models that auditors can adopt to conduct concurrent audit of smart Intrusion Detection Systems (IDSs) and log analyzers in Cyber-Physical Systems that are also founded on sound emperical claims are scarce. Recently, post-intrusion studies on the resilience of the above mechanisms and prevalence of intrusions in the above domains have shown that certain intrusions that can reduce the performance of smart IDSs can equally overwhelm log analyzers such that both mechanisms can gradually dwindle and suddenly stop working. Studies have also shown that several components of Cyber-Physical Systems have unusual vulnerabilities. These key issues often increase cyber threats on data security and privacy of resources that many users can receive over Internet of a Thing (IoT). Dreadful intrusions on physical and computational components of Cyber-Physical Systems can cause systemic reduction in global economy, quality of digital services and continue usage of smart toolkits that should support risk assessments and identification of strategies of intruders. Unfortunately, pragmatic studies on how to reduce the above problems are grossly inadequate. This chapter uses alerts from Snort and C++ programming language to practically explore the above issues and further proposes a feasible model for operators and researchers to lessen the problems. Evaluation with real and synthetic datasets demonstrates that the capabilities and resilience of smart Intrusion Detection Systems (IDSs) to safeguard Cyber-Physical Systems (CPSs) can be improved given a framework to facilitate audit of smart IDSs and log analyzers in Cyberspaces and knowledge of the variability in lengths and components of alerts warned by Smart Intrusion Detection Systems (IDSs).

Keywords Intrusion · Intrusion detection systems (IDSs) · Network intrusion detection system · Smart IDSs · IDS audit · IS auditor · Cyber-Physical Systems (CPSs)

J. O. Nehinbe (✉)
ICT Security Consultant, Lagos, Nigeria
e-mail: nehinbe@yahoo.com

A. I. Awad et al. (eds.), *Security in Cyber-Physical Systems*, Studies in Systems, Decision and Control 339, https://doi.org/10.1007/978-3-030-67361-1_5

1 Introduction

Pragmatic studies have recently shown that Cyber-Physical Systems (CPSs) must be adequately protected with security tools to reduce the rising cases of Cyber-Physical attacks and the destructive impacts of these attacks on global economy, international security, digital services and means of livelihood of many ethnic and social groups across the globe [1–3]. Further studies have shown that components of Cyber-Physical Systems (CPSs) possess individual vulnerabilities that can endanger continuous usage of Cyber-Physical Systems (CPSs) [4, 5]. The nature of the problems with different kinds of threats and cyber attacks on Cyber-Physical Systems (CPSs) can correlate to severe disasters and complex confusion that may involve different stakeholders. The motives of some intruders may be complex to understand if they simultaneously attack the seamless integration of physical components and the computational elements of Cyber-Physical Systems (CPSs) [6]. The impacts of some successful cyber attacks in this domain may corrupt or damage Cyber-Physical data [3, 7]. Some intrusions can leak sensitive information to wider audience via social media with the aims to extort and discredit victims and service providers of Cyber-Physical Systems (CPSs) [7].

The complexity and reoccurrence of threats and cyber attacks on the entire components of Cyber-Physical Systems (CPSs) have made many organizations to develop the habit of deploying several categories of Intrusion Detection Systems (IDSs) within the peripherals and gateways of their connections to the entire Cyber-Physical systems (CPSs) so that these devices can collect and analyze activities that signify evidence of intrusions against their corporate networks in real-time [8, 9]. Subsequently, analysts can quickly review the reports and respond to the attacks before they achieve the objectives of intruders that launch them [10]. These issues have inevitably generated several challenges and concerns regarding the effectiveness of IDSs and analyzers of logs of IDSs in monitoring complex architectural systems peculiar to the above domains over the years.

Figure 1 demonstrates one of the two approaches organizations can adopt to position Network Intrusion Detection System (NIDS) in relation to firewall within the peripherals and gateways that connect them to the entire Cyber-Physical Systems (CPSs) [3, 8].

Nevertheless, numerous studies often attest that Intrusion Detection Systems (IDSs) must always be upgraded to strongly help operators control the new dimensions and rising waves of intrusions against cyber-physical resources across the globe. One of the pragmatic methods to achieve this security objective is to make IDSs smarter by connecting them to the Global Systems of Mobile (GSM) communication so that the toolkits can always send alerts to remote operators such that operators can promptly respond to cyber attacks at all time [11]. Thus, smart IDSs are IDSs that are configured such that operators can receive and respond to their alerts through Short Message Services (SMS) to the GSM or email addresses of the operators of IDSs in Cyber-Physical Systems (CPSs). However, there are security and business requirements that underpin the framework upon which smart IDSs reside in private

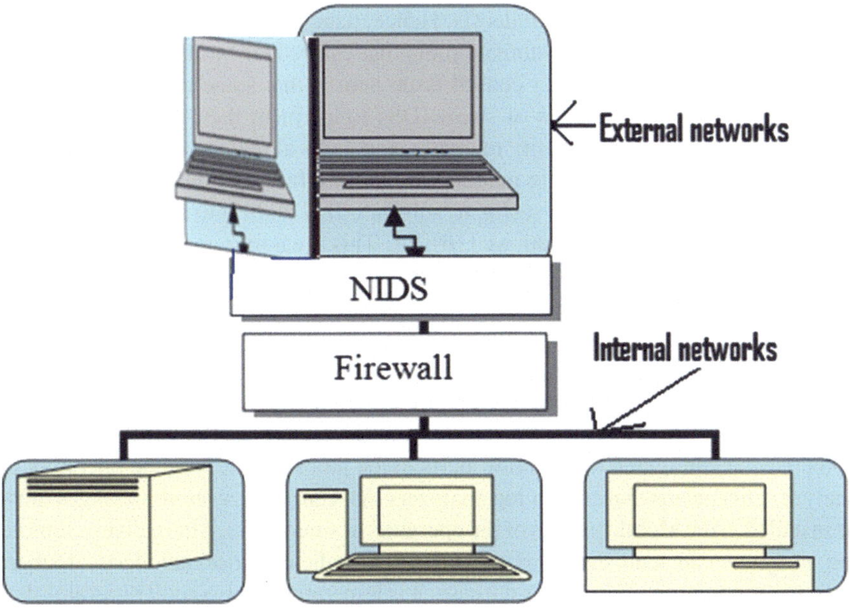

Fig. 1 NIDSs in front of firewall

and corporate settings [2, 7]. The resilience and capacities of smart IDSs can be improved if operators can combine the information they gather from audit of log analyzers with the knowledge of the variability of lengths and components of alerts that smart Intrusion Detection Systems (IDSs) in the networks have generated. This can be used to ultimately design and improve the security policy on smart IDSs in the corporate elements of Cyber-Physical Systems (CPSs) [3, 12, 13]. However, empirical studies on smart IDSs that specifically focus on audit of smart IDSs and log analyzers are inadequate over the years.

Basically, empirical studies on smart IDSs in the context of Cyber-Physical Systems (CPSs) involve pragmatic examinations of specific experiments conducted with smart IDSs to concurrently correct security concerns and audit issues. These procedures can assist operators to improve the detection of intrusions against Cyber-Physical Systems (CPSs) and cloud resources at large. The argument underpinning this chapter is that logs of smart IDSs should be concurrently audited during IDS audit. Otherwise, they may not be very useful for post-intrusion reviews. Similarly, lack of audit of logs of smart IDSs may render them ineffective for in-house training of newly recruited auditors and researchers exploring issues on identification, analysis, corroboration and mitigations of threats and security lapses in Cyber-Physical Systems (CPSs) [8, 14].

Furthermore, smart IDSs are well-known for generating large quantities of alerts whenever they are configured to detect possible intrusions against Cyber-Physical Systems (CPSs) [9, 11, 15]. It is inefficient to manually analyze massive alerts without incurring huge overheads and tradeoffs. Hence, data mining is often recommended as an underlying concept to automate tools that can reduce workload due to alerts from smart IDSs [15]. Another central issue here is that some companies use the reports obtained from the logs of smart IDSs to augment their networks security policies [8, 14. 16]. The necessity to audit smart IDSs alongside with log analyzers is not mandatory in the existing models for auditing Information Technology (IT). This generic audit framework seems to subsume IDS audit into security policy on computers and telecommunications [16–20]. This weakness may eventually lead to lack of segregation of duties among internal auditors, IDS researchers and IDS operators. The human elements of the Cyber-Physical Systems (CPSs) may place emphasize on Firewall and other forms of the Intrusion Prevention Systems (IPSs) over smart IDSs in the context of the organizational settings in the above settings. Moreover, it is plausible that some logs of regular IDSs that were archived might be relatively uninteresting details. One of the three central issues here is that the IDSs may be configured to send raw alerts to the mobile devices of the operators to analyze. This means that certain log analyzers that can analyze short messages must be installed in the Mobile phones of the operators of smart IDSs. Alternatively, remote log analyzers can send short text messages that indicate processed alerts of smart IDSs to the operators. Whichever the case, it is imperative to also audit programs that analyze logs of smart IDSs in Cyber-Physical Systems (CPSs) to regularly establish the degree of information inherent in the archived logs at each time and to ascertain the patterns of packets intended to overload smart IDSs at certain period of time in the above settings [8].

Findings suggest that suitable realistic datasets that can be used to concurrently audit smart IDSs and logs analyzers are grossly inadequate for researchers due to security issues [18, 19]. Accordingly, the above domain of IDS audit in the security of networks and other components of Cyber-Physical Systems (CPSs) continues to suffer a major setback over the years. Therefore, by using alerts from Snort and C++ programming language, this chapter presents a comprehensive review of the above research issues and further proposes a feasible model that professionals can adopt to lessen the problems. One of the significant contributions of this chapter is its ability to practically provide clear review and guidelines that experts and trainees can adopt to ensure perimeter defense of mobile and computer networks. The chapter uses four datasets to practically illustrate a new framework for concurrent auditing of smart IDSs and log analyzers within corporations in the entire Cyber-Physical Systems (CPSs). Also, the chapter broadly justifies the importance of conducting audit of log analyzers in smart phones together with IDSs audit. The remainders of this chapter are organized in the following order. Section 2 will present background research work that relates to IDS auditing. Section 3 explains the scope of IDS audit in Cyber-Physical Systems (CPSs). Section 4 discusses challenges confronting IDS auditors in auditing Cyber-Physical Systems (CPSs). Section 5 provides the proposed methodology for auditing smart IDSs and log analyzers while Sect. 6 concludes the chapter.

2 Background Information on Audit of Smart IDSs and Log Analyzers in Cyber-Physical Systems (CPSs)

Studies have shown that Cyber-Physical Systems (CPSs) are mergers of collaborative networks of automatic systems that are strongly built on sound theoretical and scientific principles and seamless integration of many disciplines [1, 2, 6]. Some of the disciplines that contribute to progressive growth and modernize Cyber-Physical Systems CPSs) over the years include informatics, computer and, mobile systems, Wireless Sensor Networks (WSNs), cyberspace, system designs, software, process, robotic, automobile and mechanical engineering [1, 2, 6, 21]. The underlying benefit of incorporating integrated components to drive Cyber-Physical Systems (CPSs) is easy connectivity of many devices and systems to Cyber-Physical systems (CPSs) across the globe. This capability has resulted into wider applications of Cyber-Physical Resources (CPRs) in the areas of medical services, agriculture, electric installations, space engineering and other notable facets of human life [6, 21].

Critical issues begin to surface with the inexhaustible growth currently recorded in this domain in recent years especially on the numbers of service users, service providers and revenue accrued from sales of products and services that relate to Cyber-Physical Systems (CPSs) [2]. Empirically, experts have argued that security, computational efficiencies and degree of helpfulness of complex architectural framework that underlying seamless integrations of physical and computation components of Cyber-Physical Systems(CPSs) are serious doubts whenever these components are evaluated on the basis of performance, quality of service, users' satisfactions and robustness to counter threats and challenges [5, 14, 22]. Yet, emphasis over the years has focused on the computational capabilities of Cyber-Physical Systems (CPSs) but less attention has been paid to the link between the computational and physical elements of this domain [5]. These flaws have raised series of technical and research issues on how to forecast traffic flow, optimize Mobile Cyber-Physical applications and how to achieve high performances of social services and healthcare facilities like wearable devices that run on Internet of a Thing (IoT) [1]. The correlations between social settings and industrial applications of cloud-based services that interact with Cyber-Physical Systems' designs; innovation and manufacturing of digital resources continue to generate new paradigms in manufacturing and design's settings [4, 6]. These necessitate the importance of measures to bridge the gap between the Cyber-Physical resources and social setting. Collaborative design of embedded systems and various algorithms that experts have designed to carry out co-modeling and co-simulation of novel innovations begin to emerge. However, majority of these algorithms often exhibit invisible flaws [21].

The above issues coupled with the alarming increase of intrusions against Cyber-Physical Systems (CPSs) have resulted in the needs for organizations to adopt Intrusion Detection Systems (IDSs) [5, 8, 10]. These toolkits can then provide automated ways to monitor, analyze all incoming and outgoing network packets in their corporate networks, trigger and log alerts on suspicious packets they observe for security and decision purposes. Nevertheless, most of these mechanisms can only detect suspicious packets [9]. They have been criticized for lacking capabilities to make dependable decisions on suspicious activities of users that may signify security breach to Cyber-Physical Systems (CPSs) [23]. Operators must carefully review alerts they generate to isolate false positives from realistic attacks. Alerts can be daunting and overwhelmingly difficult to manually analyze by operators. Series of log analyzers have been proposed over the years to compensate for these weaknesses [23, 24]. Studies have shown that significant numbers of log analyzers have limited capabilities required to categorize cyber attacks on the basis of all attributes of alerts [23]. A few numbers of researches has suggested that, the above devices should be upgraded so that they can intimate operators with alerts on real-time basis [5, 11]. The rationale is that operators should be able to remotely analyze intrusion logs and counter attacks on Cyber-Physical Systems without the need to physically .report to their offices.

These developments have led to the need to audit smart Intrusion Detection Systems (IDSs) to improve their efficacies. Audit of smart Intrusion Detection Systems (IDSs) or IDS audit involves comprehensive and thorough examination of the networking infrastructure and security controls upon which the management and operations of all smart Intrusion Detection Systems (IDSs) in an organization are established [17, 18, 25]. Ordinarily, one of the duties of IDS auditors is to thoroughly scrutinize IDSs, establish and report the efficacies of internal controls that the organization has implemented to safeguard each detector and resources related to these toolkits [16]. The evaluation and the reports of this kind of audit can go a long way to determine the level of compliance and operations of all intrusion detectors in the company with best global practices. Nonetheless, there are numerous challenges with research on audit of smart IDSs in corporate setting in the past years [18]. Studies advise that skilled intruders are common threats that are extremely disturbing corporate and private users of computer systems in Cyber-Physical systems (CPSs) [3, 7, 10]. Unfortunately, researchers habitually ignore the audit of smart IDSs that should have established exploitable pathways, audit issues and novel paradigms on network security and perimeter defense since the inception of IDS technology. This neglect has countless impacts on digital resources that connect to cyber-physical resources. This shortcoming is explicitly dangerous because it is generating warning signals service providers concerning data reliability and quality of service on local computing resources in many organizations. The impacts of some of these security concerns may appear negligible while significant numbers of them are grievous and hazardous to corporate existence considering the capabilities of demoralizing intrusions recently reported in some public media. Recently, the neglect of this aspect of IDS audit and lack of correlation of IDS audit with research findings have begun to subject sequence of findings from logs analyzers, integrity and compliance with professional standards and regulatory authorities to series of contentions [6, 22, 26].

Importantly, sudden changes in the classifications and dimensions of intrusions that often aim to attack computer and mobile services operating within the purview of Cyber-Physical Systems (CPSs) are global concerns [3, 24]. Intruders have acquired more skills such that they can launch packets that have short and long datagram to achieve different motives in cyberspace. Studies of many trace files suggest instances whereby intruders have split some inbound and outbound packets into fragments. Some studies believe that attackers on cyber-physical systems (CPSs) can suddenly varied the intensities of packets to smartly elude detections. Numerous audit and networking issues may begin to build up whenever new IDSs are installed in the perimeters of digital networks to complement existing IDSs that auditors have been previously audited. There are possibility that audit exercises may exclude auxiliary issues like log analysis on fragmented packets.

The location of IDSs relative to the firewall in an organization depends on their security policy. A growing numbers of opinions affirm that organizations can install Network Intrusion Detection System (NIDS) in the front or back of a firewall for different intentions [3, 24]. However, models that auditors can adopt to establish suitable approach to organizations are very scarce. Furthermore, current model of ICT audit restrict IDS auditors to the physical security, hardware and software components of smart IDSs [25, 27]. Auditors must use simulated attacks to investigate the initialization, configuration, interface, processing and performances of smart IDSs and to ascertain the tendency of the toolkits to dwindle after a prolonged usage. They must also evaluate the available disk spaces for both the toolkits and mobile devices that receive alerts from IDSs and log analyzers. They must assess the contingency plans in the organization to establish business continuity and preparedness of the toolkits to resume surveillance after intruders have attacked them or after downtime. Auditors must equally evaluate the internal and change controls designed to safeguard the smart IDSs from computer viruses and intruders. In addition, they will investigate the signatures, alert's mechanism, policies and possible rules that have been updated, their corresponding approvals and authorizers of the approvals to modify them [17, 25]. Nonetheless, the above procedures are flawed in the sense that both the experienced and inexperienced intruders may obfuscate and evade smart IDSs audited with the above model. Thus, intrusions on cyber components such as sensing, cyber communication mechanisms and physical resources like computer hardware, data center, employees and mobile devices that the detectors should have discerned and operators would have timely countered often achieve intruders' missions at long run.

One of the fundamental ways this chapter premises for operators and resident auditors to lessen the above problems is for both of them to periodically corroborate research with audit reports on smart IDSs in the perimeters of the organization. However, IDS audit is quite challenging nowadays because it is clearly different from the conventional IS audit process [17, 18]. Besides, IDSs audit requires the engagement of qualified IS auditors that also possess wide experience and knowledge in the above roles. Suitable IS auditors must also have practical experience on the installations of smart IDSs, logs' analyzers, reporting and countermeasures. Moreover, there are acute shortages of operators that also possess auditing skills. Besides, standard IDS audit templates and models that can serve as guiding principles to IDS auditors and operators in corporate environment in the context of Cyber-Physical Systems (CPSs) are scarce [17, 18, 25]. Consequently, most IDS operators ignore the research aspect of their jobs that should be regarded as interim audit and concentrate on IDS operations.

Furthermore, approaches that most auditors frequently adopt to conduct IDS audit with generic Information System (IS) and audit process often exclude evaluation of the significance of log analyzers in the organization [26]. The dangers of the above methods are enormous especially if both reviews are inconclusive, unreliable and unsupported by empirical claims before major infringement occurs in the digital networks of the organization. Organizations can experience infringements in critical and less critical areas of their business operations. Intruders may attack resources or areas of corporate systems that attract little or no attention of IT managers, inspection and internal control's managers with the aims to have enough time to achieve their objectives and to equally evade detection. Consequently, feelers premise that smart IDSs should be strategically installed in the segments that will make it difficult for intruders to bypass them. Smart IDSs that are located at the hearts of huge inbound or outbound traffic should be thoroughly verified by IS auditors from time to time. Traffic that migrates across spanning mode can overwhelm smart IDSs that are technically weak to compromise.

Generally, research findings and related work in the domains of IDS audit and log analyzers are novel issues in network security and Cyber-Physical Systems (CPSs) [16, 18, 25]. Conventionally, experts have justified the significance of IDS policy in the perimeter defense of networks of corporate organizations [8, 13, 26]. An empirical study that examined risk-based systems and process audit method has been carried out as a strategy to bridge the gap between auditors and architectural designs of IT resources [18]. The model was able to detect the weaknesses of the process in terms of risk of material deficiencies and thirteen control patterns. However, the research was basically a generalized audit process that has a better performance whenever the model is adopted to audit financial data. Moreover, a study on how to debug Network Intrusion Detection Systems (NIDSs) has been explored [24]. The proposed model uses detection rules to debug NIDSs and eradicate defective rules that are well-known for triggering repetitive alerts. The model can assist IDS operators to reduce workload. However, the major flaw of this model is that it has the tendency to be operationally proprietary. The model will require routinely extension and upgrade before it can broadly relevant to other categories of smart IDSs in the market.

3 The Scope of Audit of Smart IDSs and Log Analyzers in Cyber-Physical Systems (CPSs)

A systematic review of IDS audit is a methodical review or examination of the operational conditions of IDSs with the aims to ensure their protection and to guarantee efficient, effective and reliable IDS operations within the perimeters of computers; cyber-physical and sensing resources and mobile networks in an organization [13, 17, 28]. The scope of Cyber-Physical Systems (CPSs) varies from organization to organization. Algorithms are the underpinning mechanisms that control and regulate collaborative networks of theories, concepts and embedded disciplines that constitute Cyber-Physical Systems (CPSs) in each organization [21]. Audit review should reflect components of computer and mobile systems to be audited. It should state cloud resources such as networks of computers, mobile systems, Wireless Sensor Networks (WSNs), front-end and back-end of the networks, software, hardware, human element, work flow and process engineering [6]. Audit of smart IDS can be performed in conjunction with or separated from the conventional audit exercises in an organization. The audit time table, management, misgivings and repeated outbreak of intrusions can influence the necessity to conduct IDS audit and its scope of coverage. For Cyber-Physical Systems (CPSs), the scope of the audit should include the security of sensing processing, storage of large alerts, performance of hardware and software and reliability of the systems. It should also extend to validation of algorithms, automatic systems, theoretical and scientific principles and seamless integration of disciplines underlying the systems with best practices. Hence, this type of IDS audit is eventful [15, 18]. Examiners must carefully review and match the security policy of the organization with the implementations of smart IDS in the live and test environments to establish areas of compliance and noncompliance with best practice. Fundamentally, enterprise must have IDS policy. An IDS policy is a standard document stating a plan of actions an organization adopts regarding the administration and management of IDSs within their digital networks [8]. Besides, IDS policy should state IDS procedures, IDS rules and conditions that should be meant before rules can be activated, updated or deactivated [13]. The main challenge that IDS auditors often face is that most organizations do not have IDS policy [25]. Findings suggest that some companies do not isolate IDS policy from their security policies [8]. Hence, rather than separating both policies, some of them embedded a few sentences about IDSs in their security policies. Consequently, IDS audit and its ancillaries often lack exhaustive reviews over the years. Therefore, IDS auditor that wishes to conduct the above IDS audit must have well-established knowledge of IDS policy and major components of the smart IDSs within the networks.

In Snort for instance, the objectives of the audit must include critical review of IDS policy, physical security relating to the IDS (Snort in this case), the hardware component and software components of the toolkit. The audit must also include packet decoder, preprocessors, detection engine, logging and alerting system and output modules [8, 13]. Serious audit issues may arise whenever auditors lack strong knowledge of the above components and how they cooperatively work together to detect intrusions and to generate output in the required format.

4 Auditors' Challenges in Auditing Smart IDSs in Cyber-Physical Systems (CPSs)

Cyber-Physical Systems (CPSs) lack the robustness to counter threats, challenges and cyber attacks due to weaknesses genetic to individual components that form these domains. Hence, there are critical challenges that face auditors and researchers of smart IDSs regarding IDS auditing and log analyzers in these domains. This section discusses and categorizes some of these issues into two groups; namely, the challenges with smart IDSs and challenges with log analyzers.

4.1 Research and Audit Issues on Smart IDSs in Cyber-Physical Systems

Different types of smart IDSs keep different categories of logs and alerts in different formats. The default settings of parameters that coordinate alerts of smart IDSs can enable the toolkits to trigger and log wordy and more explicit warnings than the setup that customize these parameters [9, 10]. Figure 2 illustrates one of the kinds of alerts that Snort can generate. The alerts are in comma delimited format because each attribute of an alert is separated by a comma. Operators of smart IDSs can implement the formats of alerts they want during implementation and before executing IDSs like Snort. The major issue is that the preferred formats of alerts cannot be reversed while the toolkits are working. This can create series of setbacks if operators if the formats they have implemented do not convey sufficient information operators will need to decide on the security matters of Cyber-Physical Systems in the organization. For instance, it is evidence in Fig. 2 that the alerts contain IP addresses to uniquely identify computers and their domain names on the Internet. The alerts are samples of comma delimited alerts extracted from Defcon-11 traces. Some of the attributes of the alerts were Transmission Control Protocol (TCP). However, further information is still required to ascertain attributes like the names, of the attacks to understand data transmission and exchange that occurred between sources and destinations of various attacks.

Figure 3 illustrates conventional kinds of alerts that the Snort would log whenever its default parameters on logs and alerts are implemented in Cyber-Physical Systems (CPSs). This format is simple because each alert is explicit to human interpreters.

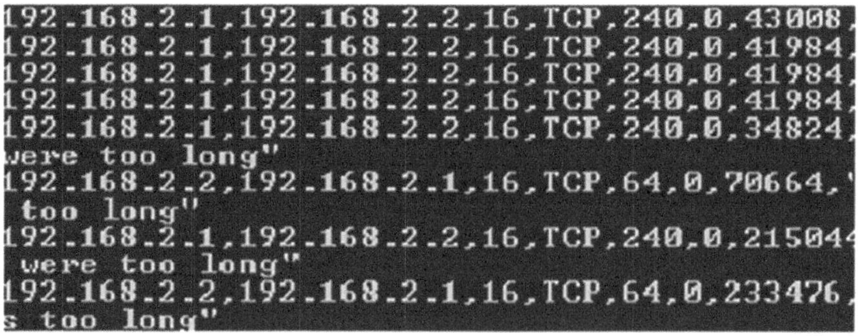

Fig. 2 A sample of alerts from Defcon11 in comma delimited format

```
[**] [125:1:1] (ftp_telnet) TELNET CMD on FTP Command Channel [**]
[Priority: 3]
08/03-22:14:26.756815 192.168.2.2:21 -> 192.168.2.1:1067
TCP TTL:64 TOS:0x10 ID:6518 IpLen:20 DgmLen:83 DF
***AP*** Seq: 0x17BA8D92  Ack: 0xFBCEEF87  Win: 0x7D78  TcpLen: 32
TCP Options (3) => NOP NOP TS: 116909 288736
[**] [125:1:1] (ftp_telnet) TELNET CMD on FTP Command Channel [**]
[Priority: 3]
08/03-22:14:26.757820 192.168.2.1:1067 -> 192.168.2.2:21
TCP TTL:240 TOS:0x10 ID:0 IpLen:20 DgmLen:66
***AP*** Seq: 0xFBCEEF87  Ack: 0x17BA8DB1  Win: 0x7D78  TcpLen: 20
[**] [125:1:1] (ftp_telnet) TELNET CMD on FTP Command Channel [**]
[Priority: 3]
08/03-22:14:26.762762 192.168.2.2:21 -> 192.168.2.1:1067
TCP TTL:64 TOS:0x10 ID:6519 IpLen:20 DgmLen:75 DF
***AP*** Seq: 0x17BA8DB1  Ack: 0xFBCEEFA1  Win: 0x7D78  TcpLen: 32
TCP Options (3) => NOP NOP TS: 116910 288736
[**] [125:1:1] (ftp_telnet) TELNET CMD on FTP Command Channel [**]
[
```

Fig. 3 A sample of alerts of Snort in a default formst

For example, the signature generator (Sig_generator) of the first attack in Fig. 3, the identification number (Sig_id) of the rule that triggered the alert and the number of times the rule has been reviewed or updated (Sig_rev) were 125, 1 and 1 respectively [9, 10]. The alerts are samples of default alerts extracted from Defcon-10 traces. The attack signified telnet's exploits. In other words, the Intrusion Detection System (Snort) detected telnet commands on the FTP command prompt or channel. The attack also indicated that someone used a computer with IP address 192.168.2.2 and port 21 to transfer file to a computer with IP address of 192.168.2.1 and port 1067 at 10:14 PM on 3rd of August. The problem with alerts that are formatted by comma delimiters is that auditors would require their documentations to properly understand them because they are not constantly explicit.

It is imperative for the auditor to establish the directory where the alerts and systems files of the IDS are kept or recorded in the hardware before the beginning of the audit. By default, shows will Snort log alerts to*/vary/log/snort/alerts*. However, the auditor begins to face further challenges if the directory is changed during implementation contrary to the conventional or documented standard. Additional challenges can occur due to the noncompliance of the organization to both the recommended disk space and accepted format for alerts in the IDS policy [8, 10]. The implication is that it will be difficult to compare the sufficiency of the information conveyed in the IDS logs and short text messages that are extracted from different segments of the perimeters of the same organization if they have heterogeneous formats. In essence, the above sample of the raw alerts explains the link between research on smart IDSs and IDS audit.

4.2 Issues with Detection Rules or Policies of Smart IDSs in Cyber-Physical Systems

Practical experience shows that the formats of detection rules vary from Intrusion Detection System to another. The rules within the detection engine of smart IDSs are many and they are mostly protected by copyright. The rules usually instruct smart IDSs to discriminate by logging and raising alerts on specific packets that migrate from specific networks into local subnets. Some rules are also designed to instruct smart IDSs to indiscriminately log and raise alerts on all suspicious packets that migrate from any network into local subnets [9]. These rules can also instruct the toolkit to always trigger an alert whenever the device observes any TCP packet that contains "USER root" in its header [8]. Rules can be localized, designed or configure such that they will report suspicious packets heading towards a computer in the subnets of Cyber-Physical Systems [10]. Several audit issues arise regarding to best strategies to audit rules or policies of IDSs. These toolkits have several inbuilt rules or policies. There may be some discrepancies between the rules or policies that have been implemented in the organization and the security policy driving the implementation of rules or policies of the smart IDSs in the system. Discrepancies can also occur if some IDSs in the networks are not configured to operate as smart toolkits. Professionalism is required in adapting framework for auditing smart IDSs to audit IDSs that are not configured as smart toolkits in other to adequately safeguard the entire components of cyber-physical resources in the organization. One of the reasons behind these challenges is that the security policy of the organization might not fully reflect the totality of the rules or policies in the detection engines of all smart IDSs in the networks. The IS auditor needs to evaluate if the IDS policy actually states specific rules or policies that should be activated or deactivated during implementations of smart IDSs. It is also necessary for auditors to establish the level of compliance of the organization with best security practice on the detection rules or policies approved by the management of the organization [8, 25].

New rules or policies can be added to the smart IDSs in other to improve their efficacies. However, some rules or policies may generate redundant alerts. Hence, it is often difficult to immediately establish the criticality of new and old rules or policies without a critical exploration of log analyzers that process alerts that correspond to these rules or policies. Also, session printable policies or rules are difficult to recommend for deactivation because they enable the detector to log everything attackers have typed [8, 10]. It is possible that all sections of the IDS policy will not fully capture the sensitivity of detection rules or policies in organizations. The chapter encourages auditors to thoroughly audit available IDS policy to ensure the policy is providing suitable standard that covers all components of cyber-physical resources adopted in the organization.

4.3 Issues with Maintenance of Smart IDSs in Cyber-Physical Systems

Smart IDSs must undergo regular maintenance so that they can adequately monitor very high traffic rates migrating into or outside the organization [20, 23, 24]. The maintenance of smart IDSs is the process of performing system tuning and routine checks on all smart Intrusion Detection Systems in the organization; the directory of each configuration file, logs, text messages; available storage size, available disk space, disk space each toolkit has already utilized and the last time each toolkit was debugged to establish their readiness to promptly report intrusions that aim to exploit features of Cyber-Physical Systems that provide opportunities for intruders to cause havoc without corrupting cyber-physical data or leaking sensitive information from cyber-physical Networks. Furthermore, constant maintenance of smart IDSs will enable their operators to correct new and past errors that were not recognized during the installations, configurations and testing phases of these devices. Usually, corrective maintenance is desirable because it will enable the operators of smart IDSs to perfect and improve the operations and performance of smart IDSs [20].

Intruders can compromise the mobile phones and email accounts of operators of smart IDSs [11, 22]. Therefore, the above maintenance will equally help operators of smart IDSs to fine-tune the toolkits so that they can effectively work in new environments and whenever the operators replace their mobile devices or renounce old email accounts. However, maintenance of smart IDSs requires extra efforts than the efforts required to configure and analyze their logs. Hence, most operators of smart IDSs often shy away from carrying out IDS porting, corrective and adaptive maintenance of these toolkits. From experience, IS auditor can perceive series of audit issues whenever the IDS policy does not recognize the significance of maintenance of smart IDSs in the enterprise networks.

4.4 Issues with Configurations of Smart IDSs in Cyber-Physical Systems

There are hardware and software requirements for each smart IDS to exhibit performance that will always conform to best security practices. For NIDSs like Snort, the toolkit works on operating System like Linux, Windows 2003 Server Enterprise Edition and Microsoft Windows XP and hardware like Compaq 1600 Pentium III with dual Processor Server and Pentium IV workstation.

Using Snort as an example [8, 10], this premises that components such as Apache, Pretty Home Page (PHP), WinPcap and Analysis Console for Intrusion Databases (ACID) must be audited to ascertain their levels of compliance to best industrial practice [9]. The combination of Snort, Apache, database and ACID enable the NIDS to log alerts into a database. Two or more toolkits can be configured to centrally log alerts to unified database. Conversely, each toolkit may be setup to log its alerts to a different database. The above components also enable analysts to visualize and analyze alerts on web interface [8, 10]. Hence, the database (back-end) that may be MySQL must also be audited. IS auditors must always refer to the IDS policy for guidance. It is a good practice to complement the audit process by referring to the security policy of the organization to gain insightful evidence on degree of compliance and conformity of both documents.

The dangers are enormous whenever intruders compromise the back-end of the toolkit. Intruders can crash the entire toolkit, alter its cryptographic keys and render it bad and unintelligent [22]. Subsequently, they can illegally reconfigure the smart IDS to log no alerts or to suppress useful alerts [22]. New waves of stealthy attacks can shutdown IDSs; enable triggers and disable or re-start the back-end databases of the detectors. In the case of Snort, attackers can suddenly shutdown the Apache upon which the smart IDS runs. Hence, auditors must establish the level of control that safeguards all the components of smart IDSs in the firm. Usually, in Snort, Apache's server uses configuration file that is stored in the/etc.*/apache2/apche2.conf* [8, 25]. Therefore, auditors must also establish the last date the configuration's file was updated. Nonetheless, the above ideas are plausible whenever the auditors possess the needed skills to critically explore them.

4.5 Issues with IDS Policy and Security Policy in Cyber-Physical Systems

IDS policy is a document that is approved by top management in an organization [8]. This document reflects and states how all IDSs in the organization are implemented and managed. The document further reveals types of IDSs and their versions, configurations, license fees and expiry date and vendors. The document defines activities that managements of the organization have agreed to be regarded as normal and intrusive activities in their Cyber-Physical Systems. It is expected to reflect the approved connectivity between log analyzers and logs of smart IDSs. It might be uneconomical to send overwhelming alerts directly to the operators of smart IDSs. Additionally, some smart IDSs can encrypt the email reports or alerts they intend to send to the operators or recipients. However, operators or recipients must install suitable tool in their mobile phones or computers to decrypt them. Thus, IDS policy should categorically state how the email addresses and mobile phones of operators of smart IDSs will receive concise and helpful alerts.

The security policy of an organization is the totality of security mechanisms that is approved by top management of the organization. This broad document usually states how the security's architecture of the organization should be deployed, monitored and managed annually. IDS policy is a segment of security policy. Auditors may find it difficult to challenge operators of smart IDSs in an organization whereby IDS policy is subsumed in security policy. In addition, the appropriateness of time that the organization must review their IDS policy will be difficult to criticize in this circumstance.

Most often, some intruders prefer to launch attacks that can probe or scan cyber networks to compensate for their inabilities to have access to the above policy's frameworks [3, 9, 10]. Information System auditors need to assess the security of the above policies in the organization to establish how they are kept, the custodian of both documents, access and procedures for granting approvals to the employees that have the rights to use and rights to know these documents.

4.6 Research and Audit Issues with Log Analyzers in Cyber-Physical Systems

The quality of information that various log analyzers can derive from different formats of alerts that smart IDSs generate depend on many factors. Some analyzers of logs that originate from smart IDSs can process specific attributes such as Transmission Control Protocol (TCP), Internet Control Message Protocol (ICMP), Type of Service (TOS) and Internet Protocol (IP) length. Intruders that compromise the TCP and IP of computer networks will distort network conversations or communications and the exchange of data through application programs [10]. The attacks will also affect apps that send packets of data from one computer to another. Similarly, the values held in the flags of parameters or attributes of alerts also differ from one attribute to another. For instance, log analyzer that analysis the parameters of ICMP in Cyber-Physical Systems intend to discover actions of intruders that have requested for certain details about the systems [29]. The intention of the intruder may be to establish computers or mobile devices that signify echo reply and destination unreachable. The attack may also reveal weaknesses in the configurations of router within Cyber-Physical Systems (CPSs). The attack can publicize details of routers, timestamp, timestamp reply; redirect message headers, domain name request, domain name reply, mobile registration request, mobile registration reply, errors in the conversion of datagram; address mask request and address mask reply. Intrusions on trace route can provide trodden paths for Distributed Denial of Service (DDoS) attacks in Cyber-Physical Systems (CPSs) [29].

In addition, TOS is designed to categorize and prioritize networks' data so that digital devices will process critical data packets before they will process data packets that of less significant. However, intruders have many ways they can check the reliability of the networks. Attacks on TOS intend to undermine the quality of services rendered by the host and routers in the networks. This category of attacks can indiscriminately affect the migrations of different kinds of inbound and outbound data within the networks of Cyber-Physical Systems (CPSs) [3, 7]. Intruders can insert fake data into the networks given the knowledge of TOS in the networks. The impacts of this attack can be severe if it occurs at the peak of operations whereby it coincidentally hinders the priority and migrations of data of higher importance than data of less importance in the networks. Moreover, attack on TOS can increase the numbers of fragmented packets that lost in transit. It can also cause significant delay of packets to complete computer and mobile communications, reassembling of fragmented packets and routing of multimedia data.

Each of the above attributes of alerts conveys different meanings to different organizations [9]. The mode that every log analyzer adopts to write their results into the output files (folders) is very important. Programs that append new records with old records would require enough disk space than programs that always clear all the content of old records in the output files during execution. For these reasons, IDS auditors often face many challenges from company to company in conducting thorough investigations on outputs of log analyzers and establish the significance of the output files in accordance to best practice.

4.7 Issues with Theoretical Frameworks for Designing Log Analyzers in Cyber-Physical Systems

There are several theoretical frameworks that programmers can adopt to design log analyzers to analyze logs of smart IDSs within Cyber-Physical networks. Studies show that Statistical techniques, subjective logic, Visualization, Artificial Intelligence (AI), Neural Networks (NNs), Ensemble techniques and data mining have been used to design log analyzers in recent years [23]. Some analyzers may adopt priority of alerts, similarity of values held in the attributes of alerts, human observations, attack scenarios, hierarchical graphs, attacks that overlap, subjective reasoning and evidence of the damage the attack has caused as underpinning philosophies to design log analyzers [23]. Auditors must be thorough in this regards because features of non-related attacks may overlap and this will lead to mismatch of intrusions [16, 26]. The maximum error of log analyzer will increase if it mismatches intrusions. In other words, reports from log analyzer that mismatches intrusions are misleading and ineffective to design strong counter measures against intrusions in progress.

In addition, it is necessary for auditors to establish how each analyzer select minimum similarity and expectation of similarity in other to establish how the toolkits merge related alerts together. Also, different algorithms and metrics can compute weighted average of related alerts in different ways. Hence, it is challenging for auditors to be vast in different algorithms for comparing overall similarity of the alerts and how various algorithms isolate patterns of alerts that are false positives from real positives.

4.8 Issues with Metrics for Designing Log Analyzers in Cyber-Physical Systems

Programmers can design log analyzers that adopt multiple metrics and different data mining concepts to analyze logs of smart IDSs [15]. It is easy to compare outputs of closely related metrics together. IDS auditors must conduct routine research to ascertain strengths and weaknesses of statistical metrics that programmers have used to support intrusion detections in corporate organization that is under review. There are different ways to interpret and improve the quality of alerts from smart IDSs. Hence, the interface between log analyzer and logs of smart IDSs must be reviewed. These will enable auditors to establish suitable metrics for cross-correlation of alerts rather than interpreting uncorrelated attacks with heuristic methods. The instant that the design will update email addresses and mobile phones of operators of smart IDSs with new alerts should immediately IDS detects every suspicious event. Security issues begin to build up whenever there are networks failures such as poor Internet connection and poor mobile signals.

Auditors must review operational logbooks to determine whether operators of smart IDSs keep track of cases of networks failures such as poor Internet connections, inability to access emails and poor mobile signals in the organization. These will give insightful evidence into the effectiveness of Internet and mobile service providers that are supporting the organization. The findings in this case may also guide the auditor in recommending to the organization to sustain or review the Service Level Agreements (SLAs) they agreed with their service providers. The new threats to Cyber-Physical resources how to mitigate intrusions that can co-occur together without sharing the same impacts on the targets. Outputs of log analyzers may indicate graphical illustrations of alerts [8]. Some operators of smart IDSs may prefer to adopt visualizations to interpret alerts in the form of histogram, pie charts, bar charts and simple correlation graphs [8]. Figures 4 and 5 demonstrate graphical illustrations of alerts from Snort whenever the valued held in the TCP and TOS are used to analyze alerts from the same dataset.

For these reasons, IDS audit must be able to establish audit issues concerning attributes and metrics the organization are adopting to differentiate sequences or patterns of alerts that have tendencies to possess different interpretations from alerts that have regular patterns even if these alerts are analyzed with different attributes. Some interpretations of alerts may not impact directly on business operations that human element of Cyber-Physical Systems (CPSs) transacts on daily basis. In addition, it is plausible that some intrusions are seasonal threats to Cyber-Physical Systems (CPSs).

A seasonal rise in successful cases of cyber attacks on corporate elements of Cyber-Physical Systems (CPSs) can co-occur with a seasonal rise in unemployment and suspension of skilled workers. Therefore, IDS audit must establish the availability of inbuilt functionalities and capability of log analyzers in the organization to enable operators of smart IDSs to timely detect and mine frequent alerts from multiple sensors. Some IDS auditors can face challenges in recommending simple methods

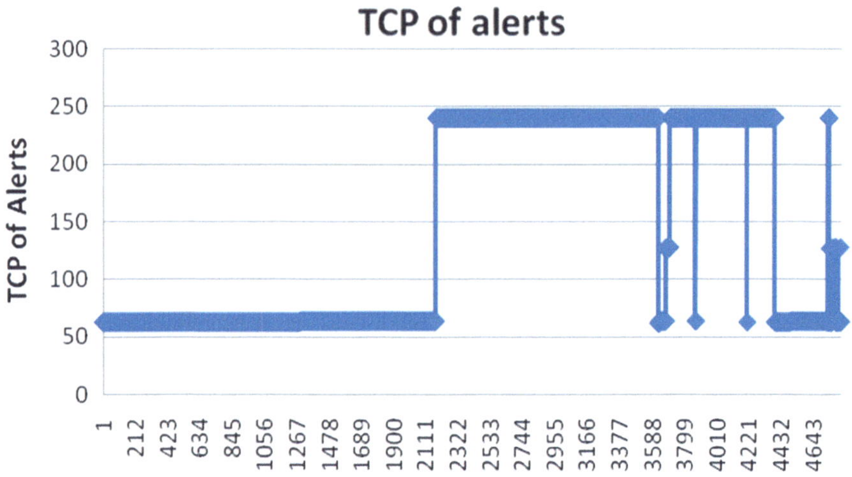

Fig. 4 Log analysis of alerts by values held in TCP of alerts

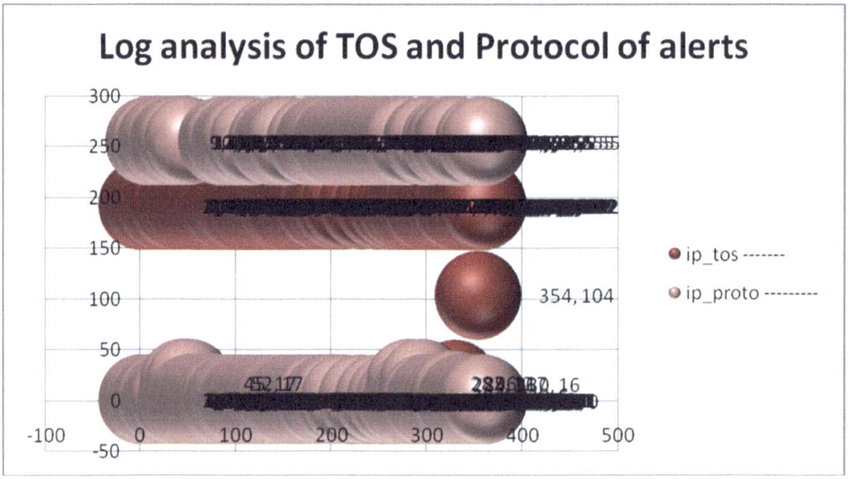

Fig. 5 Log analysis of alerts by values held in TOS and Protocol of alerts

for graphical interpretations of IDS logs to organizations that do not include methods they prefer to illustrate intrusions against their Cyber-Physical Systems in their IDS policy.

5 Methodology for Auditing Smart IDSs and Log Analyzers in Cyber-Physical Systems (CPSs)

Log analyzers are defined in this chapter as various programs that are designed to analyze logs of IDSs in a corporate setting [8]. Log analyzers have different objectives. The chapter proposes log analyzers that are interfaced with GSM to send short text messages after they have processed alerts of smart IDSs to operators. Log analyzers often have different objectives. For instance, log analyzers can be designed to debug NIDSs in the organization. There are log analyzers that determine the degree of predictability of attributes and information conveyed by attributes of alerts. Similarly, there are log analyzers that focus on correlation and aggregation of alerts. Sources of input data to each log analyzers in the same organization may also vary.

Some log analyzers may derive their input data from homogeneous logs of smart IDSs while significant numbers of them may receive input data from heterogeneous IDSs. By auditing them, operators and IDS auditors will be able to ascertain how the existing Log analyzers cluster alerts to arrive at the succinct texts they send to operators. For log analyzers that receive input data from several smart IDSs, it is necessary for the IS auditors to assess the locations of the contributing IDSs in relation to the log analyzers that aggregate or analyze their logs. Evaluators should ask questions like was the input modules of various log analyzers designed to override old alerts or append new alerts to previous ones and what programming language was used to design them? The time to upload new alerts to the input modules of the log analyzers should also be audited.

The results from the above enquiry can determine log analyzers that should be recommended for upgrade and new development that should be incorporated to improve intrusion detection in the organization. Figure 6 illustrates samples of execution of four categories of log analyzers that are designed to support the aruments raised in this chapter. These log analyzers are implemented with C++ language and they are based on the attributes of alerts from Snort IDS. The input to three of the analyzers

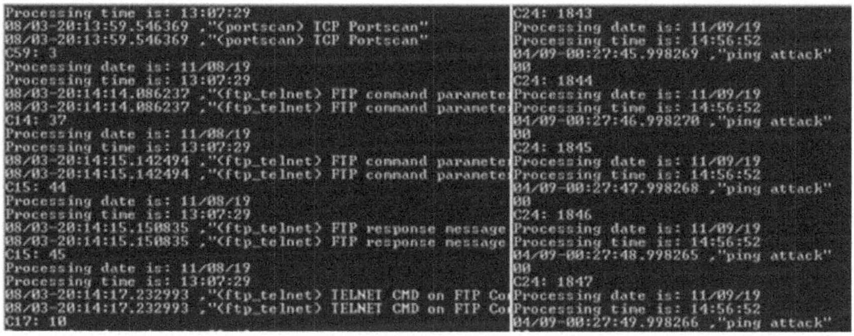

Fig. 6 A sample of execution of log analyzer of alerts

Table 1 Log analysis of online trace files

Dataset	Attribute	Number of cluster	Gini Index
DDOS-1-SIP	Source IP	408	0.998
DDOS-1-DIP	Destination IP	1	0.000
DDOS-2-SIP	Source IP	265	0.996
DDOS-2-DIP	Destination IP	1	0.000

were alerts that Snort triggered on the DATA01, DATA02 and DEFCON-10 dataset in IDS and offline modes. The input to forth analyzer was alerts that Snort triggered on DDoS datasets supplied by the DAPRA to assist research community. The IDS triggered 4,919 alerts and dropped 250 packets after analyzing the packet capture (PCAP) file of the dataset. Typical IDS research can explore many concepts with the above alerts.

The first log analyzer explores the rules that triggered the above alerts and a sample of its results is shown in Table 3. The second log analyzer explores the sources of the intrusions and all the addresses of computers they attacked and categorize them on the basis of date, time, sequence number, source IP address, source port number, destination IP address and port number of destination address. The third log analyzer explores the sources and destinations of the intrusions captured in the dataset. To ascertain the variability and quality of the alerts, the analyzer went further to compute Gini Index on the basis of sources and destnations of the attacks to further classify the alerts as shown in Table 1.

Given the probability of each cluster $[p(c_t)]$ and for each attribute (SIP or DIP), the Gini Index is expressed as [15]:

$$GIndex(SIP/DIP) = 1 - \sum_{t=1}^{n} (p(c_t))2 \tag{1}$$

The fourth analyzer uses alerts from DATA01 and DATA02 to compute the lengths of alerts and the pattern within them.

5.1 *A Model for Auditing Smart IDSs and Log Analyzers in Cyber-Physical Systems*

The auditors of smart IDSs must have audit plan and feasible audit time table. The audit time table should categorically state the annual frequency proposes for conducting audit of smart IDSs and log analyzers in the organization [12, 17, 25, 26].

The audit plans can be an annual arrangement or a short-term plan that itemize the procedures the auditors will adopt to conduct IDS audit in the organization at the due dates. Figure 7 illustrates the schematic diagram of a new framework for auditing smart Intrusion Detection Systems (IDSs) and log analyzers in this chapter.

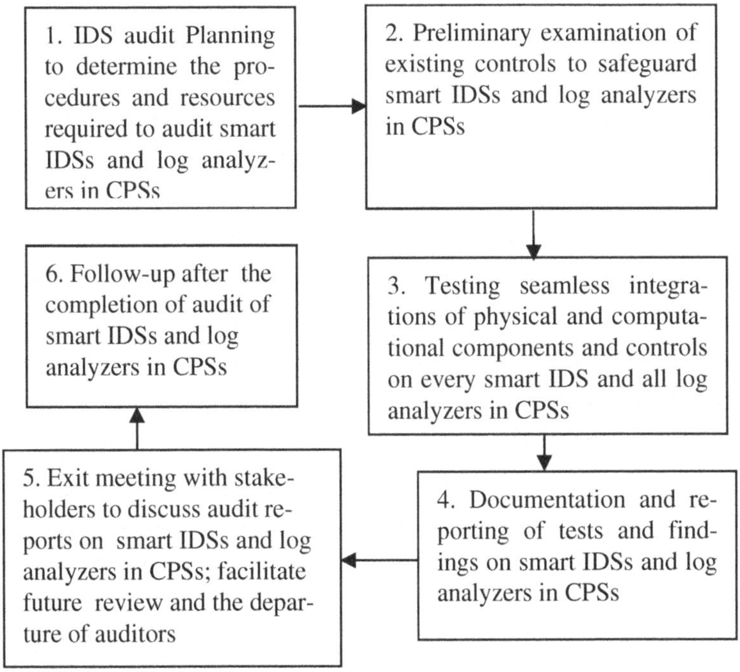

Fig. 7 A model for auditing Smart IDSs and Log analyzers in CPSs

Accordingly, IDS auditors should preview the entire processes they will follow to carry out the audit of smart IDSs and log analyzers in advance. This is called the planning phase. This is the stage at which the auditors must delineate the objectives, scope, budget and resources they would require to comprehensively accomplish the audit [12, 14, 25]. The auditors will also need to establish the methods they will adopt to carryout fact-finding; the duration or time frame they will spend on each stage and the total time they will generally spend to conduct the review. The IDS audit team should categorically state the format of the IDS audit reports, potential challenges they envisage and the period they schedule to conduct exit meetings with the management of smart IDSs in Cyber-Physical Systems (CPSs).

The second stage of this model is the preliminary examination of smart IDSs' controls and Log analyzers. In this stage, the IDS auditors ought to carry out initial assessment of the existing IDS resources, all related components of the IDS; operational procedures and the controls that were implemented in the enterprise to safeguard the smart IDSs and log analyzers. The auditors should interview or send questionnaires to main employees that are responsible for the management of different smart IDSs and all log analyzers in the organization [18, 19]. The review should cover all the IDSs in the organization together with infrastructure in the organization that relates to them, logical access and physical security of each smart IDS. The directory of each smart IDS, access to the root directory, procedure to log on to the root, permissions granted to read, write, execute and modify files and log analyzers; operating systems; hardware requirements including security, usage and available disk space; configuration files (signatures, profiles, etc.) and respective logs kept by each smart IDS and log analyzer must be requested from the dedicated IDS operators. The review of the log analyzers and other programs that interface with the logs of the smart IDS should also be carried out at this stage using simulated attacks.

Furthermore, at the third stage of this model, the IDS auditors begin to critically examine Service Level Agreement (SLA) on the smart IDSs and verify the SLA for proprietary log analyzers [18]. They will scrutinize process flow, incident reporting procedures; relevant features of physical and organizational structures; training and users manuals in the organization that is using Cyber-Physical System to support their business operations. They must test and validate the level of security and controls that have been implemented to counter likely threats and attacks on smart IDSs and related infrastructure in the networks [3, 13, 17]. Auditors must examine the seamless of the entire components of the engineered systems and quantify the level of protection smart IDSs in the organization can render to them. They must review controls and configurations of operating systems, security of smart IDSs and database access controls. The review at this stage should include various strategies the organization has implemented to hardening the host computer(s) and the networks so that auditors can establish the levels of compliance of operations of smart IDSs in the company with best practices [14, 17, 22].

In the fourth stage, proper documentation and reporting are critical elements that auditors must carryout to achieve comprehensive auditing of smart IDSs and log analyzers [12, 17, 18]. Hence, it is imperative for the IDS auditors to document key findings they observe at each stage of the audit. This chapter proposes that the IDS auditors should appoint dedicated scribers among the audit team to document tests and respective findings as the audit progresses. IDS audit reports should include executive summary, suitable headings, controls investigated during the audit and corresponding findings the team of auditors have observed in the organization [19, 25]. They must include remarks, recommendations and practical suggestions on how IDS operators and designers of existing log analyzer can fix audit issues they have identified in the review. Thus, this chapter proposes that documentation and reporting of findings should be incorporated into stage 4 of a comprehensive audit of smart IDSs and Log analyzers in Cyber-Physical System.

Exit meeting is the fifth stage for a comprehensive audit of smart IDS and log analyzers in the context of Cyber-Physical Systems. The auditors and audit team from the organization that is under review must gather together in interactive conferencing to discuss the audit reports before the audit team will exit the organization [17, 19]. The meetings are avenues for both teams to agree on the date and how various audit issues raised on the smart IDSs, log analyzers; computational and cyber-physical infrastructure in the organization will be fixed. The meetings should state the date the representatives of audit team will revisit the unit of the organization to check that issues raised in the IDS audit reports have been fixed.

Finally, follow-up is the sixth and last stage of the above framework. The representatives of the IDS audit team must revisit the organization to examine documents like visitor's diary and access log to the above resources. They need to also report on the status of all the issues that have been raised in the audit reports they recently submitted to the organization [17]. To conclude the audit, the reports of this team should categorically state audit issues on smart IDSs and log analyzers that have been fixed, pending issues and reasons behind the delay on audit issues that end-users have not fixed. We suggest that auditors must advise the organization to develop a suitable IDS policy whenever they have none.

5.2 *Results and Discussions*

The attacks illustrated with the DDOS-1 and DDOS-2 datasets in Table 1 did not vary on the basis of their respective destinations' IP addresses when compared with the sources' IP addresses of the attacks. The results sugest that the entire alerts that originate from the dataset are mostly repeated information that belongs to one group of destination's IP address. Hence, the Gini Index was 0.000.

Therefore, IDS auditors must as well audit codes and Log analyzers to establish the input data, their functions and capabilities in other to establish the strenghts and limitations of each analyzer. Such systematic review will enable the auditor to establish Log analyzers that analysts should optimize either by splitting them or by merging two or more codes together. Table 2 illustrates cumulative length of attributes that Snort has used to report 4,919 and 75,390 alerts on DATA01 and DATA02 respectively. Table 3 interprets the attacks from the above evaluation and the rules that detected them. Thus, auditors can adopt information in Tables 2 and 3 to conduct risk assessments and identify strategies of some intruders in Cyber-Physical Systems (CPSs).

Figures 8 and 9 illustrate the patterns that lengths of alerts from both datasets can generate. Thus, the chance that intruders can overload smart IDSs over time depends on the quantity of alerts the detectors can trigger on daily basis. The results further suggest that automated strategy for forecasting length of alerts smart IDSs generate is critical to auditors in conducting audit of smart IDSs in the context of Cyber-Physical Systems (CPSs). This can assist operators to forecast patterns of attacks, workload and how human aspects of security and privacy can link to Cyber-Physical Systems (CPSs).

Table 2 Log analysis of components of alerts

Dataset	Total alerts	Total attributes
DATA01	4,919	345,375
DATA02	75,390	2,893,183

Table 3 Log analysis of rules that generate alerts

Sig_generator	Sig_id	Sig_rev	Description of alert/attack	Summary of attack
119	2	1	Double decoding attack	The attack was an http exploit. The intuder inllegally inspected Hyper Text Transfer Protocol (http) to gather information about application protocol for distributing hypermedia data in the networks
119	18	1	Webroot directory traversal	The attack was an http exploit. The intuder possibly accessed data, codes, files, etc. via root directory of the web server in the networks
122	1	0	TCP portscan	The intuder inllegally scanned a computer port with intention to gather information about open ports, close ports and services running in the computer
125	2	1	Invalid FTP command	The intuder used invalid FTP command to possibly transfer files in the networks
125	3	1	FTP command parameters were too long	The attack was buffer overflow exploits with FTP client. The intruder used telnet's client to possibly transfer files that exceeded maximum length in the networks
125	4	1	FTP command parameters were malformed	The intruder used badly formed FTP command to possibly transfer files on FTP client

5.3 *Suggestions for Improving Security in Cyber-Physical Systems*

The above models have practical implementations in protecting computational, human, mechanical and physical components that are fundamental to Cyber-Physical Systems (CPSs). IDS policy must state the configurations and various types of smart IDSs in the above settings. This document should state the vendors of Network Intrusion Detection Systems (NIDs) and Host-based Intrusion Detection Systems (HIDSs) installed to safeguard all entities in Cyber-Physical Systems (CPSs). Auditors must verify whether the policy approves software-based IDSSs or hardware-based IDSs, or

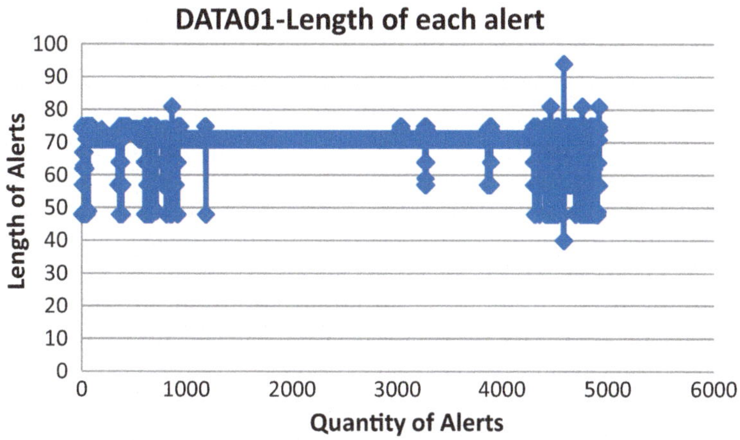

Fig. 8 Log analysis of lengths of alerts (DATA01)

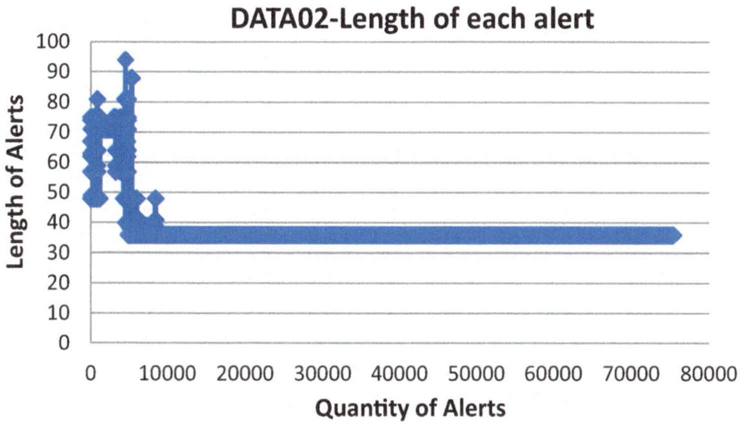

Fig. 9 Log analysis of lengths of alerts (DATA02)

combinations of both detectors. Among other things, auditors should further investigate this document to ascertain if it contains information regarding license fees, number of users, expiration date for the payment of license fees and bank accounts of the vendors of smart IDSs procured in these settings.

IDS policy must reflect operators of smart IDSs responsible for the administration and monitoring of various smart IDSs and Log analyzers in the organization. Recently, intruders keenly probe source codes to establish their limitations. Therefore, it is imperative for IDS auditors to carefully scrutinize IDS policy. The document must categorically state allowable length of time to train supervised learning algorithms as well as the acceptable level that log analyzers must reduce workload due to IDS alerts in other to undermine the generality of intrusions IDSs have warned.

What is the acceptable way to classify similar alerts and similar intrusions? Should similar intrusions be classified on the basis of temporal relationships, intrusive objectives, capabilities to support subsequent intrusions or values held in the attributes of alerts? The audit must be able to match IDS policy with the above questions for the document to be useful for mitigating problems of alert correlations that have raised serious concerns among security experts in recent time. IDS policy document should not reflect ambiguity in any aspect. The document should be simple and explicit. It should also include the incident and reporting team; processes of escalating cases of intrusions and response strategy approved by the management. In all, it is equally suggested that IDS policy should include methods for handling public awareness and lessons learnt in the case of devastated attacks that require the organization to intimate the general public.

An organization may deploy smart IDSs that run on different operating systems. The performance of smart IDSs becomes necessary whenever they run on different operating systems. For instance, experience shows that Bro usually operates in Linux/Unix, FreeBSD and Solaris' environment while Snort can run with Windows and Unix/Linux operating systems. There are different ways to hardening different operating systems. Therefore, auditors must familiar with different ways to hardening common operating systems in the industry. Some smart IDSs require installations of client software on computers in the networks of Cyber-Physical Systems (CPSs). Hence, auditors must also ensure they audit client software on computers in the networks that interface with smart IDSs. Uninteresting activities and activities that are important attacks can vary from organization to organization. Hence, auditors must be professional at all time. They should professionally handle recommendations aiming to limit the number of false positives especially while suggesting extra policy scripts that should be included with existing rules for detecting cyber attacks.

Some toolkits can express their signatures as regular expressions or as fixed strings. Audit of smart IDSs in Cyber-Physical Systems (CPSs) should establish how signatures are designed in each detector. This information is needed in recommending suitable training and professional development to operators of smart IDSs whenever audit reports suggest that operators lack sufficient knowledge to carry out their daily jobs' specifications. Auditors of smart IDSs and log analyzers should evaluate the effectiveness of training facilities that are available for conducting in-house training in the organization. In-house training can be recommended to operators in case the required facilitators are available in the organization. It is ethical for auditors to recommend training outside the organization to operators of smart IDSs whenever there are insufficient facilities to conduct in-house training in the organization [17, 26]. Operational training should include topics such as network or traffic content, false positives. false negatives, policy scripting or writing rules or signature, signature-matching, uninteresting activities, interesting activities, cyber threats and attacks, security, user privileges, front-end and back-end of smart IDSs; installation, configuration, maintenance and execution of smart IDSs and log analyzers to empower operators of smart IDSs in Cyber-Physical Systems (CPSs). Auditors should ascertain operators if smart IDSs that aware or unaware of the official websites of various smart IDSs in the organization during audit of smart IDSs and log analyzers. The audit should establish operators of smart IDSs that subscribe or unsubscribe to news update in the official websites of IDSs in the organization. The reason is that official websites of IDSs often contain helpful documentations and new tips about bugs and attacks and strategies to fix them. There should be no bandwidth limitations in the networks for most smart IDSs to be effective. Organizations should strictly adhere to the hardware requirements such as hard disk and processor of host computers; software requirement such as operating systems (Linux, Windows and Solaris) and the required versions of auxiliary tools such as libpcap, Perl and tcpdump that service providers recommend for smart IDSs to ensure high performance. Audit reports should state the location of smart IDSs in the organization; other options for location the toolkits and their respective benefits to enlighten the organization. For instance, smart IDSs can be installed behind an external firewall in the networks. This will enable the firewall to reduce numbers of suspicious packets that smart IDSs in CPSs will analyze. Some organizations may install smart IDSs before the external firewall. This method will enable smart IDSs to detect potential attacks migrating into the networks. The trade-offs is that smart IDSs will produce high number of alerts for log analyzers and operators to analyze. Smart IDSs can also be installed inside internal firewall if the human element in Cyber-Physical Systems (CPSs) aims to detect internal hosts that are vulnerable to computer worms and computer virus.

Audit reports should specify agencies that require external reports of incidents from the organization that is being audited. Statistics on incident information can suggest prevalence of security breaches of Cyber-Physical systems (CPSs) nationwide. Auditors can evaluate compliance of the organization to the various requirements of regulatory bodies by reviewing information about the frequency regulators required for submitting mandatory reports to the government and National Agency for Incident Analysis (NAIA). The formats of the reports may be summary of critical incidents or all cases of security violations on monthly, quarterly, biannual or annual basis. Interview with someone who inspects and forwards the reports to the required external recipients will appropriately establish details of how and when the reports are due for submission. The reports to agencies should be informative in case they require the reports in specific formats. Operators should express the date and time the incident begin and end. The number of each type of incident could be included in the report period for statistical purpose.

Smart IDSs and log analyzers merely detect suspicious events. They cannot make authoritative decisions if a suspicious event is an attack or not attack. These mechanisms also lack the intelligence to decide whether an attack is successful attack or a failed or unsuccessful attack. Therefore, operators and recipients of alerts from smart IDSs and log analyzers must constantly investigate the reports they receive from the above mechanisms. Furthermore, IDS audit reports and reports on log analyzers should be simultaneously made available to the IDS operators in the organizations to address all audit issues pinpointed in the reports.

Above all, the above audit model is an integral part of the information security of an organization. Host machines, hardware-based IDSs and repository for storing reports on smart IDSs should be regularly protected from intruders like burglars. For software-based IDSs, the logical security of databases of the IDSs; web servers and various infrastructural components on the networks such as router, firewall and location of the smart IDSs in relation to the firewall should be thoroughly reviewed to ascertain their levels of compliance with best security standards. Segregation of duties among network engineers, Database Administrators (DBAs), internal control and operators of smart IDSs in Cyber-Physical Systems (CPSs) is highly recommended. It is disastrous if the logs are deleted while the toolkit is running. Auditors should recommend enforcement of strong access controls to restrict illegal logging into the configurations and logs of smart IDSs as panacea to information leakages and attacks on smart IDS through the back-end of applications in Cyber-Physical Systems (CPSs) [2].

The root causes of intrusions are dynamic security and privacy issues in Cyber-Physical Systems (CPSs). Broad audit should be able to reveal how log analyzers adopt classification rules to segment logs of smart IDSs in Cyber-Physical Systems (CPSs) and classify alerts into normal and abnormal events. Without sound understanding of data mining procedures, IDS auditors might face difficult challenges to audit association and episode rules necessary to expose hidden relationship among alerts that are not obviously related. Research has discovered that sequence of the intrusions on cyber-physical resources in an organization can occur within different timestamp. Practically, it is difficult to find the mean of categorical datasets that have no numerical attributes. Instances whereby the designers of log analyzers have adopted weighted values to transform alerts in the logs of smart IDSs must be clearly reviewed during audit. The reports will enable end users to establish limitations of algorithms that adopt concepts like k-nearest-neighbor (KNN) classifiers and how to improve on the underpinning concepts for transposing alerts into human readable form in the organization. Auditors should establish types of Security Information and Event Management (SIEM) and other threat solutions in Cyber-Physical Systems (CPSs).

The above results submit that auditors must audit log analyzers irrespective of whether they are locally designed or they are proprietary models in the organization. The reports should reveal expert rules that are used to process events' logs and their characteristics. Auditors should strongly recommend proper documentations for log analyzers and other threat solutions in Cyber-Physical Systems (CPSs). Essentially, the above audit model should establish the existence or absence of audit team in the organization. Reports obtained from the audit should be submitted to the unit in charge of monitoring smart IDSs in the organization. Thereafter, auditors should notify them and management with written reports stating past audit issues that have been suitably addressed [16, 24]. Otherwise, a terminal date to ensure that all pending audit issues must be addressed and potential impacts of noncompliance must be issued to the above stakeholders as well.

6 Conclusion

This chapter shows that pragmatic studies on audit of smart IDSs in the context of Cyber-Physical Systems (CPSs) are erroneously taken lightly over the years. This gap has generated negative impacts in the security of computational components, cyber and physical resources of Cyber-Physical Systems (CPSs) over the years. Manufacturers of smart IDSs can design rules or policies that are deactivated by default because they are not immediately needed to protect Cyber-Physical Systems (CPSs). Such rules or policies can be completely useless if smart IDSs are not periodically audited. Operators can waste huge resources to redesign inactive rules or policies due to lack of information about possible threats and cyber attacks in Cyber-Physical Systems (CPSs) and ignorance of the existence of similar rules or policies in the detection engines of smart IDSs. Consequently, the chapter demonstrates that log analyzers can serve diverse objectives in a corporate setting. It has also been stated that series of intrusions can elude smart IDSs whenever the periodic audit of smart IDSs in Cyber-Physical Systems (CPSs) is not based on empirical findings. The idea is that smart IDSs and all log analyzers in a corporate setting must be specially audited and their readiness for packets processing must be routinely verified to ascertain their compliance with best security practices.

There are several concerns that may arise if the computers hosting smart IDSs are weakly protected or if they are not protected at all. The toolkit can be compromised by intruders, thereby under-reporting or over-reporting security breaches in Cyber-Physical Systems in the organization. Intrusions that overpower hosts of smart IDSs can suddenly shutdown the toolkits without the awareness of operators. The smart IDSs can begin to generate series of false alerts. These devices can suddenly stop to trigger alerts if intruders cleverly re-configure them without the awareness of dedicated employees. Experienced intruders may modify rules or policies of smart IDSs and compromise the passwords for logging to the root directories of smart IDSs in Cyber-Physical Systems (CPSs). They may delete logs, modify alerts and other related components of these toolkits. Some intruders may disable smart IDSs in Cyber-Physical Systems (CPSs) before they will attacks the networks. The integrity of the log analyzers that analyze logs of compromised smart IDSs in these circumstances will also be subjective. Therefore, smart IDSs and log analyzers in Cyber-Physical Systems (CPSs) must be periodically audited to establish lapses or hidden faults in the validity and the strength of the protection that the internal controls offered to the detectors and to help the company to settle on the cost of ownership of their smart IDSs.

This chapter has proposed an audit model that should contain significant and explicit information necessary to guide human elements in Cyber-Physical Systems (CPSs). The chapter also substantiates the importance of smart log analyzers in the security of Cyber-Physical Systems (CPSs). These groups of log analyzers are configured to remotely send brief statements that present the main points about alerts/attacks and in the form of short text messages to the operators of smart IDSs in Cyber-Physical Systems (CPSs). The message may include "source IP, destination IP, short descriptions and time of occurrence of the attacks". The above model has also suggested that audit reports should contain executive summary on audit of smart IDSs and log analyzers in Cyber-Physical Systems (CPSs); objectives or purpose and scope of the audit. The reports must also include all proprietary and locally developed log analyzers that relate to smart IDSs in the review. The reports will be informative if they convey information about the available resources, challenges and date of the audit. Columns that outline the serial number (S/N); control tests that auditors have carried out, findings, risk assessment of each problem, suggestions that can mitigate the problems; human elements in Cyber-Physical Systems (CPSs) that should fix the problems and remarks or explicit comments (that will state whether the problem has been fixed or is still a pending issue) should be incorporated in the audit reports. Useful explanations regarding the entire phases of the audit, signatories to the reports and annotations should be included in the reports to clarify and substantiate the validity of the reports to stakeholders in Cyber-Physical Systems (CPSs).

Furthermore, auditors must periodically verify that logs of smart IDSs and log analyzers in Cyber-Physical Systems (CPSs) are regularly archived and operators strictly adhere to the modality for maintaining them. This chapter has further provided a new pathway on how to investigate the sufficiency of IDSs intelligence and log analyzers and the degree at which they conform to IDS policy and best security practices in a real-life environment and in the context of Cyber-Physical Systems (CPSs). Since empirical studies have shown that IDS policy is a well-established fact in IDS manuals, similarly, future studies should provide best standards and frameworks for concurrent auditing of smart IDSs and log analyzers in Cyber-Physical Systems (CPSs) using non-statistical metrics. Finally, strong cooperation between organizations, GSM operators and research community can help to lessen issues and challenges in Cyber-Physical Systems (CPSs) that have been identified in this chapter.

References

1. Colombo, A.W., Bangemann, T., Karnouskos, S., Delsing, J., Stluka, P., Harrison, R., Jammes, F., Lastra, J.: Towards the next generation of industrial cyber-physical systems In: Industrial Cloud-Based Cyber-Physical Systems: The IMC-AESOP Approach, pp. 1–22 (2014). ISBN 9783319056234
2. George, L.: Cyber-Physical Attacks: A growing invisible threat. Oxford, UK. Elsevier Science (2015). ISBN 9780128012901
3. Phatak, T., Isal, P., Kadale, O., Nalage, A., Bhongle, S.: Smart intrusion detection system. Int. Res. J. Eng. Technol. **4**(04) (2017)
4. Wu, D., Rosen, D.W., Wang, L., Schaefer, D.: Cloud-based design and manufacturing: a new paradigm in digital manufacturing and design innovation. Comput. Aided Des. **59**, 1–14 (2014)
5. Ciprian-Radu, R., Olimpiu, H., Ioana-Alexandra, T., Gheorghe, O.: Smart monitoring of potato crop: a cyber-physical system architecture model in the field of precision agriculture. Agric. Agric. Sci. Procedia **6**, 73–79 (2015)
6. Stallings, W.: Network Security Essentials: Applications and Standards, 4th edn. Prentice Hall (2011)

7. Murray, W.H.: Data security management: principles and applications of key management. Auerbach publication (1999)
8. Rehman, R.U.: Intrusion detection systems with snort: advanced IDS techniques using snort, apache, MySQL, PHP, and ACID. Library of Congress, New York (2003)
9. Buchanan, W.: The Handbook of Data and Networks Security, 1st edn. Springer-Verlag New York, Inc. Secaucus, NJ, USA (2007)
10. Alder, R., Baker, A.R., Carter, E.F., Esler, J., Foster, J.C., Jonkman, M., Keefer, C., Marty, R., Seagren, E.S.: Snort: IDS and IPS Toolkit. Syngress publishing, Burlington, Canada (2007)
11. Kumar. T.S., Radivojac, P.: Introduction to data mining:- lecture notes (2017)
12. Epstein, J.: Security lessons learned from société générale. IEEE Secur. Priv. **6**(3) (2008)
13. Rainer, R.K., Cegielski, C.G., Splettstoesser-Hogeterp, I., Sanchez-Rodriguez, C.: Introduction to Information Systems: Supporting and Transforming Business, 3rd Canadian edn. (2013). ISBN: 9781118476994
14. The National Science Foundation-US: Cyber-Physical Systems (CPS) (2020)
15. Snort Users Manual 2.9.11:The Snort Project; Cisco and/or its affiliates (2017)
16. Adams, D., Maier, A.: Confidentiality Review & Audit of GoldBug-Encrypting E-Mail-Client & Secure Instant Messenger (2016)
17. ISACA: Information Systems Auditing: Tools and Techniques Creating Audit Programs (2016)
18. Julish, K., Suter, C., Woitalla, T., Zimmermann, O.: Compliance by design–bridging the chasm between auditors and IT architects. Computers & Security, vol. 30 (6–7). Elsevier (2011)
19. Bitterli, P.R., Brun, J., Bucher, T., Christ, B., Hamberger, B., Huissoud, M., Küng, D., Toggwhyler, A., Wyniger, A.: Guide to the Audit of IT Applications. ISACA (2009)
20. Gubb, P., Takang, A.: Software Maintenance. World scientific Publishing, New Jersy, USA (2003)
21. Fitzgerald, J., Larsen, P.G., Verhoef, M. (eds.): Collaborative Design for Embedded Systems: Co-modelling and Co-simulation. Springer Verlag (2014). ISBN 9783642541186
22. The Global Information Assurance Certification (2003) Snort Intrusion Detection System Audit: An Auditor's pers-pective, GSNA practical version 2.1 (2007)
23. Nehinbe, J.O.: Methods for reducing workload during investigations of Intrusion Logs, PhD Thesis, University of Essex, Colchester, London (2011)
24. Nehinbe, J.O.: Automated Technique for Debugging Intrusion Detection Systems, 1st International Conference on Intelligent Systems, Modelling and Simulations (ISMS2010), proceedings of IEEE Computer Society's Conference Publishing Services (CPS), London (2010)
25. Baker, W.H., Hutton, A., Hylender, C.D., Novak, C., Porter, C., Sartin, B., Tippett, P.: Data Breach Investigations Report, Verizon Business (2009)
26. Robert, D.E.: IT auditing: an adaptive process. Mission Viejo: Pleier Corporation (2005)
27. Cascarino, R.E.: Auditor's Guide to Information Systems Auditing. John Wiley & Sons publication (2007)
28. Senft, S., Gallegos, F.: Information Technology Control and Audit. Auerbach Publications (2009)
29. IANA: Internet Control Message Protocol (ICMP) Parameters. https://www.iana.org/assignments/icmp-parameters/icmp-parameters.xhtml

Model-Based CPS Attack Detection Techniques: Strengths and Limitations

Surabhi Athalye, Chuadhry Mujeeb Ahmed, and Jianying Zhou

Abstract Given the intricate interactions between the physical and cyber components found in urban cyber-physical systems (CPSs), the detection of attacks in such infrastructure has been approached in various ways. This work presents an exhaustive study that compares different kinds of attack detection mechanisms and evaluates them using a set of defined metrics. Model-based attack detectors are investigated in this report, which use mathematical system models with the input and output as the sets of actuators and sensors of the underlying physical processes, respectively. The detection methods comprise statistical change monitoring procedures (CUSUM and bad-data detectors) and a device fingerprinting technique. The case studies of two research facilities, a smart water treatment plant (SWaT) and a water distribution plant (WADI), have been used to assess these security measures. These testbeds represent the diversity of CPS infrastructures found in cities today. Several types of attacks have been simulated on the plants to experimentally analyse the performance of the detection methods.

Keywords Cyber-physical systems · Water treatment systems · Water distribution systems · Model-based attack detection · NoisePrint · Statistical detectors

1 Introduction

Comprised of physical infrastructure that is controlled by computation and communication frameworks, a cyber-physical system (CPS) includes a combination of several interconnected elements. Programmable Logic Controllers (PLCs), sensors, actuators, a Supervisory Control and Data Acquisition (SCADA) workstation, and a

S. Athalye (✉) · C. Mujeeb Ahmed · J. Zhou
Singapore University of Technology and Design, Singapore, Singapore
e-mail: surabhi_athalye@sutd.edu.sg

C. Mujeeb Ahmed
e-mail: chuadhry@mymail.sutd.edu.sg

J. Zhou
e-mail: jianying_zhou@sutd.edu.sg

A. I. Awad et al. (eds.), *Security in Cyber-Physical Systems*, Studies in Systems, Decision and Control 339, https://doi.org/10.1007/978-3-030-67361-1_6

Human Machine Interface (HMI) are some of these components, which interact using a network. The PLCs are responsible for implementing the necessary control actions based on the present state of the system, which is checked through the SCADA. This directly affects the progress and normal functioning of the plant's sub-processes.

The network and physical elements of the CPS directly influence the physical processes, and for their normal operation, it is essential that these units work in tandem. Such systems comprise information technology (IT) and operational technology (OT) components that interact with each other to ensure proper functionality overall. Communication within these industrial IoTs also introduces vulnerabilities that could be abused by malicious entities [1, 2]. Furthermore, since CPSs could be hampered via both, the cyber and physical domain, the design of their security measures becomes more complicated when compared to that of strictly IT systems [3]. The physical infrastructure of a CPS could be targeted and sabotaged by cyber attacks. The wireless communication incorporated in an interconnected CPS makes it susceptible to remote breaches and also attacks [4, 5]. This poses a direct danger to the communication among the nodes of a CPS, and is detrimental to the infrastructure. Physical attacks usually involve damaging the sensors or other components, which compromises the integrity of the data. Consequently, faulty readings get forwarded to the PLCs, thus rendering the resulting control actions incorrect or even harmful. The conventional focus of security research is on anomaly-detection in the communication network component of a CPS [6]. However, attacks in the physical domain could be more challenging to detect, for they may not reflect in the system's network [7, 8].

In recent years, many different CPSs such as nuclear enrichment plants, power grids, and water treatment systems have been victims of successful cyber and physical attacks [9, 10]. This chapter examines water treatment and distribution plants. Lately, water distribution systems have been reported to be targeted on several instances. The attacks on the Maroochy Shire Water Services [11] and at the Kemuri Water Company [12] are well known.

Due to the interdependence of network and physical elements, CPS security measures can adopt many different methodologies as compared to traditional IT security approaches. Standard security solutions like verification, access control and message integrity are not sufficient for adequate protection in the case of a CPS [13]. This is because such methods are incapable of correlating the sensor measurements with the respective physical process or control action, implying that they cannot be used to identify aberrations in the physical dynamics of the systems [13]. Such security measures typically focus on securing the communication network itself, while others are based on validating the integrity of data from the physical components of the system and the interactions among them. However, a majority of these works have been done using simulated data or theoretical examples [14, 15]. Hence, a thorough process-level security study of a real system is necessary.

This work performs case studies on a water treatment and a distribution plant and takes into consideration model-based approaches for attack detection. These plants are run under normal operation, and then the corresponding sensor dataset is used to generate Linear Time-Invariant (LTI) system models. The residual, which is the difference between the estimated and actual sensor values, is obtained and the attack detection methods are then applied on it. A control theoretic-approach is used to create these models, thereby allowing the dynamics of the physical processes to be mathematically analysed. This way, anomalies arising due to either cyber or physical attacks can be identified by comparing the modelled and the actual behaviour of the plant. The dynamic system models representing the two plants in this study have been created using two ways, (1) using system identification techniques, and (2) from fundamental physics principles. The evaluation of the models obtained using both these approaches is presented.

This chapter evaluates the detection performances of three attack detection techniques. The first method is a standard statistical mechanism called Cumulative Sum (CUSUM) that is used to detect attacks by monitoring changes in the expected values. The second one is called bad-data detector and it identifies instances of abnormal data using empirically determined thresholds. The last technique, called *NoisePrint*, is a device identification method that is based on fingerprinting the sensor and process noise [3].

When accounting for the performances of the attack detection techniques, the sensitivity of the method is also an important consideration, besides precision. This implies that the instances of false alarms under normal operation of the plants must be reasonably low. This is crucial while securing actual industrial systems that comprise of a large number of physical devices, whereby checking each component could be tedious. Thus, the evaluation of the detection mechanisms is done under normal operating conditions as well as when the plants are under attack, in order to review their performances comprehensively.

The motive behind this work is to carry out exhaustive testing of detection techniques for CPSs on different testbeds, and then compare their efficiency. This chapter is builds on the preliminary results presented at a conference [16] and gives some insight on the issues as highlighted below:

1. Effect of Noise on the System Models: One of the problems faced while implementing and verifying the models was that of process noise for each individual run. The effect of the noise from environmental disturbance causes erratic deviations from the modelled behaviour.
2. Faults in Sensors: Even when the plants were operating normally, the presence of some hidden faults in the sensors was noted, thereby hindering the creation of valid system models. Hence, the components have to be carefully checked to ensure proper functionality during the data collection.
3. Availability and Reliability of Data: The design and performance of an anomaly detector are directly affected by data availability. In order to obtain accurate models, sufficient data must be procured that (a) captures the components' complete performance cycles, and (b) represents all the possible modes of operation of the Industrial Control System (ICS) without temporary glitches and/or outliers. In this study, the plants were run continuously under normal operating conditions and the datasets obtained were used to create the models. However, unexpected results were observed when the models were tested on the plants when they were not running.

4. Attack Detection Speed: The anomaly detection speed is a vital consideration for the safety of the plant, but is often neglected while assessing the performance of a detection technique [17]. The faster the anomaly is detected, the better the impact can be mitigated, as it allows the required protective measures to be taken earlier. Hence, one of the performance attributes that has been evaluated and emphasised in this study is Time Take for Detection (TTD).

Organization: The chapter henceforth is organized as follows. Literature related to the work reported in this chapter is reviewed in Sect. 2. The research facilities used to test the performance of the attack detection mechanisms are described in Sect. 3. The physical systems in these facilities have been modelled in two different ways, which are explained in Sect. 4. Section 5 explains the attack detection framework as well as the three detection techniques that this work focuses on. Subsequently, the attacker profile is defined in Sect. 6, which also elaborates the likely attack cases and their execution. Section 7 presents the performance evaluation of the detection mechanisms under normal and attack scenarios. Lastly, the results obtained above are analysed, and the conclusions mapping the above contributions are put forward in Sect. 8.

2 Related Work

In this section, the research done in CPS security is highlighted. One of the earlier works on the security of power systems against data injection attacks is detailed in [18]. It has been shown that a bad-data detector could catch an attack resembling a fault, but it would miss a stealthy attack. This has also been noted in study on a smart water distribution in [19]. In these two studies, the models of underlying process have been generated using simulations and the real testbed, respectively.

Active Defense Some detection techniques combine the modelling of the physics of the system with active detection mechanisms. The authors of [7] present a sensor attack detection technique based on challenge-response. This method has been tested on active sensors in vehicles. Physical watermarking is also an active technique, which is proposed in [20].

Control Theory/State Estimation The detection techniques that are based on physics originate in control theory because of the ample literature on the modelling of the physical processes. The past half-century has seen the extensive study of the detection of faults in control systems. Many works have reported the use of models for a physical process [21–24]. Fault detection literature provides ideas for these works, which have also highlighted the limitations of fault detectors when it comes to attack detection. Secure state estimation has been exhaustively analysed, towards that end. The recent research in [25] proposes a search algorithm that is based on Satisfiability Modulo Theory (SMT) to accelerate the search of potential sets of sensors, followed by an extended work to model the noisy systems [26].

Physical Authentication Authors of [27, 28] demonstrate interesting approaches for authenticating the control logic in a PLC by making use of the physics of the process. Recently, the authors of [29–31] were able to uncover an insider threat by exploiting the physics of the process.

Device Fingerprinting A closely related work [32] proposed to fingerprint sensors based on the measurement noise. However, the technique works only in specific states, for example, when the water in the tank is constant. For the extraction of sensor noise for certain components (e.g. level sensors), the process has to be in its static mode. In a particular type of ICS (automotive industry), researchers have tried to fingerprint devices in Controller Area Networks (CAN) bus [33].

Attacks on Water Systems Attack detection in canal networked systems using hydrodynamic models has been discussed in [34, 35]. An investigation into the challenges in the security of control systems when sensor and actuator data are compromised has been reported in [36]. The general approach in the literature is to study the effect of specific attacks on a particular system. These attacks include denial-of-service and deception attacks against a networked control system. A denial-of-service attack refers to obstructing the availability of resources by jamming communication channels [37, 38]. In deception attacks, the integrity of the sensor and actuator data is compromised. Specific types of deception attacks include false data injection, replay, and stealthy attacks. In [39, 40], false data injection attacks on power networks are studied while assuming that the attacker has perfect knowledge of the system model. In [41], authors have demonstrated the effect of replay attacks on the sensor readings and have proposed a methodology to detect such attacks. Various cyber attacks on networked control systems are studied in [42] using a quadruple tank testbed. An anomaly-based methodology for the detection of a wide range of integrity attacks in cyber-physical critical infrastructure is reported in [19, 43].

Most of the aforementioned related work provides theoretical models and examples for CPS security. However, research on cyber-security of CPSs through empirical analysis is lacking. Recently, in [19, 22, 44], researchers reported an investigation into the effectiveness of another detection scheme that uses the Kalman filter on an operational water treatment testbed. Experimental study on these kinds of real testbeds provides insight on how attacks can be launched on such systems. However, the work presented in this chapter is the first of its kind as it demonstrates the security testing on CPSs using two different testbeds. Also, it is a novel attempt to compare the different types of attack detection techniques on the basis of their performance in terms of accuracy and the detection speed.

3 Testbeds: Our Playground

This work has made use of two operational testbeds, a secure water treatment plant (SWaT) and a water distribution plant (WADI), that serve as exhibits of urban cyber-physical systems. The security measures have been implemented on these plants and their capabilities have been tested.

3.1 SWaT: A Secure Water Treatment Testbed

The Secure Water Treatment (SWaT) testbed is an operational water treatment plant scaled down to produce doubly filtered water at the rate of 5 gallons/minute. This testbed simulates the larger plants found in cities today. In addition to the following brief overview of this testbed, further details about SWaT can be found in [45].

Water Treatment Process As seen in Fig. 1, the water is passed through six distinct stages in order to undergo the complete treatment. Each of the stages is controlled by a dedicated PLC, and the necessary control actions are taken based on the data from sensors.

A motorised valve in stage P1 controls the inflow of the untreated water into the raw water tank. From this tank, the water is sent to the next stage P2 for chemical dosing where it undergoes chlorination. The water is then pumped to the ultra-filtration (UF) feed water tank in stage P3. The UF feed pump in this stage then transfers the water to the Reverse Osmosis (RO) feed water tank in stage P4 via the UF unit. In this stage, an RO feed pump sends the water through an ultraviolet dechlorination unit, which is controlled by the PLC. If necessary, sodium bisulphite ($NaHSO_3$) is added to the water to regulate the Oxidation Reduction Potential (ORP). Following this, the chlorine-free water is sent to a two-stage RO filtration unit in stage P5. The permeate tank stores the filtered water from the RO unit while the reject is accumulated in the UF backwash tank. The membranes of the UF unit are cleaned by activating the UF backwash pump when required in stage P6.

Communication Network and Vulnerabilities The PLCs in this testbed control the pumps and valves within their domain by obtaining data from the sensors associated with their corresponding stages. Communication among the PLCs occurs over a separate network. The connections between the sensors, actuators and PLCs can be either wired or wireless. Attackers could exploit the vulnerabilities of the protocol used to compromise the communication links among the PLCs and the sensor-actuator set. The PLC firmware, and consequently the PLCs themselves, could be the subjects of attack as well. Gaining control over such vital links, the attacker could use several strategies to insert fake data to the PLCs, adversely affecting the control actions taken based on it.

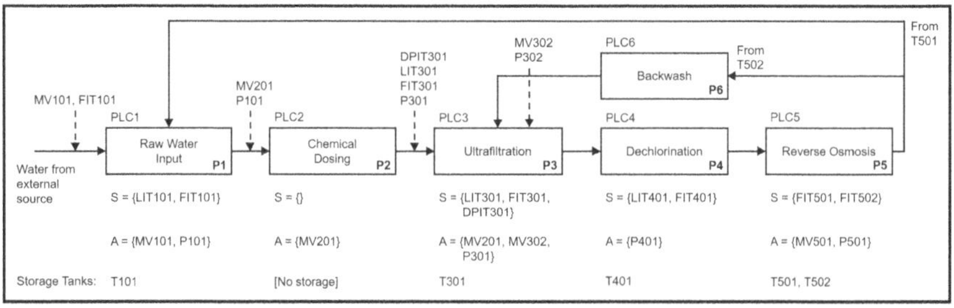

Fig. 1 Architecture of the SWaT Testbed

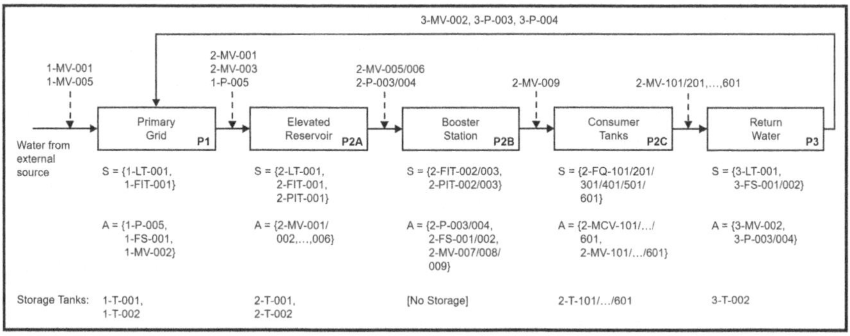

Fig. 2 Architecture of the WADI Testbed

3.2 *WADI: A Water Distribution Plant*

The Water Distribution (WADI) plant is an operational testbed supplying 10 US gallons/minute of filtered water [46]. It serves to represent the large distribution networks in cities that provide water to consumers.

Water Distribution Network As seen in Fig. 2, WADI is a three-staged plant with a primary grid (P1), a secondary grid (P2), and a return water grid (P3). In order to simplify the architecture, the stage P2 has been divided into three parts, P2a, P2b and P2c. The return water grid pumps water to the primary grid for recycling. Water quality analyzers have been installed in the return water grid to check water quality before pumping it into the primary grid.

Communication Network and Vulnerabilities All of the three stages in WADI have an individual PLC controlling them. These PLCs use National Instruments Compact RIO as Remote Input Output (RIO) devices. The communication network comprises three layers namely, layer-0 (L0), layer-1 (L1) and layer-2 (L2). Being at the process level, layer L0 connects the actuators/sensors and input/output (I/O) modules via the RS485-Modbus protocol. Layer L1 forms the plant control network and has a central node to which all the PLCs are connected in a star topology. The communication between the plant control network and the touch panel serving as the Human-Machine Interface (HMI) occurs via the third layer L2. Penetrating any of the layers gives the attacker access to manipulate the data and control actions.

4 System Models

The two testbeds are treated as multi-input, multi-output systems, applying the model-based approach. This section covers the creation and validation of their system-models.

A system model is used to represent the dynamics of a physical process as a mathematical formulation. The example of an Industrial Control System (ICS) is considered in this chapter. The state of the system is measured using sensors. Actuators are controlled by the Programmable Logic Controllers (PLCs) to affect the process state and their control actions are determined by the design of the system. For example, in a water tank, the level of water is taken as its state and is measured by a level sensor. Such a state in a water distribution process is dynamic and is governed by the actuators at the inlet and outlet of the tank, in this case. These actuation signals are considered as the inputs to the control system while the sensor measurements act as its outputs. This way, the systems and their underlying processes can be described using system models.

The models created largely pertain to the physical processes and do not represent the chemical behaviour of the plants. Hence, the components accounted for are pumps (Ps), motorised valves (MVs), electromagnetic flow meters (FITs), level sensors (LITs, LTs) and pressure sensors (PITs), while chemical sensors are excluded.

As explained below, the system models are derived using (a) sub-space system identification, and (b) basic physics principles. The first method is data-driven while the second approach is design-centric. When the data-driven procedure is used, obtaining the model can be generalised for any type of plant, making the model-based approach applicable to several industrial systems. Deriving the models from physical principles might not be as straightforward, based on the complexity of the system involved, but this approach can also be made generic by dividing the plants into simpler sub-systems. By having system models, the plants' behaviour can be represented in a quantitative manner, allowing their operation to be estimated. Analysis of the deviation from the expected and actual performance gives insight on possible anomalies and/or attacks in the ongoing processes.

4.1 System Modelling Using Sub-space System Identification

In this approach, the goal is to obtain a model of the following form at any time instance k for a system with p control inputs (actuators) and m outputs (sensors):

$$\begin{cases} x_{k+1} = Ax_k + Bu_k + v_k, \\ y_k = Cx_k + \eta_k \end{cases} \tag{1}$$

where $x \in \mathbb{R}^n$ is the system state vector of n states, $A \in \mathbb{R}^{n \times n}$ is the state space matrix, $B \in \mathbb{R}^{n \times p}$ is the control matrix, $y \in \mathbb{R}^m$ is the vector of the measured outputs, $C \in \mathbb{R}^{m \times n}$ is measurement matrix, and $u \in \mathbb{R}^p$ denotes the system control input. The system dynamics are captured by the state space matrices A, B and C, which can then be used to find a specific system state from an initial state. The noise vectors for the sensor and process are represented by η_k and v_k, respectively.

Such a Linear Time-Invariant (LTI) model can be obtained using sub-space system identification algorithms. The objective of system identification is to find these matrices using the input and output data from the plant. The system state for a physical process can be modelled with a precise analytical model. The system model can be used to predict the normal operation of a dynamic physical process, and deviations from the expected system behaviour can be observed in the case of anomalies. Such 10-state ($n = 10$) models have been created to represent both the testbeds.

4.2 System Modelling Using First Principles

In this section, a model for the level sensor outputs of two tanks in the WADI testbed is derived based on fundamental principles of physics. The change in the level of water in a particular tank across a time interval, as measured by level sensors, can be estimated to be a linear function of the net flow into that tank using the readings of the inflow and outflow sensors (Fig. 3).

The net flow into a particular tank (measured as the rate of volume) is given by the difference in the values of the flow sensors at the inlet and outlet, f_{in} and f_{out}, respectively:

$$\Delta f = f_{in} - f_{out} \tag{2}$$

Hence, the estimated net flow into a particular tank, $\hat{y}$, is given as follows:

$$\Delta \hat{y} = K_a \Delta f \tag{3}$$

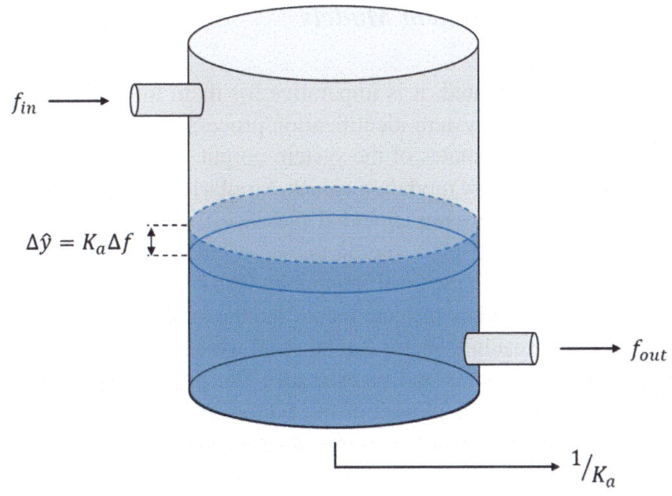

Fig. 3 Deriving a model for the level of water in a tank using first principles

Table 1 Models based on first principles for the two tanks in WADI testbed

Sensor	$\hat{y}_{k+1} = \hat{y}_k + K_a \Delta f_k + L(y_k - \hat{y}_k)$	
	K_a	L
1-LT-001	0.01503	0.5
2-LT-002	0.02356	0.5

where K_a is an empirically determined constant with a value approximately equal to the reciprocal of the cross-sectional area of the tank. Therefore, at any time step, $(k + 1)$, the level sensor estimate for a particular tank, $\hat{y}_{k+1}$, can be given as follows:

$$\hat{y}_{k+1} = \hat{y}_k + K_a \Delta f_k \tag{4}$$

To further improve this model, the error between the actual measurement and estimate at the current time instance, y_k and $\hat{y}_k$, respectively, can also be taken into account while computing the level sensor estimate at the succeeding time instance:

$$\hat{y}_{k+1} = \hat{y}_k + K_a \Delta f_k + L(y_k - \hat{y}_k) \tag{5}$$

where L is a linear filter that takes the weighted average of the actual and estimated values (similar to the Kalman filter-based gain matrix described Sect. 5). The constants in Eq. (5) for the models for level sensors 1-LT-001 and 2-LT-002 in WADI are listed in Table 1.

4.3 Validation of the System Models

After all the models are created, it is imperative for them to be validated. For the models generated using the system identification process, the state-spaces matrices are applied to obtain the estimates of the system output. Next, the real-time sensor readings are compared to these modelled values. Similarly, the first principles-based models for the level sensors are implemented and the estimated and actual values are compared.

The comparison for one of the level sensors in SWaT is illustrated in Fig. 4. The top pane shows the actual sensor measurements and the model estimate of the sensor. In the middle pane, the residual vector can be seen, which is the difference between the actual sensor measurements and its estimate. The probability density function (PDF) of the residual vector is plotted in the bottom pane. The estimate obtained from the model converges with the actual sensor measurements to a large extent, since the PDF of the residual has a very small variance.

The PDF plots for the models of the level sensors in the WADI testbed are shown in Fig. 5. It can be seen that the overall variance of the residuals from both, the

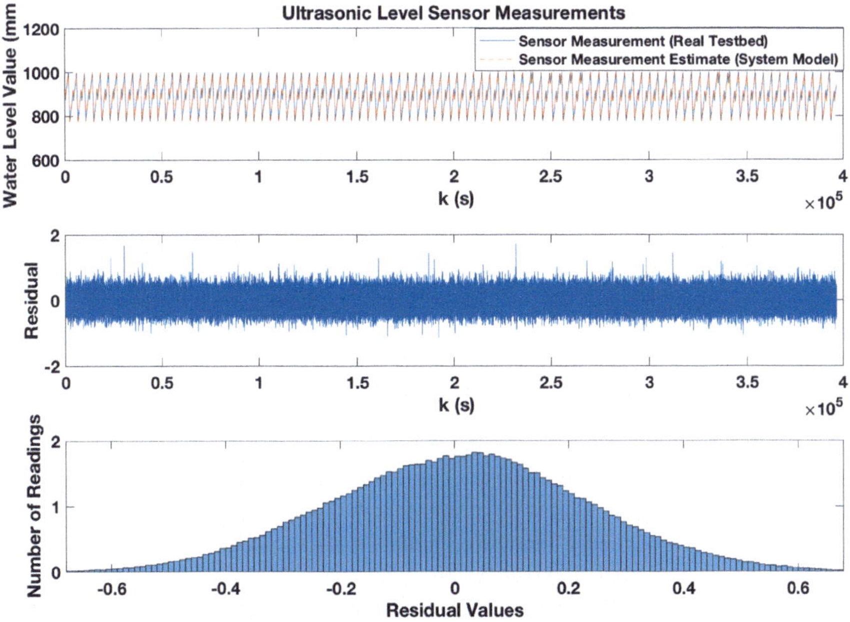

Fig. 4 Validating system model obtained using sub-space system identification method for a level sensor in SWaT testbed

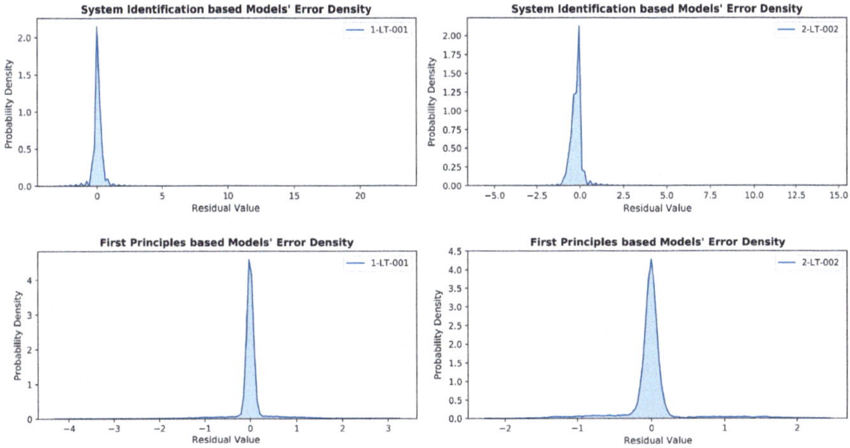

(a) PDF plots for 1-LT-001.

(b) PDF plots for 2-LT-002.

Fig. 5 Probability density function plots for the level sensors in WADI for the errors obtained from both models

system identification and first principles-based models, is small. There are some stray values that are very few in number and are considered to be outliers caused by unstable measurement instances. Comparing the models for both the sensors in Fig. 5, the model based on system identification has smaller error variance than the one derived from first principles. This is a significant observation regarding model accuracy and its effects would be seen on the attack detection accuracy discussed in Sect. 7.

Besides the visual validation, the root mean square error (RMSE) is used as a metric of estimation accuracy. The RMSE value for N readings is given as follows:

$$\text{RMSE} = \sqrt{\frac{\sum_{i=1}^{N} \left(y_i - \hat{y}_i\right)^2}{N}} \tag{6}$$

where y_i is the actual i-th sensor reading, and $\hat{y}_i$ is the i-th estimate obtained from the model. The RMSE value essentially indicates how far the estimated value is from the real measurement, and $(1 - \text{RMSE})$ represents the model accuracy as a percentage. According to control theory literature, models with an accuracy of even 70% are considered to approximate the real system dynamics accurately enough [47].

Table 2 shows the RMSE values of the models obtained using system-identification for 6 sensors in the SWaT testbed. It can be seen from these results that the model has high accuracy. In the case of WADI, the two models have been tested, the first obtained by following the system identification process, while the other is derived from first principles. As shown in Table 3, the system identification model reports a higher accuracy than the first principles-based models in the case of both the level sensors. Comparing the models for the level sensors in SWaT and WADI, it can be seen from the two tables that the model for SWaT is more accurate. This can be attributed to noisy measurements in WADI, which are caused by the higher sensitivity of its components in comparison to SWaT.

Table 2 Validating SWAT model obtained from sub-space system identification

Sensor	FIT101	LIT101	LIT301	FIT301	LIT401	FIT401
RMSE	0.0363	0.2867	0.2561	0.0200	0.2267	0.0014
(1-RMSE)*100%	96.3670	71.3273	74.3869	98.0032	77.3296	99.8593

Table 3 Validating both types of models for the level sensors in WADI

Sensor	**System identification model**		**First principle model**	
	RMSE	(1 - RMSE)*100%	RMSE	(1 - RMSE)*100%
1-LT-001	0.2623	73.7729	0.3946	60.5394
2-LT-002	0.2506	74.9446	0.4111	58.8854

Based on the RMSE values and the PDF plots, it can be safely concluded that the models obtained for both the plants are reasonably accurate, and can therefore be implemented on the testbeds to form the basis for the model-based attack detection methods.

5 Attack Detection Framework

The attack detection framework is based on validating the live sensor measurements to detect attacks. The involves (1) applying the models created in the earlier section to obtain the expected values using Kalman filter-based state estimation, (2) comparing these to the actual sensor measurements to get the error or residual vector, and (3) analysing the residual vector to check the source of the sensor readings. The following attack detection techniques are applied on this vector to determine the occurrence of an attack:

1. CUSUM detector
2. Bad-data detector
3. *NoisePrint*.

5.1 *Kalman Filter*

The state of the system with m outputs at a time instance k is estimated based on the model matrices obtained from system identification and the available real-time output y_k using a linear filter of the following structure:

$$\hat{x}_{k+1} = A\hat{x}_k + Bu_k + L_k(\bar{y}_k - C\hat{x}_k) \tag{7}$$

with the estimated state given by $\hat{x}_k \in \mathbb{R}^n$ for an n-state system, $\hat{x}_1 = E[x(t_1)]$, where $E[\cdot]$ is used to denote expectation, and gain matrix $L_k \in \mathbb{R}^{n\times m}$. Equation (7) depicts the system model where the Kalman filter is being employed for estimation. Using the previous readings up to x_{k-1} and the current sensor reading y_k, an estimate for the current state variable $\hat{x}_k$ is made. In Eq. (7), $\bar{y}_k$ represents a generic form for sensor measurements that might contain the data spoofed by an attacker. Thus, an error vector can be defined as follows:

$$e_k = x_k - \hat{x}_k \tag{8}$$

where $\hat{x}_k$ denotes the optimal estimate for x_k given the readings for all the system outputs until the $(k-1)$th time step. Let P_k denote the error covariance, $Cov(e_k) = E[(x_k - \hat{x}_k)(x_k - \hat{x}_k)^T]$, and $\hat{P}_{k|j}$ be the estimate of P_k given $y_1, ..., y_j$. The prediction equation for the state variable using the Kalman filter can be written as follows:

$$\hat{x}_{k+1|k} = A\hat{x}_{k|k} \tag{9}$$

$$P_{k+1|k} = AP_{k|k}A^T + Q \tag{10}$$

where $\hat{x}_{k|k}$ is the estimate at time step k (using measurements up to the time step k), and $\hat{x}_{k+1|k}$ is the $(k + 1)$th estimation based on previous k measurements. Similarly, $P_{k|k}$ is the error covariance estimate until time step k. The covariance matrix for the process noise is given by Q. The time step is updated using the Kalman gain, L_k:

$$L_k = P_{k|k-1}C^T(CP_{k|k-1}C^T + R)^{-1} \tag{11}$$

$$\bar{x}_{k+1|k} = \hat{x}_{k+1|k} + L_k(y_k - C\hat{x}_{k+1|k}) \tag{12}$$

$$\bar{P}_{k+1|k} = (I - L_kC)\hat{P}_{k+1|k} \tag{13}$$

where $\bar{x}_{k+1|k}$ and $\bar{P}_{k+1|k}$, are the updates for the $(k + 1)$th time step using measurements y_i from the ith sensor, and Kalman gain L_k. The measurement noise covariance matrix is denoted by R.

Assuming that the system is stable, the initial state for estimation can be selected arbitrarily, since it eventually manages to converge. For e.g.., the initial estimated state can be x_0 with $P_0 = E[(\hat{x}_0 - x_0)(\hat{x}_0 - x_0)^T]$. The Kalman gain L_k is updated at every time step, but after a few iterations, it converges and subsequently operates in a stable manner. The Kalman filter is an iterative estimator and $\hat{x}_{k|k}$ in Eq. (9) is obtained from $\bar{x}_{k-1|k}$ in Eq. (12). It is assumed that the system is in a steady state before attacks are launched. The Kalman filter gain is represented by L in steady state.

5.2 Residuals and Hypothesis Testing

Model-based fault detection is commonly found in related literature [47]. Use of the residual vector for fault detection concentrates on a particular structure of the faults. However, an intelligent adversarial attack is challenging to detect. This study focuses on the performance of the residual-based anomaly detection schemes against strategically designed attacks, such as, replay attacks [20, 22]. The performance of the detection procedures has been assessed for a variety of such attacks.

These procedures are dependent on a state estimator, such as the Kalman filter, to predict system state. The estimated state is used to obtain the expected output (sensor value) $\hat{y}_k$ and this is then compared to the actual sensor measurements $\bar{y}_k$, which may have been compromised. The sensor measurement $\bar{y}_k$ for the kth time step is given as follows:

$$\bar{y}_k = y_k + \delta_k \tag{14}$$

where y_k is the actual sensor value and δ_k represents the attack signal that may or may not be present at that time instance. Under normal operation, the difference between this value and the estimate should stay within a defined threshold, else, an alarm would be triggered. From the state-space matrices, the sensor estimate $\hat{y}_k$ for the kth time step is obtained as follows:

$$\hat{y}_k = C\hat{x}_k \tag{15}$$

Using this, the residual random sequence $r_k, k \in \mathbb{N}$ is defined as:

$$\begin{aligned} r_k &= \bar{y}_k - \hat{y}_k \\ &= (y_k + \delta_k) - \hat{y}_k \\ &= (Cx_k + \eta_k + \delta_k) - C\hat{x}_k \\ r_k &= Ce_k + \eta_k + \delta_k \end{aligned} \tag{16}$$

When there are no attacks ($\delta_k = 0$), the mean of the residue can be computed:

$$E[r_{k+1}] = CE[e_{k+1}] + E[\eta_{k+1}] = \bar{r}_{m\times 1} \tag{17}$$

where $\bar{r}_{m\times 1}$ denotes an $m \times 1$ matrix composed of the mean of the residuals under normal operation, and the co-variance is given by:

$$\Sigma := E[r_{k+1}r_{k+1}^T] = CPC^T + R_2 \tag{18}$$

For the residual, the hypothesis testing is for $\mathcal{H}_0$, the *normal mode* (no attacks), and $\mathcal{H}_1$, the *faulty mode* (with attacks). In this study, the system output is taken as the flow at the nodes and the water level in the tank. From this data and the state estimates, the residuals are obtained. Thus, the two hypotheses are state as follows:

$$\mathcal{H}_0 : \begin{cases} E[r_k] = \bar{r}_{m\times 1}, \\ E[r_k r_k^T] = \Sigma \end{cases} \text{or } \mathcal{H}_1 : \begin{cases} E[r_k] \neq \bar{r}_{m\times 1}, \\ E[r_k r_k^T] \neq \Sigma \end{cases}$$

In the later sections, these hypotheses are tested using commonly used change detection techniques based on the statistics of the residues.

5.3 Cumulative Sum (CUSUM) Detector

The CUSUM procedure, also known as the stateful detector, takes in the residual vector as an input to its algorithm. This input can be considered as a *distance measure* of how far the estimate is from the real measurements. For this work, all the sensors and actuators are considered together while creating a system model, implying that these together form the system output. Hence, the Kalman filter gain is presented in form of a matrix. A dedicated detector for each sensor is designed.

The index i denotes the sensor/detector, $i \in \mathfrak{I} := \{1, 2, \ldots, m\}$, where m is the number of sensor outputs. Thus, the attacked output vector at the kth time instance can be partitioned as $\bar{y}_k = \text{col}(\bar{y}_{k,1}, \ldots, \bar{y}_{k,m})$, where $\bar{y}_{k,i} \in \mathbb{R}$ denotes the i-th entry of $\bar{y}_k \in \mathbb{R}^m$. At any time step k, the individual attacked output for the i-th sensor, $\bar{y}_{k,i}$ is given as:

$$\bar{y}_{k,i} = C_i x_k + \eta_{k,i} + \delta_{k,i}, \tag{19}$$

where C_i denotes the i-th row of C, and $\eta_{k,i}$ and $\delta_{k,i}$ denote the i-th entries of η_k and δ_k, respectively. The residual vector for each sensor can be given as follows:

$$r_{k,i} = C_i e_k + \eta_{k,i} + \delta_{k,i}. \tag{20}$$

CUSUM is a standard procedure [48], and is explained using the following equations.

CUSUM: $S^-_{0,i} = 0,\ \ S^+_{0,i} = 0,\ \ \tilde{k}^+_i = 0,\ \ \tilde{k}^-_i = 0,$

$$\begin{cases} S^+_{k,i} = \max(0, S^+_{k-1,i} + r_{k,i} - \bar{T}_i - \kappa_i), & \text{if } S^+_{k-1,i} \leq \tau^+_i, \\ S^+_{k,i} = 0 \text{ and } \tilde{k}^+_i = \tilde{k}^+_i + 1, & \text{if } S^+_{k-1,i} > \tau^+_i. \end{cases} \tag{21}$$

$$\begin{cases} S^-_{k,i} = \min(0, S^-_{k-1,i} + r_{k,i} - \bar{T}_i + \kappa_i), & \text{if } S^-_{k-1,i} \geq \tau^-_i, \\ S^-_{k,i} = 0 \text{ and } \tilde{k}^-_i = \tilde{k}^-_i + 1, & \text{if } S^-_{k-1,i} < \tau^-_i. \end{cases} \tag{22}$$

Design parameters: bias $\kappa_i > 0$ and threshold $\tau_i > 0$.
Output: $alarm(s) = \tilde{k}^+_i + \tilde{k}^-_i$.

From (21) to (22), it can be observed that the CUSUM values $S^+_{k,i}$ and $S^-_{k,i}$ accumulate the distance measure $r_{k,i}$ over time to measure how far are the values of the residual from the target mean ($\bar{T}_i$). This window for error can be tuned using the slack variable κ. This is chosen to be $\frac{1}{2} * \sigma_i$ in this study, where σ_i is the standard deviation of the residual vectors for the i-th sensor. The CUSUM threshold τ_i is computed as follows:

$$\tau_i = \pm \Gamma * \sigma_i \tag{23}$$

where Γ is a multiplier to scale the standard deviation (σ) and is usually assigned a value between 3 and 5 [48].

An alarm is raised when this accumulation exceeds the chosen threshold τ_i in its magnitude. The sequence $S_{k,i}$ is reset to zero every time it becomes negative or greater than τ_i. If $r_{k,i}$ is tightly bounded and κ_i is not adequately large, the CUSUM sequence $S_{k,i}$ indefinitely grows till the threshold τ_i is reached, regardless of how large τ_i is set to be. To prevent such drifts, the slack variable κ_i must be appropriately selected based on the statistical properties of the distance measure. Once κ is chosen, the threshold τ_i must be selected to achieve a required false alarm rate $\mathcal{A}_i^*$. The *false alarm rate* for the CUSUM procedure is denoted by $\mathcal{A}_i \in [0, 1]$ and is defined as the expected proportion of observations which are incorrect alarms raised in the absence of an attack [49, 50].

The CUSUM values are expected to stay within the thresholds when the system is operating normally, since the estimates would mostly conform to the actual sensor values. Whereas, attack conditions would tend to cause the residual values to change unexpectedly, which would be reflected in the CUSUM values and consequently raise alarms to indicate the presence of an adversary.

5.4 Bad-Data Detector

The bad-data detector is known to be widely used in the CPS security literature [51]. It has been implemented on the residual sequence $r_{k,i}$ of the sensor i at the time instance k given by Eq. (16) in the following way:

Bad-Data Procedure:

$$\text{If } |r_{k,i}| > \alpha_i, \quad \tilde{k}_i = k, \quad i \in \mathfrak{I}. \tag{24}$$

Design parameter: threshold $\alpha_i > 0$.

Output: alarm time(s) $\tilde{k}_i$.

Using the bad-data detector, an alarm is reported if the distance measure, taken as $|r_{k,i}|$, exceeds the threshold α_i. Analogous to the CUSUM procedure, the parameter α_i is selected to satisfy a required false alarm rate $\mathcal{A}_i^*$.

Similar to the CUSUM detection, the alarms triggered using the bad-data detector indicate the presence of an anomaly or attack, since the residual values would be expected to remain within the threshold under normal conditions.

5.5 NoisePrint (Residual and Noise Fingerprint)

In this section a sensor fingerprinting technique called *NoisePrint* is presented, which is based on the sensor and process noise [3]. The working principle of this method is the intuition that when the system is in steady state [52], the residual vector obtained by applying the system model is a function of sensor and process noise. As seen earlier, at the kth time step, the residual vector is given as $r_k = Ce_k + \eta_k$, and the error vector can be expressed as $e_k = \sum_{i=0}^{k-2} (A - LC)^i (v_{k-i-1} - L\eta_{k-i-1})$, where $v_k \in \mathbb{R}^n$ is the process noise and $\eta_k \in \mathbb{R}^m$ is the sensor noise [53]. Using the system model to obtain the state estimates, the characteristics of the sensor and process noise for any given ICS can be extracted. This way, the acquired residual vectors can now be fingerprinted using pattern recognition techniques like machine learning.

5.6 Design of NoisePrint

The first step of the proposed scheme is the collection of data, which is then segmented into smaller chunks for feature extraction in the time and frequency domains. A sensor ID is assigned to the combined set of features. Following this, a machine learning algorithm is applied for classifying the sensors on the basis of their noise profiles.

***Residual Collection*:** After obtaining the system model for the ICS of the plant, the residual vector is calculated, as explained in the previous sub-sections. The residuals are collected for the complete group of sensors in SWaT and WADI testbeds, and are then inspected to extract sensor and process noise. When the plant is running normally, the error in sensor measurement occurs due to the presence of sensor and process noise (for e.g., due to the splashing of water, etc.). The collected residual is analysed in the time and frequency domains to examine the patterns of noise, and it has been found that they follow a Gaussian distribution. Applying a machine learning algorithm on fresh readings, which is test-data from the plants, unique profiles of sensors and processes can be generated using their variance along with the other statistical features acquired from the noise vector. Noise fingerprints can be created at the commissioning phase of the plant or over the course of its usage. Whether the process is dynamic or static does not make a difference, since these noise fingerprints are acquired using the system model.

Feature Extraction: Data is collected from sensors at a sampling rate of one second. Since data is collected over time, the time domain features can be extracted from the raw data itself. The Fast Fourier Transform (FFT) algorithm [54] has been used to convert the domain from time to frequency in order to extract the spectral features. In total, eight features are used to construct the fingerprint, as listed in Table 4.

Data Chunking: After residual collection, the dataset is divided into smaller segments. In the following sections, it will be seen that for both the testbeds, experiments have been conducted using datasets collected over a period of several days. Data chunking is essential for addressing (1) *what is the required sample size for training a machine learning model that performs well*, and (2) *how much data is sufficient to reliably confirm the presence or absence of an attack.*

The whole residual dataset (with a total of N readings) is divided into m chunks (each chunk of the size $\lfloor \frac{N}{m} \rfloor$). The feature set $< F(C_i) >$ is calculate for all the data chunks. For each sensor, there are m sets of features $< F(C_i) >_{i \in [1,m]}$. It has been empirically found that one-class SVM produced highest accuracy for the chunk size of $\lfloor \frac{N}{m} \rfloor = 60$ readings. This indicates the size of the data required for a machine learning algorithm and implies that attack detection decisions cannot be made instantly.

Table 4 List of features used. Vector x is time domain data from the sensor for N elements in the data chunk. Vector y is the frequency domain feature of sensor data. The vector of bin frequencies and the magnitude of the frequency coefficients are given by y_f and y_m, respectively

Feature	Description
Mean	$\bar{x} = \frac{1}{N}\sum_{i=1}^{N} x_i$
Std-Dev	$\sigma = \sqrt{\frac{1}{N-1}\sum_{i=1}^{N}(x_i - \bar{x}_i)^2}$
Mean Avg. Dev	$D_{\bar{x}} = \frac{1}{N}\sum_{i=1}^{N} \lvert x_i - \bar{x}\rvert$
Skewness	$\gamma = \frac{1}{N}\sum_{i=1}^{N}(\frac{x_i-\bar{x}}{\sigma})^3$
Kurtosis	$\beta = \frac{1}{N}\sum_{i=1}^{N}(\frac{x_i-\bar{x}}{\sigma})^4 - 3$
Spec. Std-Dev	$\sigma_s = \sqrt{\frac{\sum_{i=1}^{N}(y_f(i)^2) * y_m(i)}{\sum_{i=1}^{N} y_m(i)}}$
Spec. Centroid	$C_s = \frac{\sum_{i=1}^{N}(y_f(i)) * y_m(i)}{\sum_{i=1}^{N} y_m(i)}$
DC Component	$y_m(0)$

6 Attacker and Attack Model

The cyber-attacks taken into consideration in this work are on sensor measurements, as shown in Fig. 6. For the purpose of this study, certain assumptions have been made about the attacker. The types of attacks carried out on the SWaT and WADI testbeds are introduced in this section. Following this, the attacker model is discussed, which outlines the attacker's intentions and capabilities. The attacker can have a wide range of motives, which include performance degradation, interfering with a physical property of the system, or harming a component.

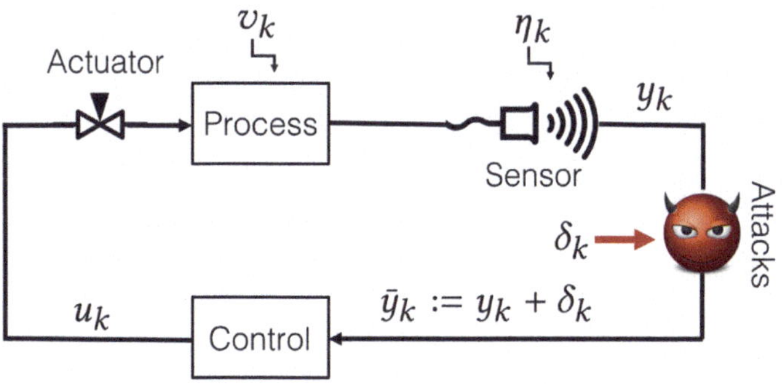

Fig. 6 CPS under attack

6.1 Attacker Model

Given the particular types of attacks considered in this work, the following assumptions have been made regarding the attacker:

1. The attacker is able to access $y_{k,i} = C_i x_k + \eta_{k,i}$ (i.e., the i-th sensor reading at the time step k).
2. The attacker knows the dynamics of the system, the state-space matrices, the control inputs and outputs, and the detection method that has been implemented.

Attacks on a CPS can be launched in several different ways. The sensors, actuators and PLCs in an ICS interface with each other using communication networks. These links of interaction can be compromised by a classic *Man-in-The-Middle (MiTM)* attack [55, 56]. Besides injecting false data into sensor measurements through the cyber domain, an attacker may also physically tamper with a component so as to send the ICS into an unstable state. Hence, authenticating the sensor readings being transmitted to the controller is crucial.

An attacker with physical access to the plants and the components within it is known as a *malicious insider*. However, physical damaging of sensors need not be done by an insider, as critical infrastructure systems like water treatment and power plants are often distributed over large areas [57, 58]. It is possible for an *outsider*, such as the end user, to execute a physical attack on sensors like the smart energy monitors found in urban regions.

6.2 Attack Scenarios

Data Injection Attacks In these attacks, the real sensor measurement is modified or fake data is injected by the attacker in order to push the system into an undesirable state. In the experiments conducted on the testbeds, the following two types of data injection attacks are considered:

- *Bias Injection Attack*: The attacker's motive in this type of attack is to mislead the control system by falsifying the sensor readings. For e.g., the level sensor measurements could be increased by the attacker while the actual level in tank level is stagnant. Considering the attacked values to be true sensor readings, the controller would keep the pump working till the tank empties itself, causing the pump to burn out. The attack vector in this scenario is defined as:

$$\bar{y}_k = y_k + \delta_k, \tag{25}$$

  where δ_k is the biased value injected by the attacker.
- *Stealthy Attack*: This attack is extremely difficult to discover using model-based methods because it is designed to remain undetected by statistical detectors such

as bad-data thresholds or CUSUM change detectors. The attacker chooses an attack vector δ_k for Eq. (25) in a way that makes it inconspicuous when statistical detectors are used. The residual vector in such attacks may not show substantial changes or cross the thresholds, which is required for detectors to recognise an attack. The impact of such attacks has been studied in literature [19]. Such attacks enable the attackers to hide their data injection while still influencing the system covertly, therefore earning their name. However, *NoisePrint* is capable of detecting such attacks, as will be seen in Sect. 7. An example of such an attack would be modifying a level sensor measurements by a small value, or spoofing a constant fake reading which might be same as the real reading from the last instance.

The attacks mentioned above have been simulated on the SWaT and WADI testbeds and the three detectors have been implemented to analyse their performance.

6.3 Attack Execution

Cyber Domain Data traffic between the sensors and the PLCs is intercepted using a *Man-in-The-Middle (MiTM)* approach. These packets are then inspected, and their payload (sensor measurement) is altered to implement the attack. The false reading could be injected by the adversary to execute either a bias injection attack or a stealthy attack.

The attacks carried out on SWaT and WADI for testing the attack detection techniques are shown in Tables 5 and 6. These attacks are of the following types, based on the manner of their execution:

- *Single-point Attack*—these kinds of attacks target one point in the system, manipulating its reading and/or severing its communication link.
- *Multi-point Attack*—in such attacks, several points are simultaneously targeted.
- *Stealthy Attack*—these are the attacks wherein minute alterations are made to the data value of a sensor, making it challenging to detect the anomaly.

The multi-point attacks could be single-stage or multi-stage, based on how spread out the target points are across the different stages in the plant. In the real scenario, these attacks are usually dependent on the competence, extent of access and intentions of the perpetrator.

Such operational technology (OT) attacks threaten the normal operation of the plant by manipulating states of sensors and/or actuators. Using the option provided by the SCADA system for the SWaT and WADI testbeds to manually alter the sensor/actuator readings being sent to the PLC, simple bias injection attacks have been simulated on the plants. Custom-developed Python programs are used that gradually modify the attack vector to imitate more complicated stealthy attacks. The researchers at the iTrust Labs have developed custom-coded modules [59] that are

Table 5 List of Attacks (SWaT)

Attack	Description (Initial state/attack state)
Stage 1	
Atk-1-s	LIT101 = 659 mm/change level +1 mm/s
Atk-2-s	LIT101 = 659 mm/LIT101 = 850 mm
Atk-3-s	LIT101 = 659 mm/LIT101 = 210 mm
Atk-4-s	LIT101 = 679 mm/LIT101 = 700 mm
Atk-5-s	LIT101 = 1029 mm/LIT101 = 700 mm
Atk-6-s	LIT101 = 789 mm/LIT101 = 789 mm
Atk-7-s	LIT101 = 784 mm/LIT101 = 600 mm
Stage 3	
Atk-8-s	L < LIT301 < H/LIT301 = HH+
Atk-9-s	L < LIT301 < H/change level −1 mm/s
Atk-10-s	L < LIT301 < H/change level −0.5 mm/s
Atk-11-s	FIT301 = 0 m^3/h/FIT301 = 2 m^3/h
Atk-12-s	L < LIT301 < H/water leakage attack
Stage 4	
Atk-13-s	FIT401 = 0.48 m^3/h/FIT401 = 0m^3/h
Atk-14-s	LIT401 < 1000 mm, P401 = ON/LIT401 = 1000 mm and P401 = ON
Atk-15-s	L < LIT401 < H, P301 = ON/LIT401 = 600 mm and P301 = ON
Atk-16-s	L < LIT401 < H/LIT401 < L
Atk-17-s	LIT401 = 1005 mm/LIT401 = 1005 mm

Table 6 List of attacks (WADI)

Attack	Description (Initial state/attack state)
Atk-1-w	1-FIT-001 = 1.71 m^3/h/1-FIT-001 = 1.5m^3/h
Atk-2-w	2-FIT-001 = 0 m^3/h/2-FIT-001 = 1.5m^3/h
Atk-3-w	2-FIT-003 = 0 m^3/h/2-FIT-003 = 1m^3/h
Atk-4-w	1-LT-001 = 55%/1-LT-001 = 80%
Atk-5-w	1-LT-001 = 40.21%/1-LT-001 = 40.21%
Atk-6-w	2-LT-002 = 46%/2-LT-002 = 65%
Atk-7-w	2-LT-002 = 71.2%/ 2-LT-002 = 71.2%

able to communicate with the LabVIEW-based[1] SCADA interface. These have been used to execute the stealthy attacks.

[1]Laboratory Virtual Instrument Engineering Workbench (LabVIEW) is a system-design software developed by National Instruments. For attack tool see: https://gitlab.com/gyani/NiSploit.

7 Performance Evaluation

This section introduces the metrics that have been used to assess the performance of the three attack detection methods. Next, the testing and efficiency of these techniques on the testbeds under normal operation and attack conditions are analysed.

7.1 Performance Metrics

A practically efficient attack detection method must be able to achieve both precision and sensitivity. The detection techniques have been assessed based on the following metrics, which form the criteria for effectiveness:

- True Positive Rate (TPR)—this is the number of instances whereby the technique correctly raises alarms (predicts an attack) over the course of the attack.
- False Positive Rate (FPR) or False Alarm Rate (FAR)—this is the number of instances when the method triggers alarms in the absence of any attack.
- Time Taken for Detection (TTD)—this is the time taken by the method to raise an alarm after the attack has been launched.

The TPR of the detection mechanism must be as high as possible as it is an obvious indication of its detection accuracy. The FPR provides insight on how likely the method is to raise false alarms, which are very inconvenient in practical applications. Hence, the FPR of the technique should be satisfactorily small. It is rather counter-intuitive to have a high TPR if the method cannot detect the attack soon enough. In a realistic scenario, the CPS is responsible for performing vital, large-scale processes which could be influencing the neighbouring economy directly or indirectly. If the attack stays unmitigated for too long, it could not only prove detrimental for the system, but also for its end-users. Therefore, the TTD of the detection method should be reasonable.

Practical implementations of the detection techniques often aim to achieve a balance between having a high TPR and a low FPR. The method could end up having a high FPR but manage to reach a good TPR. Conversely, the design may have a low FPR, but it could miss some attacks, which would lead to a low TPR. Usually, there is often a trade-off between these two rates such that a feasible FPR is maintained while attaining a satisfactory TPR.

7.2 Normal Operation

As highlighted previously, the design of the detection methods must ensure that too many false alarms are not raised. In order to test this, the detection techniques were applied on the plants when they were running normally, and their performance was studied.

Table 7 False positives under normal experiments on SWaT

Sensor	FIT101	LIT101	FIT301	LIT301	FIT401	LIT401
CUSUM detector						
Threshold	0.0149	3.1168	0.2209	0.5529	0.0156	0.5674
κ	0.0074	0.3117	0.0276	0.1382	0.0028	0.1135
FAR	5.54%	5.19%	5.34%	4.65%	4.02%	4.03%
Bad data detector						
Threshold	0.0205	1.4100	0.1184	0.4887	0.0108	0.4178
FAR	4.29%	5.32%	4.84%	4.56%	5.41%	5.42%
NoisePrint						
FAR	0%	1.29%	8.3%	2.44%	0%	0%

In the case of both the testbeds, the CUSUM and bad-data thresholds have been tuned to permit an FPR of 5% (or lesser). This allows the brief abnormalities (due to technical glitches and/or extrinsic disturbances) that occur in actual industrial plants to be accounted for.

Tables 7, 8 and 9 present the design parameters and thresholds for the detectors for each individual sensor. These results show that for both the plants, the false alarms raised were within a reasonable limit around the desired value.

In Fig. 7a, the residue from the system identification-based model for the level sensor (2-LT-002) in WADI can be seen to mostly remain beneath the bad-data threshold when the plant is operating normally. Likewise, it can be seen in Fig. 7b that the CUSUM values do not cross the threshold for 2-LT-002 under normal operation. Figure 8a, b show that similar results are achieved for the model for a level sensor (LIT-101) in SWaT when the CUSUM and bad-data detectors are used, respectively. From these figures, it can be said that the thresholds of the bad-data and CUSUM

Table 8 False positives under normal experiments on WADI (using the system identification based-model)

Sensor	1-LT-001	2-LT-002	2-PIT-001	2-PIT-002	1-FIT-001	2-FIT-001	2-FIT-002	2-FIT-003
CUSUM detector								
Threshold	1.109	0.6534	8.6809	0.2107	0.2964	0.0995	0.311	1.2972
κ	0.3466	0.2042	0.8681	0.3511	0.0823	0.0829	0.0389	0.1081
FAR	4.61%	3.76%	5.01%	3.47%	4.29%	4.13%	4.93%	5.01%
Bad data detector								
Threshold	1.122	0.7674	3.5104	0.7239	0.2063	0.3018	0.1548	0.487
FAR	4.40%	4.19%	4.08%	3.89%	4.64%	3.49%	4.56%	4.80%
NoisePrint								
FAR	13.04%	6.95%	21.74%	6.95%	6.08%	11.30%	4.34%	11.30%

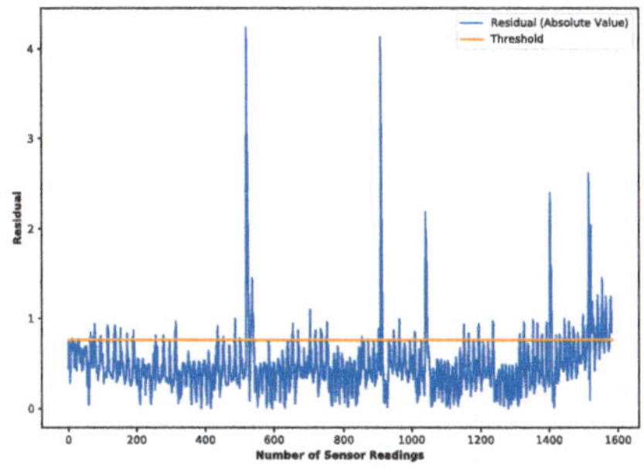

(a) Bad-Data detection

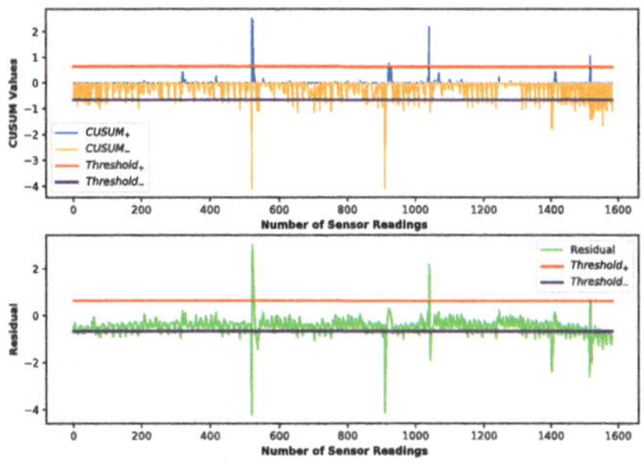

(b) CUSUM detection

Fig. 7 Statistical attack detection methods applied on the residual for level sensor (2-LT-002) estimates from WADI under normal operation

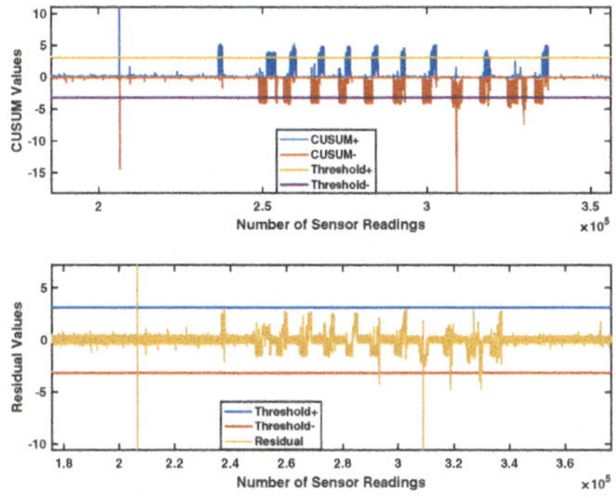

(a) Bad-Data detection

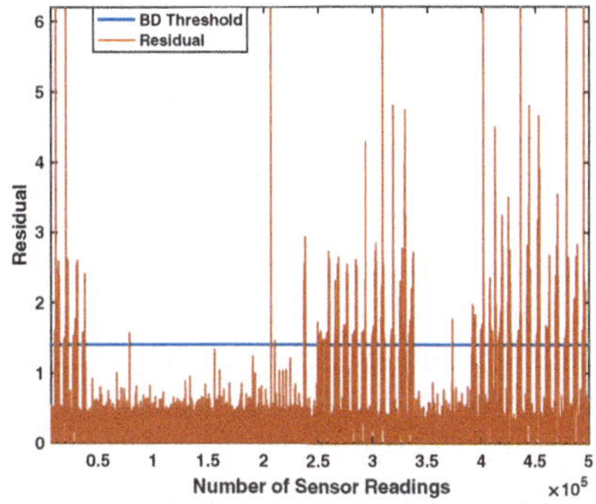

(b) CUSUM detection

Fig. 8 Statistical attack detection methods applied on the residual for level sensor (LIT-101) estimates from SWaT under normal operation

detectors have been designed to meet the requirements and can therefore be feasibly implemented on the plants when they are being run normally.

As seen in Table 7, the performance of *NoisePrint* is good when tested on SWaT, and it reports low or zero FPRs for nearly all the sensors. In the case of WADI, however, the FPR for majority of the sensors exceeded the desired 5%, as shown in Tables 8 and 9. This can be attributed to the fact that the components in WADI are known to be highly sensitive to environmental disturbances, which introduce faults to the sensor readings.

These tables and figures show that under normal operation, detection techniques perform adequately well in the case of both the testbeds (Tables 7, 8 and 9).

Table 9 False positives under normal experiments on WADI (using the first principle-based model)

Sensor	1-LT-001	2-LT-002
CUSUM detector		
Threshold	1.5315	1.6286
κ	0.2945	0.2036
FAR	3.95%	4.32%
Bad data detector		
Threshold	1.1359	0.797
FAR	3.84%	3.89%
NoisePrint		
FAR	5.21%	9.56%

7.3 Attack Detection

The SWaT and WADI plants were subjected to different attacks and the three detection methods were tested under these scenarios. The attacks executed on the plants are listed in Tables 5 and 6. The system identification- and first principle-based models were applied on the testbeds to obtain the residuals for all the sensors when the plants were under attack, and these were then examined using the detection methods. Tables 10, 11 and 12 present the metrics to gauge the performances of the detectors for each of the attacks.

Table 10 shows that when applied on SWaT, the performance of the CUSUM and bad-data detectors is good for many of the bias injection attacks, such as Atk-11-s, Atk-4-s and Atk-5-s. However, the stealthy attacks Atk-17-s and Atk-6-s are not caught by these detectors. *NoisePrint*, on the other hand, succeeds in detecting all of the attacks, including the stealthy attacks, and its TPR is comparable for other cases. The attacks that demonstrate poor TPR when CUSUM and bad-data thresholds are used can be detected more accurately using *NoisePrint*. But despite its superior performance, *NoisePrint* is not able to achieve a fast speed of detection. The CUSUM and bad-data detectors have a low TTD, which means that they are able to catch attacks in significantly lesser time as compared to *NoisePrint*.

The residual of the level sensor (LIT-101) in SWaT under stealthy attack is illustrated in Fig. 9a, b. In this attack, the sensor reading is spoofed by the attacker to remain at the same value as the last known normal measurement, therefore tricking the controller, while the actual process state keeps progressing differently. It can be seen in Fig. 9a that this residual never crosses the CUSUM thresholds. Likewise, Fig. 9b shows that the residual remains below the bad-data threshold during the stealthy attack. This implies that neither of the two detectors were able to detect the presence of this stealthy attack. However, *NoisePrint* catches this attack, as seen in Table 10.

In the case of WADI, when the residues obtained from the system identification-based model are examined using the CUSUM detector, it fails to perform well under

Table 10 Attack detection performance on SWaT testbed

Attack	NoisePrint			CUSUM			Bad data		
	TPR	FNR	TTD (s)	TPR	FNR	TTD (s)	TPR	FNR	TTD (s)
Single point attacks									
Atk-8-s	85.72%	14.28%	121.22	17.46%	82.54%	2	16.75%	83.25%	2
Atk-9-s	14.50%	85.50%	179	88.15%	11.85%	2	93.18%	6.82%	2
Atk-10-s	80.64%	19.35%	130.09	56.30%	43.70%	5	58.48%	41.52%	3
Atk-11-s	87.50%	12.50%	89.59	100%	0%	1	100%	0%	1
Atk-12-s	63.63%	36.37%	117.83	95.42%	4.58%	6	96.64%	3.36%	6
Atk-1-s	88.88%	11.12%	32.48	91.16%	8.83%	2	91.34%	8.66%	1
Atk-2-s	67.56%	32.44%	46.90	85.08%	14.92%	1	78.02%	21.98%	1
Atk-3-s	90.91%	9.09%	35.25	98.92%	1.08%	1	99.08%	0.92%	1
Atk-7-s	88.24%	11.76%	57.35	77.58%	22.42%	1	60.62%	39.38%	1
Atk-13-s	55%	45%	44.43	32.82%	67.18%	2	13.94%	86.06%	2
Atk-16-s	86.21%	13.79%	56.26	6.21%	93.79%	1	6.32%	93.68%	1
Multi-point attacks									
Atk-14-s	81.82%	18.18%	125.59	16.32%	83.68%	1	6.76%	93.24%	1
Atk-15-s	77.78%	22.22%	105.3	54.68%	45.32%	2	99.64%	0.36%	2
Atk-4-s	94.73%	5.26%	35.59	99.66%	0.34%	1	100%	0%	1
Atk-5-s	90.47%	9.53%	44.50	99.68%	0.32%	1	100%	0%	1
Stealthy attacks									
Atk-17-s	80%	20%	67.03	0%	100%	**ND**	0%	100%	**ND**
Atk-6-s	75%	25%	174.84	0%	100%	**ND**	0%	100%	**ND**

Table 11 Attack detection performance on WADI (System Identification Model)

Attack	NoisePrint			CUSUM			Bad data		
	TPR	FNR	TTD (s)	TPR	FNR	TTD (s)	TPR	FNR	TTD (s)
Single point attacks									
Atk-1-w	25%	75%	100	7.89%	92.11%	1	21.74%	78.26%	1
Atk-2-w	100%	0%	50	51.28%	48.72%	2	91.11%	8.89%	2
Atk-3-w	100%	0%	50	22.22%	77.78%	1	13.16%	86.84%	1
Atk-4-w	20.51%	79.49%	150	1.81%	98.19%	1	3.59%	96.41%	1
Atk-6-w	56.25%	43.75%	100	17.67%	82.33%	1	32.49%	67.51%	1
Stealthy attacks									
Atk-5-w	19.44%	80.56%	200	1.40%	98.60%	2	2.51%	97.49%	1
Atk-7-w	100%	0%	50	45.79%	54.21%	3	94.02%	5.98%	1

Table 12 Attack detection performance on WADI (First Principle Model)

Attack	NoisePrint			CUSUM			Bad data		
	TPR	FNR	TTD (s)	TPR	FNR	TTD (s)	TPR	FNR	TTD (s)
Single point attacks									
Atk-4-w	20.51%	79.49%	150	1.55%	98.45%	1	2.84%	97.16%	1
Atk-6-w	25%	75%	100	0%	100%	**ND**	0 %	100%	**ND**
Stealthy attacks									
Atk-5-w	25%	75%	100	1.12%	98.88%	2	2.51%	97.49%	2
Atk-7-w	12.24%	87.76%	250	0%	100%	**ND**	0%	100%	**ND**

all the attacks, as seen in Table 11. The bad-data detector achieves reasonable TPRs for attacks Atk-2-w and Atk-7-w, while *NoisePrint* is able to reach 100% detection accuracy for the attacks Atk-2-w, Atk-3-w and Atk-7-w. These two techniques report poor TPRs in detecting the other attacks. As is the case for SWaT, *NoisePrint* has a considerably higher TTD than that of bad-data detector even for WADI.

The bad-data and CUSUM detection when the system identification-based model is used on the level sensor 2-LT-002 in WADI is seen in Fig. 10a, b, respectively. Figure 10a shows that the residual is well above the bad-data threshold during the attack, implying the detector's good performance. However, as illustrated in Fig. 10b, the CUSUM detector does not catch the attack as the CUSUM values stay below the thresholds most of the time over the course of the attack.

As seen in Table 12, none of the detection mechanisms performed well when they were used on the residuals obtained from the first principle-based models for the level sensors in WADI. Attacks Atk-6-w and Atk-7-w remain entirely undetected by both the CUSUM and bad-data detectors.

The bad-data and CUSUM detection of Atk-6-w on 2-LT-002 in WADI using the residuals obtained from the first principles-based model is illustrated in Fig. 11a, b, respectively. The residual of the sensor under attack does not cross the bad-data threshold, as seen in Fig. 11a. Similarly, the CUSUM values of the residual when the sensor is under attack remain within the threshold, as shown in Fig. 11b, depicting that the attack has passed undetected.

From these results, it can be seen that the statistical detectors, bad-data and CUSUM, succeed in detecting simple bias injection attacks. But these detectors fail in catching the complicated stealthy attacks. This behaviour is not unusual, because stealthy attacks tend to ensure that the residuals obtained from the models do not change noticeably, therefore preventing the thresholds that confirm the presence the attack from being crossed. Whereas, *NoisePrint* shows the capability of identifying these types of attacks, because the attacker might not be able to replicate the sensor and process noise, which are the basis of detection for this technique. But the added accuracy of *NoisePrint* comes at the cost of its slow detection speed.

The applicability of these detection mechanisms in practical critical infrastructure systems can be questioned, considering the performance and nature of the techniques.

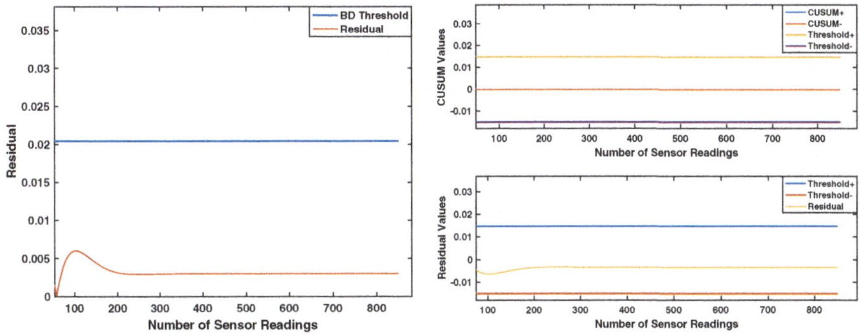

(a) Bad-Data detection (b) CUSUM detection

Fig. 9 Statistical attack detection methods (Bad-Data and CUSUM) applied on the residual for level sensor (LIT-101) estimates from SWaT under stealthy attack

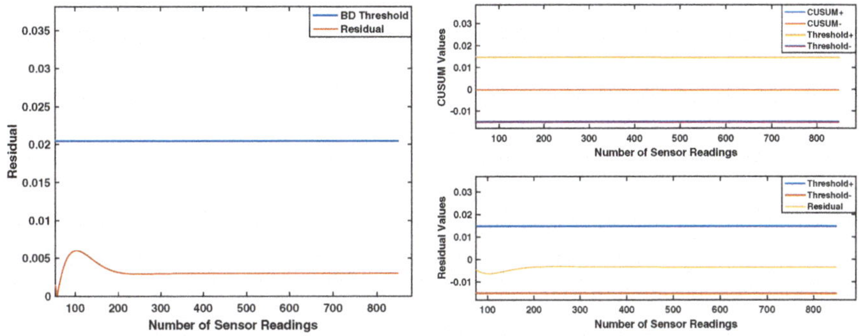

(a) Bad-Data detection (b) CUSUM detection

Fig. 10 Statistical attack detection methods applied on the residual values from system identification-based model for level sensor (2-LT-002) from WADI under stealthy attack

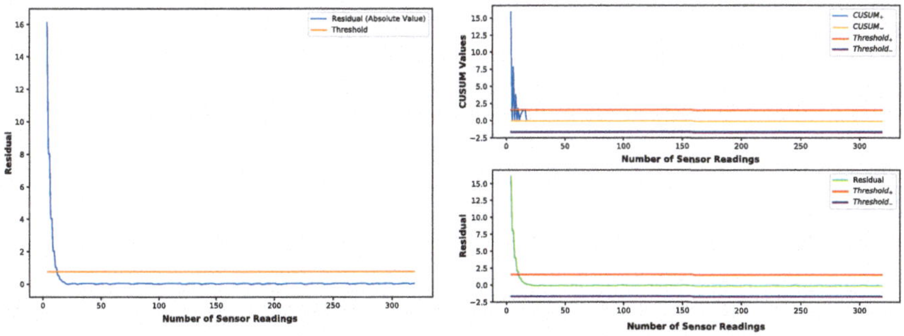

(a) Bad-Data detection (b) CUSUM detection

Fig. 11 Statistical attack detection methods applied on the residual values obtained using the first-principle-based model for level sensor (2-LT-002) under attack (Atk-6-w)

Unlike industrial CPSs, the testbeds studied in this work are scaled down, making it feasible to obtain their system models. In the case of larger plants, this could be addressed by dividing them into multiple sub-stages based on the processes taking place in them, and having individual models for each sub-system.

The longer detection time of *NoisePrint* is not favourable for CPSs requiring prompt response during attacks or anomalies, such as power grids. But its accuracy makes it an efficient detection technique for plants with large number of sensors, and it is still applicable to CPSs wherein the physical damage due to attacks can take a longer time to manifest.

8 Conclusions

From the results of the model validation, it can be understood that the models obtained by employing established system identification algorithms perform satisfactorily well as compared to the models generated using the heuristic or analytical methods, that involve the approximation of the process dynamics using first principles. An important insight is that for the infrastructure and sensors prone to environmental disturbances (for instance, the WADI testbed used in this study), obtaining a normal reference model is a non-trivial task.

It can be concluded that the statistical techniques, such as bad-data and CUSUM detectors, are easily able to catch basic bias injection attacks that resemble sensor faults. However, in order to detect advanced stealthy attacks, sophisticated methods such as *NoisePrint* are required.

From the tests conducted on the plants, it can be concluded that while detection methods should be able to demonstrate accuracy, the speed at which they detect the attack is also an essential metric for evaluating its utility in actual critic infrastructure systems. For assessing the performance of novel security techniques, the time taken for detection should be given more importance.

Acknowledgements The authors thank the reviewers for their comments. This work has extensively made use of the research facilities offered by the iTrust research centre at Singapore University of Technology and Design for which, the authors would like to express their gratitude.

References

1. Cardenas, A., Amin, S., Lin, Z., Huang, Y., Huang, C., Sastry, S.: Attacks against process control systems: Risk assessment, detection, and response. In: 6th ACM Symposium on Information. Computer and Communications Security, pp. 355–366 (2011)
2. Ahmed, C.M., Zhou, J.: Challenges and opportunities in cps security: a physics-based perspective. IEEE Secur, Priv (2020)
3. Ahmed, C.M., Ochoa, M., Zhou, J., Mathur, A.P., Qadeer, R., Murguia, C., Ruths, J.: Noiseprint: attack detection using sensor and process noise fingerprint in cyber physical systems. In: Proceedings of the 2018 on Asia Conference on Computer and Communications Security, ser.

ASIACCS '18, pp. 483–497. ACM, New York, NY, USA (2018). http://doi.acm.org/10.1145/3196494.3196532
4. Adepu, S., Mathur, A.: Distributed detection of single-stage multipoint cyber attacks in a water treatment plant. In: Proceedings of the 11th ACM on Asia Conference on Computer and Communications Security, pp. 449–460. ACM (2016)
5. Prakash J., Ahmed, M.: Can you see me on performance of wireless fingerprinting in a cyber physical system. In: 2017 IEEE 18th International Symposium on High Assurance Systems Engineering (HASE)
6. Krotofil, M., Gollmann, D.: Industrial control systems security: what is happening? In: 2013 11th IEEE International Conference on Industrial Informatics (INDIN), pp. 664–669, July 2013
7. Shoukry, Y., Martin, P., Yona, Y., Diggavi, S., Srivastava, M.: Pycra: physical challenge-response authentication for active sensors under spoofing attacks. In: Proceedings of the 22Nd ACM SIGSAC Conference on Computer and Communications Security, ser. CCS '15, pp. 1004–1015. ACM, New York, NY, USA (2015). http://doi.acm.org/10.1145/2810103.2813679
8. Ahmed, C.M., Prakash, J., Zhou, J.: Revisiting anomaly detection in ICS: Aimed at segregation of attacks and faults (2020)
9. Humayed, A., Lin, J., Li, F., Luo, B.: Cyber-physical systems security—a survey. CoRR (2017). arxiv:abs/1701.04525
10. Tahsini, A., Dunstatter, N., Guirguis, M., Ahmed, C.M.: Deepbloc: a framework for securing cps through deep reinforcement learning on stochastic games. IEEE Conference on Communications and Network Security (CNS) **2020**, 1–9 (2020)
11. Slay, J., Miller, M.: Lessons learned from the maroochy water breach. In: Goetz, E., Shenoi, S. (eds.) Critical Infrastructure Protection, pp. 73–82. Springer, Boston, MA, USA (2008)
12. Hemsley, K., Fisher, R.: A history of cyber incidents and threats involving industrial control systems. In: Staggs, J., Shenoi, S. (eds.) Critical Infrastructure Protection XII, pp. 215–242. Springer International Publishing, Cham (2018)
13. Pasqualetti, F., Dörfler, F., Bullo, F.: Attack detection and identification in cyber-physical systems. IEEE Trans. Autom. Control **58**(11), 2715–2729 (2013)
14. Pasqualetti, F., Dorfler, F., Bullo, F.: Cyber-physical attacks in power networks: Models, fundamental limitations and monitor design. In: Proceedings of the 50th IEEE Conference on Decision and Control and European Control Conference (2011)
15. Mo, Y., Sinopoli, B.: Secure control against replay attacks. In: 47th Annual Allerton Conference on Communication, Control, and Computing, Allerton (2009)
16. Athalye, S., Ahmed, C.M., Zhou, J.: A tale of two testbeds: a comparative study of attack detection techniques in cps. In: Rashid, A., Popov, P. (eds.) Critical Information Infrastructures Security, pp. 17–30. Springer International Publishing, Cham (2020)
17. Mitchell, R., Chen, I.-R.: A survey of intrusion detection techniques for cyber-physical systems. ACM Comput. Surv. (CSUR) **46**(4), 1–29 (2014)
18. Liu, Y., Ning, P., Reiter, M.K.: False data injection attacks against state estimation in electric power grids. In: Proceedings of the 16th ACM Conference on Computer and Communications Security, ser. CCS '09. Association for Computing Machinery, New York, NY, USA (2009). https://doi.org/10.1145/1653662.1653666
19. Ahmed, C.M., Murguia, C., Ruths, J.: Model-based attack detection scheme for smart water distribution networks. In: Proceedings of the 2017 ACM on Asia Conference on Computer and Communications Security, ser. ASIA CCS '17, pp. 101–113. ACM, New York, NY, USA (2017). http://doi.acm.org/10.1145/3052973.3053011
20. Mo, Y., Weerakkody, S., Sinopoli, B.: Physical authentication of control systems: designing watermarked control inputs to detect counterfeit sensor outputs. IEEE Control Syst. Mag. **35**(1), 93–109 (2015)
21. Bai, C.-Z., Gupta, V.: On kalman filtering in the presence of a compromised sensor: fundamental performance bounds. In: American Control Conference, vol. 2014, pp. 3029–3034. IEEE (2014)
22. Ahmed, C.M., Adepu, S., Mathur, A.: Limitations of state estimation based cyber attack detection schemes in industrial control systems. In: 2016 Smart City Security and Privacy Workshop (SCSP-W), pp. 1–5, Apr 2016

23. Murguia, C., Ruths, J.: Characterization of a cusum model-based sensor attack detector. In: 2016 IEEE 55th Conference on Decision and Control (CDC), vol. 12, pp. 1303–1309 (2016)
24. Qadeer, R., Murguia, C., Ahmed, C.M., Ruths, J.: Multistage downstream attack detection in a cyber physical system. In: Katsikas, S.K., Cuppens, F., Cuppens, N., Lambrinoudakis, C., Kalloniatis, C., Mylopoulos, J., Antón, A., Gritzalis, S. (eds.) Computer Security, pp. 177–185. Springer International Publishing, Cham (2018)
25. Shoukry, Y., Chong, M., Wakaiki, M., Nuzzo, P., Sangiovanni-Vincentelli, A., Seshia, S.A., Hespanha, J.P., Tabuada, P.: Smt-based observer design for cyber-physical systems under sensor attacks. ACM Trans. Cyber-Phys. Syst. **2**(1), 1–27 (2018)
26. Mishra, S., Shoukry, Y., Karamchandani, N., Diggavi, S.N., Tabuada, P.: Secure state estimation against sensor attacks in the presence of noise. IEEE Trans. Control Netw. Syst. **4**(1), 49–59 (2016)
27. Roth, T., McMillin, B.: Physical attestation in the smart grid for distributed state verification. IEEE Trans. Dependable Secur. Comput. **15**(2), 275–288 (2016)
28. Chen, Y., Poskitt, C.M., Sun, J.: Learning from mutants: Using code mutation to learn and monitor invariants of a cyber-physical system. In: IEEE Symposium on Security and Privacy (SP), vol. 2018, pp. 648–660. IEEE (2018)
29. Agrawal, A., Ahmed, C.M., Chang, E.-C.: Poster: physics-based attack detection for an insider threat model in a cyber-physical system. In: Proceedings of the 2018 on Asia Conference on Computer and Communications Security, pp. 821–823 (2018)
30. Ahmed, C.M., Prakash, J., Qadeer, R., Agrawal, A., Zhou, J.: Process skew: fingerprinting the process for anomaly detection in industrial control systems. In: Proceedings of the 13th ACM Conference on Security and Privacy in Wireless and Mobile Networks, ser. WiSec '20, pp. 219–230. Association for Computing Machinery, New York, NY, USA (2020). https://doi.org/10.1145/3395351.3399364
31. Ahmed, C.M., Mathur, A.P.: Hardware identification via sensor fingerprinting in a cyber physical system. In: 2017 IEEE International Conference on Software Quality, Reliability and Security Companion (QRS-C), pp. 517–524, July 2017
32. Mujeeb, A., Mathur, A., Martin, O.: NoiSense: detecting data integrity attacks on sensor measurements using hardware based fingerprints. ArXiv e-prints, Dec. 2017
33. Choi, W., Jo, H.J., Woo, S., Chun, J.Y., Park, J., Lee, D.H.: Identifying ecus using inimitable characteristics of signals in controller area networks. CoRR (2016). arxiv: abs/1607.00497
34. Amin, S., Litrico, X., Sastry, S.S., Bayen, A.M.: Cyber security of water SCADA systems 2014-Part II: attack detection using enhanced hydrodynamic models. IEEE Trans. Control Syst. Technol. **21**(5), 1679–1693 (2013)
35. Amin, S., Litrico, X., Sastry, S., Bayen, A.M.: Cyber security of water SCADA systems 2014Part I: Analysis and experimentation of stealthy deception attacks. IEEE Trans. Control Syst. Technol. **21**(5), 1963–1970 (2013)
36. Cárdenas, A.A., Amin, S., Sastry, S.: Research challenges for the security of control systems. In: Proceedings of the 3rd Conference on Hot Topics in Security, ser. HOTSEC'08, pp. 6:1–6:6. USENIX Association, Berkeley, CA, USA (2008)
37. Amin, S., Cárdenas, A., Sastry, S.S.: Safe and secure networked control systems under denial-of-service attacks. In: proceedings of the 12th International Conference on Hybrid Systems: Computation and Control (HSCC), vol. 5469, pp. 31–45. LNCS, Springer (2009)
38. Gupta, A., Langbort, C., Basar, T.: Optimal control in the presence of an intelligent jammer with limited actions. In: 49th IEEE Conference on Decision and Control (CDC), pp. 1096–1101, Dec. 2010
39. Liang, G., Zhao, J., Luo, F., Weller, S.R., Dong, Z.Y.: A review of false data injection attacks against modern power systems. IEEE Trans. Smart Grid **8**(4), 1630–1638 (2017)
40. Deng, R., Xiao, G., Lu, R.: Defending against false data injection attacks on power system state estimation. IEEE Trans. Industrial Informatics **13**(1), 198–207 (2017)
41. Mo, Y., Sinopoli, B.: Secure control against replay attacks. In: 2009 47th Annual Allerton Conference on Communication, Control, and Computing (Allerton), pp. 911–918 (2009)

42. Teixeira, A., Pérez, D., Sandberg, H., Johansson, K.H.: Attack models and scenarios for networked control systems. In: Proceedings of the 1st International Conference on High Confidence Networked Systems, ser. HiCoNS '12, pp. 55–64 (2012)
43. Ntalampiras, S.: Detection of integrity attacks in cyber-physical critical infrastructures using ensemble modeling. IEEE Trans. Ind. Inf. **11**(1), 104–111 (2015)
44. Palleti, V.R., Tan, Y.C., Samavedham, L.: A mechanistic fault detection and isolation approach using kalman filter to improve the security of cyber physical systems. J. Process Control **68**, 160–170 (2018)
45. SWaT: Secure Water Treatment Testbed (2015). https://itrust.sutd.edu.sg/wp-content/uploads/sites/3/2015/11/Brief-Introduction-to-SWaT_181115.pdf
46. Ahmed, C.M., Palleti, V.R., Mathur, A.P.: WADI: a water distribution testbed for research in the design of secure cyber physical systems. In: Proceedings of the 3rd International Workshop on Cyber-Physical Systems for Smart Water Networks, ser. CySWATER '17, pp. 25–28 (2017)
47. Wei, X., Verhaegen, M., van Engelen, T.: Sensor fault detection and isolation for wind turbines based on subspace identification and kalman filter techniques. Int. J. Adapt. Control Signal Process. **24**(8), 687–707 (2010). http://dx.doi.org/10.1002/acs.1162
48. Montgomery, D.: Introduction to Statistical Quality Control. Wiley (2009)
49. Adams, B., Woodall, W., Lowry, C.: The use (and misuse) of false alarm probabilities in control chart design. Front. Stat. Qual. Control **4**, 155–168 (1992)
50. van de Dobben, C.: Bruyn: Cumulative Sum Tests: Theory and Practice. Griffin, London (1968)
51. Liu, T., Gu, Y., Wang, D., Gui, Y., Guan, X.: A novel method to detect bad data injection attack in smart grid. In: 2013 IEEE Conference on Computer Communications Workshops (INFOCOM WKSHPS), pp. 49–54. IEEE (2013)
52. Aström, K.J., Wittenmark, B.: Computer-controlled Systems, 3rd edn. Prentice-Hall Inc, Upper Saddle River, NJ, USA (1997)
53. Ahmed, C.M., Zhou, J., Mathur, A.P.: Noise matters: using sensor and process noise fingerprint to detect stealthy cyber attacks and authenticate sensors in cp. In: Proceedings of the 34th Annual Computer Security Applications Conference, pp. 566–581 (2018)
54. Welch, P.: The use of fast fourier transform for the estimation of power spectra: a method based on time averaging over short, modified periodograms. IEEE Trans. Audio Electroacoust. **15**(2), 70–73 (1967)
55. Urbina, D.I., Giraldo, J.A., Cardenas, A.A., Tippenhauer, N.O., Valente, J., Faisal, M., Ruths, J., Candell, R., Sandberg, H.: Limiting the impact of stealthy attacks on industrial control systems. In: Proceedings of the 2016 ACM SIGSAC Conference on Computer and Communications Security, ser. CCS '16, pp. 1092–1105. ACM, New York, NY, USA (2016). http://doi.acm.org/10.1145/2976749.2978388
56. Amin, S., Litrico, X., Sastry, S., Bayen, A.: Cyber security of water SCADA systems; Part I: Analysis and experimentation of stealthy deception attacks. IEEE Trans. Control Syst. Technol. **21**(5), 1963–1970 (2013)
57. Formby, D., Srinivasan, P., Leonard, A., Rogers, J., Beyah, R.: Who's in control of your control system? device fingerprinting for cyber-physical systems. In: NDSS, Apr 2016
58. Sridhar, S., Hahn, A., Govindarasu, M.: Cyber physical system security for the electric power grid. Proc. IEEE **100**(1), 210–224 (2012)
59. Adepu, S., Mishra, G., Mathur, A.: Access control in water distribution networks: a case study. In: 2017 IEEE International Conference on Software Quality, Reliability and Security (QRS), pp. 184–191, July 2017

Security of Cyber-Physical Monitoring and Warning Systems for Natural and Technological Threats

Mirosław Hajder, Piotr Hajder, and Mariusz Nycz

Abstract This work describes the use of cyber-physical architectures for the construction of regional environmental monitoring and alert system for residents to the natural and technological threats. Regional systems are structures covering an area of several km^2. The work focused on ensuring the security of collected and processed information as well as the continuity of system operation. The base tool to improve its parameters is the use typical solution for cyber-physical systems, including component of Internet of Things. Nationwide, environmental monitoring in most countries is still based on IT solutions from the early 1980s. In the case of slow-changing global threats covering large parts of country, this solution works well. However, for dynamically changing regional threats, using them for early preparation of accurate warnings is impossible. As a solution for such problem, we are using the data that comes from weather stations owned by the residents of the monitored area, that was designed using Internet of Things technology. This approach creates new, unprecedented threats to information security and system continuity. In addition to threats resulting from the use of the Internet of Things, there are new, resulting from the properties of algorithms for threats forecasting, which require, among others input dataflow continuity. In work the functional organization of environmental monitoring system and their architecture are presented. On their basis the methods and means of counteracting security and accessibility threats are evaluated. This work describes also the implementation of environmental monitoring system in a town adjacent to a larger agglomeration.

M. Hajder
University of Information and Management, Rzeszow, Poland
e-mail: Miroslaw.Hajder@gmail.com

P. Hajder (✉)
AGH University of Science and Technology, Krakow, Poland
e-mail: phajder@agh.com; phajder@agh.edu.pl

M. Nycz
Rzeszow University of Technology, Rzeszow, Poland
e-mail: Mar.Nycz@gmail.com

A. I. Awad et al. (eds.), *Security in Cyber-Physical Systems*, Studies in Systems, Decision and Control 339, https://doi.org/10.1007/978-3-030-67361-1_7

1 Introduction

The last few years have been a period of a noticeable increase in the frequency of natural hazards, such as floods, atmospheric and ground drought, hurricane winds, icing, hail, etc. These phenomena are also noticed in countries where their occurrence was rare until now. The increasing disaster frequency should be associated with permanent environmental changes occurring because of human activities.

In everyday language, a disaster is a sudden change in the characteristics of the surrounding world [1]. It can positively transform the environmental or, on the other hand, destructive, negatively affecting on the environment and society. Disasters are natural phenomena and their appearance is unpredictable. When we are not prepared, they can result in a serious consequences for human and their environment [2–4]. The scientific definition of a disaster should first be sought in the area of mathematical disaster theory (MDT) described by, among others Thom and Zeeman [2, 5, 6]. MDT is concerned with analysis of spatio-temporal models and development of catastrophes occurring in complex systems and structures, without detailed distinction of types of objects and phenomena. Unfortunately, even within the MDT itself, there is a lack of terminological uniformity. According to the definition given in the works of V. I. Arnold, a catastrophe is a rapid qualitative change of an object because of uniform quantitative change of its parameters. In turn, the definition, in accordance with the works of H. Poincare, treats catastrophe as the loss in stability of the system's harmonic movement and its abrupt transition to a new state of equilibrium, with current parameters of that movement.

The most important reasons for the increase in the frequency of disasters [1, 3, 7, 8] are:

1. Increased sensitivity to external factors of the environmental and technological aspects, resulting among others from violation of the natural balance, through the massive use of technology compromising the natural environment;
2. Obstructing the natural regeneration of the biosphere: anthropogenic transformation of the natural environment, expanding the technological sphere, and also massive exploitation of the pristine areas of the Earth;
3. High sensitivity of the social sphere of life for the natural and technological disaster, occurring by intensification of the negative effects of disasters.

Increasing frequency of weather disasters of unprecedented intensity indirectly results in growing number of technological events, especially in the case of older facilities, designed and built at a time, when in the given area the climatic threats did not occur, or their intensity was significantly lower. Disasters have been described from the beginning of historiography, at least for several years scientific research have been conducted to detect them early. However, existing research and available solutions mainly refer to global catastrophes, covering large areas and causing huge damage to the infrastructure and psyche of people. In the range of global threats, the responsibility of measuring and forecasting threats in most countries lies with state authorities. Research in this area is undoubtedly important, but their importance for local communities is limited. The authors- interests include regional disasters, not sufficiently presented in contemporary scientific research. Local disasters are characterized by high intensity of occurring phenomena, dynamic course and short duration, and their detection is possible immediately before their occurrence. Methods used to analyze global phenomena (statistical methods [9, 10]; analogy methods [11, 12]; method of comparing the rate of change [13–15]; energy analysis methods [13, 14]; critical level methods [13, 14]; method of dimensionless relative coefficients [13, 16]; instrumental forecasting methods [8, 17]; risk analysis methods [17, 18]; mathematical methods of catastrophe theory [5, 6, 19] et al.) are a little use for research phenomena occurring in a regional scale. Similarly, the methods and means used to design global systems are ineffective in creating solutions functioning in regional scale. Architectures obtained using them poorly respect local conditions, and the forecasts appear too late and significantly differ from the occurring phenomena. This work contains of 5 paragraphs. The introduction discusses the purpose of the research subject, presented expectations of research results and showed a structure of work. In the next paragraph the specific features of regional monitoring system in the range of software and hardware equipment and information security are analyzed. Paragraph 3 presents the organizational and architectural conditions of monitoring systems; three generations of systems have been distinguished. Paragraph 4 presents information security threats occurring in environmental monitoring systems. It discusses the impact on safety by the basic organizational components of monitoring systems, classifies the risks and the specific effects of their occurrence. Paragraph 5 describes the system architecture, and 6 the characteristics of the residents' information subsystem, taking into account a wide age spectrum of alert recipients. The last paragraph is a summary, in which presented the basic results of work and describe aim of future research.

Further, describing about environmental monitoring we will use replacement the term monitoring, but consider the security of environmental monitoring systems we will use also term security.

The chapter describes the results of research ended the real implement the environmental monitoring systems in mainly address to regional threats in a town adjacent to a larger agglomeration with extensive and development potential.

2 Monitoring Information Conditions

2.1 *The Concept of Environmental Monitoring*

The term monitoring appeared in the second half of the 20th century and defined a system of repetitive, directed observations of one or more elements in the surrounding nature. Most researchers defines the concept of monitoring system as a set of elements that creates a structure intended for collection and process information about the state of surrounding environment. These elements are: object and entities of monitoring, its

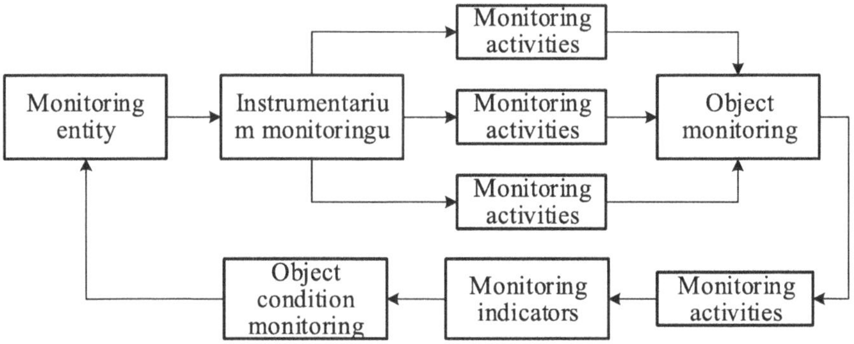

Fig. 1 Elements of monitoring system and their relationship

instruments, a set of monitoring indicators and monitoring activities. The relationship between these elements is shown in Fig. 1.

Generally, the monitoring objects are complex systems and phenomena. The common feature of all monitoring objects is the high dynamics of changes occurring in them. Only in this case it is appropriate to monitor them. Objects whose behavior is static can be observed by classical methods. Monitoring entities are most often carriers of the monitoring function, i.e. organizations, organizational structures and people who perform above activities. The entity not only make a monitoring disposition, but it is also interested in their results. A set of monitoring indicators is a collection of measured values providing a comprehensive description of the state of the environment and data about its quantitative and qualitative changes. The monitoring instrumentarium creates a set of hardware and software measures necessary to perform measurements, their statistical processing, forecasting, inform and warning people about the state of environment and potential threats. It is used in its activities by monitoring entities. Monitoring activities are a set of inclusive functional procedure: collection and processing information, its visualization as well as preparing proposition for necessary actions that are a response to the state of environment, include changes in the operation of the monitoring system itself.

We distinguish the three basic types actions perform in monitoring systems:

1. Organization and performing monitoring, for which is a measured subsystem used;
2. Collection the measures results performed by communication subsystem;
3. Processing of measurement data along with recommendations regarding their use, which the information and analytical subsystem deals with.

The mutual relationship between above actions and their implementing components are shown in Fig. 2.

Monitoring as a sequence of interrelated activities can be divided into three successive stages:

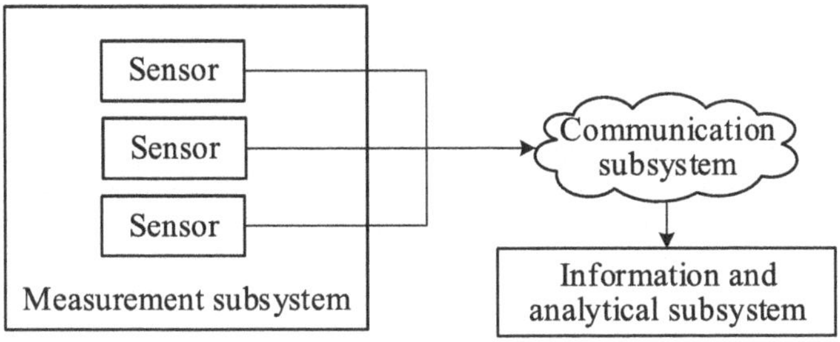

Fig. 2 Interconnection of monitoring components

1. Preparation for the legal and normative organization of monitoring [20, 21];
2. Executive, during which measurements are conducting and their results are sent to the node dealing which their further processing [22–24];
3. Analytical and decision-making, where the results of monitoring are processed and then used in the management process [24, 25].

2.2 A Formal Description of the Monitoring Process

Since the monitored environment is a complex system, the proposed theory contributes to the development of the theory of such systems. We will call a complex system an object composed of many elements, each of which can be considered as a system. As a rule, these elements according to certain, specific principles, are combined into one integral whole or are connected by appropriate functional relations. At any time, the elements of any complex system are in one of the possible states. The transition between them is made under the influence of the internal or external factors. The dynamics of the behavior of a complex system is manifested in the fact that the state of the element and its output signals are at any time determined by its previous states and input signals coming from other elements of the system or its external environment. In theory complex system, the term of external environment is described as a set of objects that are not objects of a given system, interaction with which is considered in the process of its study. Elements of complex system works in mutual connection: the properties of each element depend on the conditions set by other elements of this system. The properties of complex system are determined not only by the properties of its components, but also by the nature of the interaction between them [26].

The basic method of studying the complex system described by theory is a mathematical modeling. To carry it out, the processes of the system functioning, i.e. present it in the form of a sequence of specific events, phenomena, or procedures should be formalized, and then its mathematical description is created. According to modeling theory, to formalize the representation of any object O, first it is required to specify all their attributes creating a static object model. Then, the process Q of describing changes in their value in time, which is the result of various factors, which creates a model of object behavior under given conditions, is defined. The further consideration we assume that: K is an identificatory of the object contained to classification space; A is a description of the invariant of K attributes, V is a description of the properties, relations and functions determining the behavior of the examined object. Since this model have a static character, it always describes a system state in a moment t. Considering above findings, we can define the static object O as:

$$O \rightarrow (K, A, V, t) \tag{1}$$

The process of changing the state of an object over time, under the influence of a set of internal and external factors, is called the behavior of an object. This process can be described by the following expression:

$$Q \rightarrow (K, G, F, T) \tag{2}$$

where F–space of factors influencing objects behavior.

From the point of view of the behavior of the object, time plays a particular role, time T is beyond the set of F factors. In turn, G is a set of all object attributes, divided into two subsets: A–subset invariant attributes; X–subset of parametric attributes that change over time under the influence of internal or external factors contained in the set F, while $G = A \cup X$. The subset A includes such attributes as: the name of the object, its identification number, geographical location, etc. In turn, the subset X contains characteristics that are the parameters of the object, which are functions of time and factors from the set F affecting the object O.

The set of values of all object attributes at the moment t is called the state of the given object. The set of attributes $(A_1, A_2, \ldots, A_s, X_1, X_2, \ldots, X_n, t)$ creates the state space of object O, and the set of values of this variables is called its state coordinates. The sequential change in the state of monitoring objects, expressed by means of monitoring indicators, is called the monitoring process. According to the introduced markings:

$$Q = f(K, A, X, T) \tag{3}$$

is a mathematical description of the monitoring object state change process. For the functioning of the monitoring system to provide reliable data on the state of the environment with minimal costs of its implementation, it is necessary to introduce hierarchically linked levels of information generalization on monitoring objects. If objects are subject to generalization and they will also be its result, we can distinguish three basic types described by argument K from formula (3). These are the objects of observation, generalization and monitoring, relations between which were shown in Fig. 3.

Observation objects are objects that are subject to continuous tracking of their selected characteristics, which is performed by measuring them directly. The object of generalization will be the set of observed objects, grouped using thematic, spatial, or temporal criteria, on the basis of which the analysis of the environment state

Fig. 3 Relationship between classes of objects

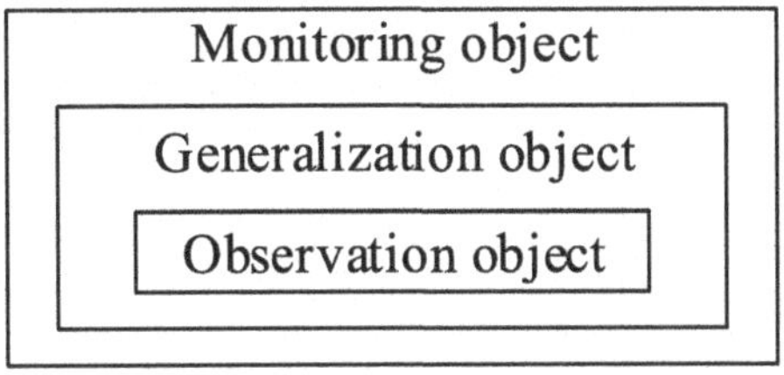

and forecasting its changes is performed. While observation objects are described by parameters whose significance is determined by measurement, generalization objects are presented as calculation parameters determined based on mathematical or statistical formulas. Monitoring objects are complex system objects, whose condition is described by means of integral assessments, which allow to present overall quantitative and qualitative changes in the state of the examined system. So we can write that: $K = \{K_o, K_u, K_m\}$, where: K_o–subset of observation objects; K_u–subject of generalization objects; K_m–subset of monitoring objects. In the description based on the expression (3) A denotes the attributed of the objects and their most important properties. At the same time, $A = \{\cup A_{K_o}, \cup A_{K_u}, \cup A_{K_m}\}$. In turn, X describes parametric properties characterizing the state of the object, which are determined by internal or external factors. As in the case of attributes, we assume that $X = \{\cup X_o(t), \cup X_u(t), \cup X_m(t)\}$. The argument T of the expression (3) defines the periodicity of recording the dynamics of changes in the state of objects, i.e. obtaining measurement results, their generalization and obtaining monitoring results, as well as the moment t_0 to start the observation process. The argument T have the form: $T = \{T_o, T_u, T_m, t_0\}$.

3 Organizational and Architectural Conditions of Monitoring

3.1 Classic Environmental Monitoring Systems

Although environmental measurements have been carried out regularly since the 16th century, contemporary methods of measurements differed significantly from those of today. Currently used solutions can be divided into three generations. The first of these includes, above all, a nationwide monitoring system built and exploited by governments in its framework of their statutory obligations. Measuring nodes are built based on equipment calibrated in accordance with applicable standards. In Poland, there are over 1000 measuring nodes of this type connected in a common structure. From an IT point of view, they take the form of centralized systems with distributed collection of information that is processed. Their organization was shown in Fig. 4.

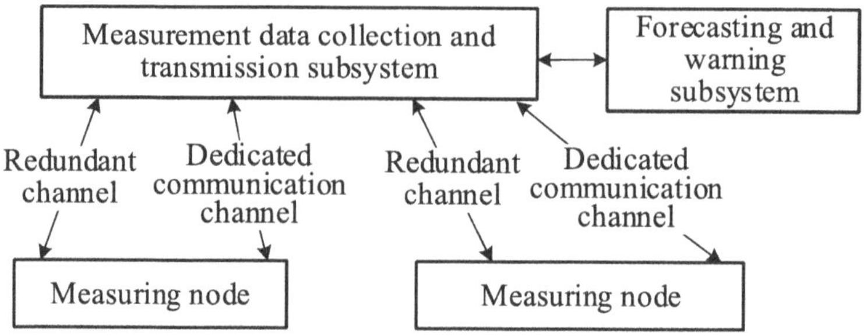

Fig. 4 Classic environmental monitoring system—the first generation of systems

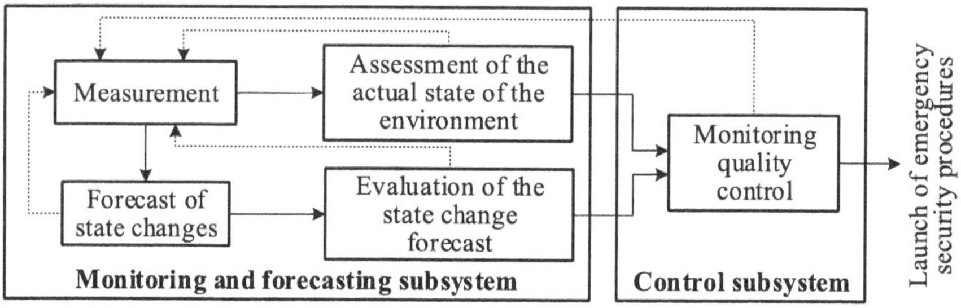

Fig. 5 Functional architecture of monitoring systems

An important factor from the security of the classic system is the measurement data collection and transmission subsystem integrating measuring nodes using encrypted dedicated communication channels. Each node is equipped with an additional redundant channel that also uses dedicated links. The preparation of forecasts and the formulation of warnings are performed centrally in the forecasting and warning subsystem especially designed for this purpose. The functional architecture of this generation environmental monitoring system is presented at Fig. 5.

In addition to permanent tracking and assessment of the state of the environment, the system in Fig. 5 forecasts possible changes and estimates emerging threats. The frequency and precision of measurements are constantly adapted to the current level of threats. The increase in accuracy is usually initiated by the components of the monitoring and forecasting subsystem. It can also be forced by the control subsystem, whose additional task is to run emergency procedures.

First generation systems should not be called cyber-physical: they do not use the solutions offered by the Internet of Things, and their functioning is not based on the extensive use of the public Internet. The architecture characteristic for modern (second generation) environmental monitoring systems is presented in Fig. 6. It highlights the components of data acquisition, processing and storage, which are crucial for modern systems operating practically in real time.

In order to guarantee the monitoring consistency, the measurement nodes, apart from vertical connections connecting them with the communication subsystem, are

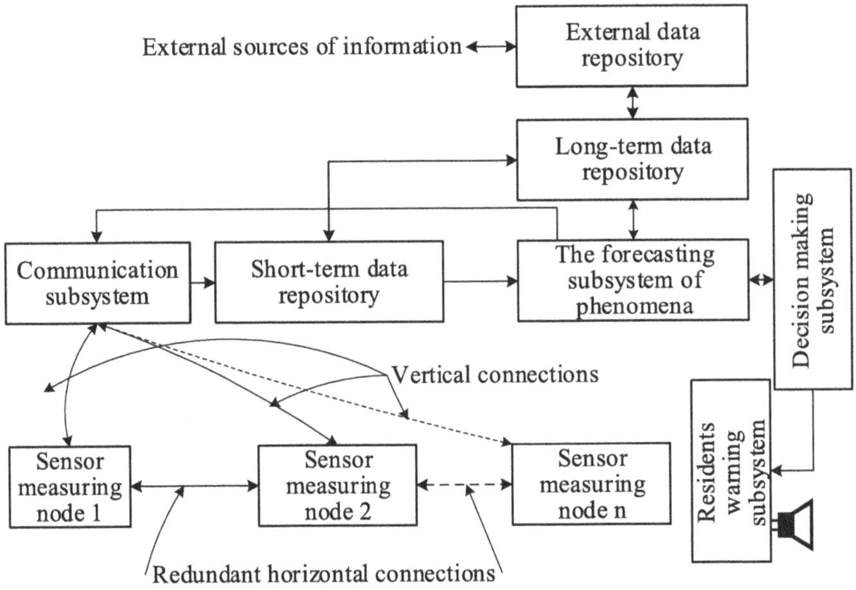

Fig. 6 Hardware and software architecture of monitoring systems

equipped with horizontal channels connecting them to adjacent measurement nodes. Typically, these connections are made using a different communication technology which further increases the service life of the system. Information used in the forecasting process is stored in three independent data repositories: short-term, long-term and external. The prepared forecasts are additionally stored in the long-term data repository.

3.2 Multi-level Sensor Monitoring Networks

From an IT point of view, the second generation of environmental monitoring systems are built on the base of sensor networks standards. These networks are derived from cellular wireless networks, currently built as multi-level hierarchical architectures, consisting of 3–5 communication levels [27–30]. In addition to hierarchy, these networks are characterized by heterogeneity, resulting mainly from the variety of requirements, posed to traffic management at each level of the hierarchy. Theoretically, the use of a flat homogeneous network for this purpose, but such a solution is currently not justified. Heterogeneity of communication has many advantages, among others ensures a wide range of available ways of user integration with the network and the possibility of construction at its higher levels, multi-protocol, multimedia communication systems. In turn, hierarchy puts them in the main direction of the development of information systems [31–35]. As the reason for the widely spread

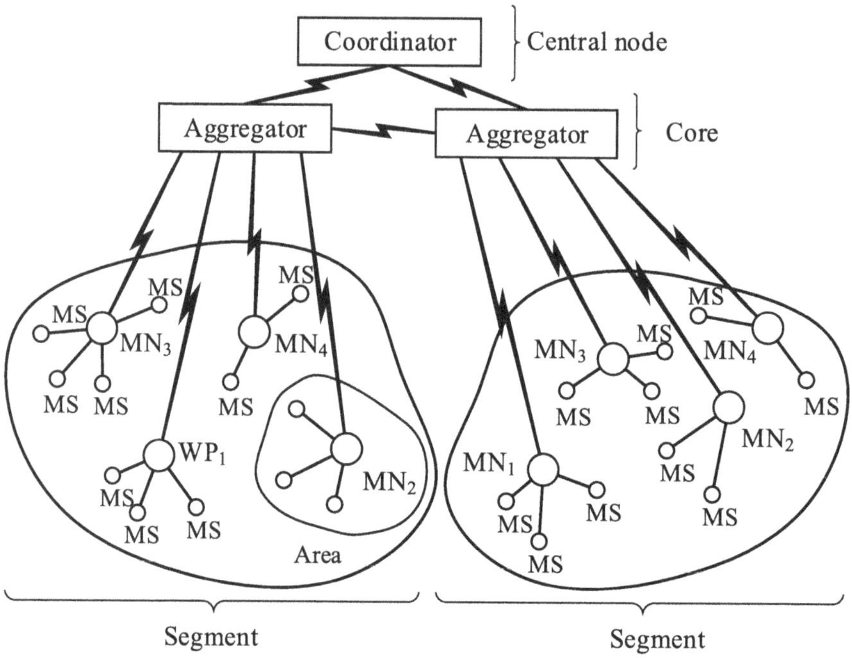

Fig. 7 Hierarchical sensor network architecture. MN—measuring node, MS—measuring sensor

hierarchical systems, usually are indicated [32, 36–38]: easier analysis of complex systems divided into smaller components; high specialization of functional blocks; simplification of operation, maintenance and servicing. An example of a network operating in accordance with the above model is shown in Fig. 7.

At the lowest level of the model, there are measuring sensors, most often connected to the measuring node using wired methods. Using RFID technology, sensors can be integrated into the network wirelessly, however, due to the power consumption during measurement and communication, this solution is used less often. The location of the measuring device is called the measuring point. The number of sensors attached to the node depends on its architecture and usually does not exceed a few. An additional limitation on the number of sensors may also be the power supply available to the node. The area where the sensors connected to the same node are located is called the measurement area. The nodes are connected into measuring segments using segment aggregators. These devices, in addition to greater computing power, have a wide set of external interfaces used to build the network core. The network coordinator is an extensive unit whose basic task is to provide external communication for the entire measurement network. The coordinator's task may also be the integration of aggregators within the core of the measurement networks.

Two alternative concepts clash in the design of monitoring systems of this class [27, 39, 41]. The first of them assumes the location stability of environmental conditions, whose measurement is the purpose of building the system. This means that the sources of potential contaminants are known and invariable. Therefore, the basic criterion for the distribution of measuring nodes are the location of pollution sources and the ways of their movement in the environment (air, water, or soil). The solution of the design task comes down to locating the nodes in places with the maximum concentration of harmful factors. Connecting such nodes can be a complex task, which is a consequence of the territorial heterogeneity of the location of the nodes, and the system itself can be heterogeneous in communication. Despite the above difficulties, due to the small number of components, such systems are relatively cheap in design, construction, and subsequent operation. However, they do not provide tracking of the state of the environment outside the previously designated areas.

The second strategy assumes that the appearance of harmful factors is equally likely throughout the entire monitored area and requires the distribution of measuring nodes in the entire protected space. Thanks to this, not only stationary sources of pollution are tracked, but also routes of harmful substances or contamination resulting from criminal activities. From the point of view of communication design, this task is simpler than the previous one—usually a homogeneous mesh network is built.

Although the second generation of monitoring networks may contain elements of the Internet of Things, its functioning is based on private computer networks. Therefore, they should not be included in the class of cyber-physical networks.

3.3 Regional Monitoring Systems

From an IT point of view, the third generation of environmental monitoring systems create systems designed to handle regional threats. Such systems are built by local governments, which favors the use of modern solutions, including the Internet of Things. In these systems are widely used a wireless sensor network. Partially giving up the use of communication methods typical for these networks in favor of LTE and 5G technologies. Correctly designed and constructed regional systems can be a great example of cyber-physical systems. They are complex distributed systems, managed by computer algorithms, closely integrated with the Internet and its users. The software and hardware components of the regional monitoring system are closely related, each component operates on different spatial and time scales, exhibits many different behavioral modalities, and interacts with each other in many ways that change depending on the context. Unlike the Internet of Things, regional monitoring systems are characterized by higher coordination between physical and computational elements.

In the presented classification, regionality is understood as territorial. Systems of this class cover an area of several km^2. In the conditions of the authors' country, it is an area occupied by a village council or a small town. The regional nature of threats translates into the functioning of the monitoring system, which results from the conditions described below.

First, medium or long-term forecasts are generated based on measurements in classic monitoring systems (first generation). The models used for this purpose use data obtained at relatively large time intervals and due to the size of the input data they are relatively resistant to distortions appearing in them. In regional systems, based on the quantitative and qualitative set of measuring sensors and frequent measurements (every few or several minutes), only short-term forecasts are prepared. Due to the rapid change in local weather processes (usually several dozen minutes pass from the first symptoms of danger to the end of its course) and the limited number of measuring nodes, the sensitivity of forecasts to measurement distortions is high. Repeated, manual size measurement is not possible.

Secondly, the recipients of hazard warnings are residents of a relatively small area, and the operators are usually local authorities. Due to the above dependence, recipients consider them to be more reliable than those generated at the national level, all the more so because of their local nature it is always possible to verify their relevance [42]. Possible imperfections in forecasts can damage confidence in local forecasting and warning systems.

Thirdly, regional catastrophic phenomena have a very large area of activity. For example, torrential rains or hailstorms travel along a 400–700 m wide [25, 42]. Therefore, the density of the location of the measuring nodes in the monitored area should be high, which is usually not allowed by the modest investment budget. Advance forecasting of hydrological phenomena occurring in local watercourses of small width and depth and a significant slope of the riverbed requires information on current precipitation. Otherwise, the forecasts will be based not on hydrodynamic models, but on current analysis of the river current and forecasting its changes based on machine learning methods. In the analyzed cases, these methods allow generating alerts less than an hour before an emergency occurs [42].

The solution to the problem of low-density sensor placement proposed and tested by the authors is the use of weather stations at the disposal of the residents of the monitored area. In an area of over 5 km^2, in private hands there are 12 weather stations connected to the Internet, of which 3 are equipped with a rain gauge and anemometer. The actual number of devices is higher but connecting them to the public Internet is not practiced. Usually these devices operate within private networks. However, the proposed solution raises several new threats to the functioning of the monitoring system, in particular in broadly understood information security. We note that:

1. The measuring devices will be managed by persons not directly related to the forecasting service provider. Usually, the use of measurement data is based on the rules relating to the public Internet domain: user has no legal guarantee of the continuity of measurements and their correctness. Such guarantees require the signing of an appropriate contract, which usually does not occur;
2. Equipment used by residents is calibrated at the factory, calibration to the standard is rare. Therefore, there is no guarantee that the measurements will be accurate. In addition, even devices from the best manufacturers require periodic maintenance, which is not practiced for this class of equipment. The system provides verification of measurement results. On the base of fuzzy algorithms and machine learning

methods, the reliability of the measurement was assessed based on a comparison with the nearest measuring nodes. If the measurement value raises doubts, it is corrected. Correction decisions are never made if the measurement results cannot be compared with the results of an adjacent measuring station;

3. An additional measuring device may become a point through which the user's network or the service provider's network will be compromised, often with hostile intentions. To avoid the above threats, only external measuring stations equipped with the option of user authentication and communication channel encryption are allowed to be used in the system;
4. For prepared forecasts to be accurate and appear maximum in advance, it is necessary to have a permanent set of measuring nodes and data collected by them. Therefore, the measuring network is subject to specific requirements for consistency, which further complicates the construction of the communication network used in monitoring. When designing the regional monitoring network, it was assumed that the main reason for the loss of network coherence are communication channel failures. Therefore, the algorithms used to design the connection network theoretically allow to obtain the indicated level of communication redundancy. The practical use of these possibilities is very limited. The available telecommunications infrastructure as well as the costs of constructing a redundant network stand in the way. Therefore, the system provides the following functions:

   a. Each remote node has been equipped with a local measurement archiver ensuring collection of results during node sleep or damage to external communication channels;
   b. If the data needed to obtain the forecast cannot be provided to the computational node that develops the forecast, the results are approximated based on the measurements of the set of nearest measuring stations;
   c. If the degree of network damage prevents approximation (also when adjacent nodes are too distant from the damaged measurement point), the calculation procedure begins based on a less accurate, simplified model.

The organization of the regional environmental monitoring system has been shown in Fig. 8.

The tasks performed by individual system components are classified into three basic classes: data collection, processing and communication, and information and warning. In practical solutions, the range of input data used is very limited. This results in the need to reduce the operating costs of the monitoring system to the necessary minimum. The use of commercial historical data is limited, satellite observation systems are practically unused.

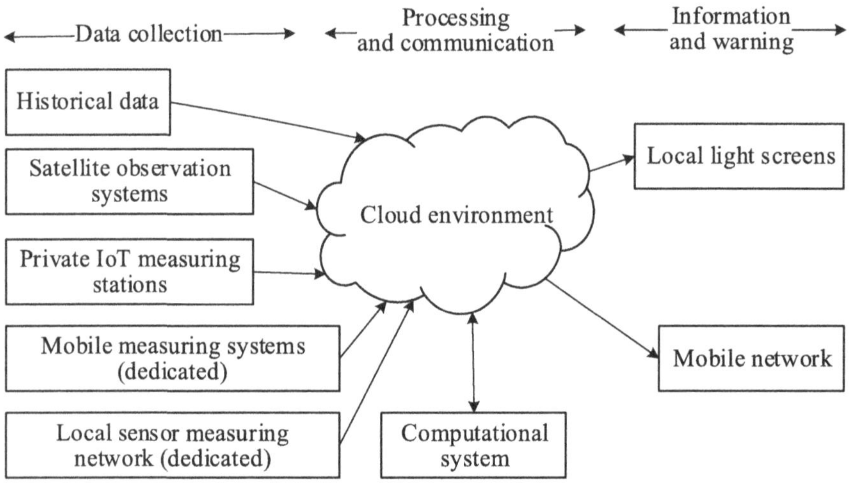

Fig. 8 Components of a cyber-physical monitoring and warning system

4 Threats to Information Security in Monitoring

4.1 Technical and Organizational Conditions for Monitoring

One of the most used definition of the threats of information security assumed, that they are a variety actions that can lead to violation integrity storing, processing and sharing information in information system or to damage itself or used by him infrastructure. The security of monitoring systems is a concept especially interesting mainly due to their organization and architecture, including technical solution developed in the recent past and unprecedented operating requirements.

The first new solution significantly affecting the monitoring systems is the use of Cyber-Physical Systems (CPSs) for their construction. The source of dynamic development of CPS should be sought in the discussed below Internet of Things (IoT). According to the popular definition, CPS is a complex system controlled or supervised by computer algorithms. According to different definition they are a complex system consisted with the set of natural or artificial systems and control controllers, allowing to integrate this organization in one whole. In CPS, there is a close relationship between computing and physical resources. Information technologies monitors and manage physical processes by feedback loop. Thanks for this all what take place in physical systems has an impact for the computing resources the whole system. In fact, CPS is a computer system that uses a set of physical components, that are closely related with their program elements.

Cyber-Physical Systems can function with various time and spatial scales, its behavior can be manifold, and the way they interact depends on the current context. The examples of successful use of CPS can be intelligent networks, autonomics car, monitoring of patients—vital functions in homes, clinics and hospitals, flexible production systems, avionics systems etc. Especially means assign CPS concept in Industry 4.0, which they form the basis.

The organization of monitoring as a Cyber-Physical System is ambivalent for its security. On the one hand, it brings new types of threats, unprecedented in information systems. On the other hand, it improves the security of the system as a whole thanks to the effect of scale and the complementarity of the measuring nodes. An example of ambivalence can be the solution massively used in monitoring systems based on CPS -application for the preparation of forecasts of measurement nodes owned by private persons. This creates new threats related to the possibility of external attacks on measuring devices, incorrect measurements, penetration attempts the whole systems through them. However, it the situation of a failure of dedicated measuring stations, their indications may be used to recreate their possible indications, necessary to develop an accurate forecast.

The above-mentioned measuring stations of private persons are more and more often made in Internet of Things (IoT) technology. Contrary to popular belief, IoT is not a recent discovery. His idea appeared in 1982, when vending machines were first connected to the vast computer network. The result was the ability to remotely control the parameters of the distributed products (e.g. the temperature of drinks, the state of containers with semi-finished products, etc.). The term of Internet of Things was used for first by Kevin Ashton in 1999 in presentation research concerned improve the supply chain commissioned by the global corporation Procter and Gamble.

The fundament of action of IoT was ability to exchange information between real devices in commonly used, described the unique addressing. The implementation of necessary functions has become possible after equipping the devices with sets of measuring sensors and the ability to independently make a number of basic decisions regarding their operation. The ability of IoT technology in the area of integration subjects surrounding world excellent describe the data prepared by corporation Cisco. They estimate, that the summary value of devices connected with Internet is now 14,4 billion USD. According to their research, nearly 99% of physical devices are not yet connected to the network, which opens up great opportunities for the development of this concept. According to these forecasts, the maximum number of items assigned to one person that can be connected to the network is currently from 3,000 to 5,000. The gradual blurring of the boundaries between people and machines will lead to exacerbation of social problems related to the loss of privacy of human life. Most likely, deep personal relationship and the trust on which they are built will become rare. The impact on the security of use in monitoring systems of IoT solutions is also twofold. It is mainly due to the IoT components having their own computing power, which can be used both for attacks on monitoring and for its protection.

A particular deep impact on the organization and architecture of monitoring results from its qualification to critical infrastructure systems (CI). According USA Patriot Act the critical infrastructure is a "a collection of physical or virtual systems and devices, important for the country to such an extent that their damage or destruction may cause catastrophic consequences in the field of defense, economy, public health and national security". In practice, we divide the critical infrastructure into:

infrastructure ensuring national security; objects necessary to its functioning, not being its part; objects necessary for solving national tasks, the failure of which will result in a deterioration of the security level, weakening of the economy or a negative impact on the prestige and credibility of the state. They are the critical infrastructure: government information systems, national security systems, health care, energy management, food delivery, transport, municipal economy, communication, civil defense, banking, etc.

The environmental monitoring systems must absolutely be qualified for CI. From the point of view end user, the action of monitoring should be characterizing two basic features: continuity functioning and an accuracy of forecasts. Both require using the solution in which partial inefficiencies will not significantly affect the functioning of the monitoring.

Summing up, let us note the multidimensional nature of the problems of ensuring the security of monitoring systems [47, 48]. It requires taking into account safety issues at every stage of the facilities life cycle: from design to disposal.

4.2 General Threats Classification

The classification of threats, typical for environmental monitoring systems, is presented in Fig. 9. Threats are divided into three main categories: natural, technological and informational. Such an extensive classification of threats results from the specific operating conditions and functional requirements of monitoring. It is the sum of threats characteristic for: traditional information systems, Internet of Things and critical infrastructure systems. Due to the recipients of information created by monitoring, regardless of the source, addresses or the essence of the threat, the consequences of refusal can be very serious. For the first group includes phenomena that

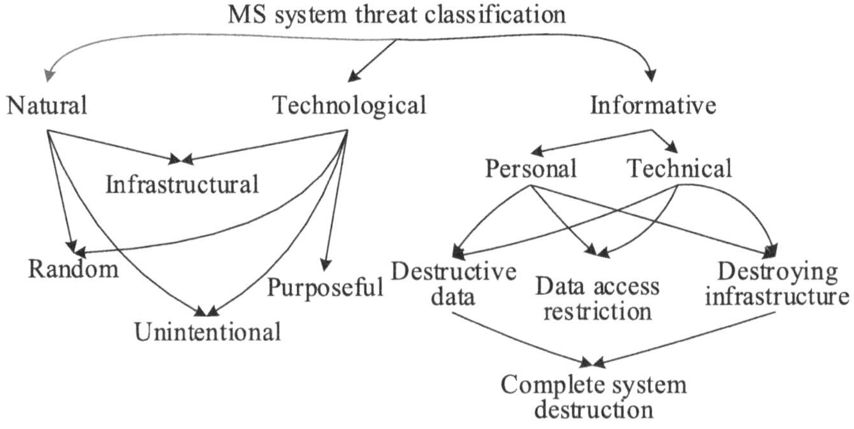

Fig. 9 Classification of threats to the safety of environmental monitoring systems

run independently of human activity. They include natural catastrophic phenomena such as hurricanes, floods, hailstorms, fires, etc. The appearance of this type of threat may be influenced by a person without any influence or directly or indirectly. In example, the fire can be a result of natural self-ignition, lightening, accidental ignition fire, but also ignition or self-ignition as a result of improper exploitation of the equipment or other neglect or omission. In the first case we discuss about natural catastrophe, in second about technological. In the situation of natural self-ignition or lightening the threat is not a result of technology created and operated by human. In none of these and similar situations the threat is the result of human actions directly related to data processing.

Natural and technological threats have infrastructural character -in the result of its appearance the communication and measurement EM components are damage. Information resources and processing are less at risk due to redundancy and their deployment in safe locations. From the point of view of the causes of occurrence, the natural and technological threats may be random or unintentional. In both cases they are consequences of carelessness, inattention, ignorance or ignorance of users. An example of such threats may be the accidental installation of programs that are not required for work, and that significantly interfere with the functioning of the system: for example, causing the loss of access to information, and even information itself. Intentional threats characteristic for the group of technological threats, in contrast to the previous ones, are caused consciously. They include the attacks also from external and internal of company. Their appearance may result in system stoppage, inaccessibility of data, and if the interference is only an intermediate step in a wider attack, also loss of funds or intellectual property of the organization or private person. Information threats have been separated from the group of technological threats. They form an independent group of dangers directly related to the collection, processing and sharing of information used, for example, in the information and warning procedure. Although similar like in classical information systems we divide these threats on person and technical, the role of human facilities is in monitoring systems much smaller, even symbolic. Access to information resources of monitoring is very limited at the stage of their design. End users obtain solely access to processed output data in read mode, without ability modify it. Person and technical threats in both cases are aimed at damage the collected, processed and shared data. Less frequently, they may result in limiting access to information, and very rarely in destroying the communication or computing infrastructure. The least common effect of information threats may be the complete destruction of the system associated with damage or destruction of infrastructure and loss of information.

4.3 Special Types of Threats

In the professional literature, the threats related to organized attacks on information resources or processing infrastructure are most often considered. Less frequently the consequences of failure of the communication infrastructure are analyzed, without the functioning of which measurement, forecasting and warning are impossible. In many cases when building the monitoring system, computing resources are redundant, while forgetting about common threats to measurement nodes or connection channels. In part, this approach is irrational: in the last 20 years, computing infrastructure components (servers, storage) have significantly improved their reliability, similar as electronic communication components. During this period, the threats to measurement nodes, transmission cabling and wireless transmission channels operating in the public space have not decreased, but even increased. The regularity applies in particular to the communication and measurement infrastructure of environmental monitoring systems. To put it simply, it consists of three types of objects: intermediate or terminal communication nodes, two-point or broadcast communication channels and measurement nodes often integrated with terminal communication nodes. From the practice are known the cases of aimed damage, destroy or theft infrastructure components. Not infrequently, damage to one node, or communication channel excluded the monitoring system from operation.

The impact of component reliability on the effectiveness of monitoring systems has been analyzed so far based on the experience of traditional computer networks. In analysis and synthesis of such networks until recently were dominated by methods successively developed in the years 1960–2000, the effectiveness of which for environmental monitoring systems is far from satisfactory. Therefore, in the design methodology developed and verified in practice, a number of solutions have been proposed to minimize the impact of even several-fold damage on the functioning of monitoring. To the most important we can include:

1. As far as technical, financial and organizational possibilities are concerned, the use of redundancy of measuring elements as well as communication channels and nodes;
2. Possibility of dynamic reconfiguration of the network organization by changing the connections used, including the replacement of the main monitoring network node;
3. The function of supplementing the unavailable measurement values necessary to develop an accurate forecast with data obtained using mathematical methods. The forecasting uses information from active measurement nodes and historical data.

The authors analysis confirmed, that research of reliability and lifetime of monitoring systems is necessary, and the possible refusals of the components of the communication infrastructure are as dangerous for security as hacker attacks [49]. This problem is particularly relevant in cases of monitoring about regional character, for which uses redundancy the nodes and channels is very limited, and the meager information resources make it difficult to use mathematical methods for generating missing measurements.

5 Monitoring System Architecture

Due to the type of tasks performed, the hardware and software architecture of environmental monitoring systems must be characterized by the maximum level of availability. This means that regardless of the extent of the catastrophic phenomena taking place, the system must fulfill its basic tasks, i.e. forecast environmental changes, and alert people in the affected area. From the point of view of the theory of technical systems, the availability of systems is characterized by: service life, fault tolerance and durability. Service life is the ability of the system to perform its basic functions despite the damage. Damage resistance is most often understood as a reflection of the service life properties in normal system operation mode. Unlike lifetime, immunity has its numerical expression. In technical systems, it is defined as the ability of the system to perform its basic functions, despite the occurrence of damage to one or more of its components. Damage resistance depends on the number of each subsequent individual component failure, after which it maintains the efficiency of the entire system. The basic level of resistance to damage means protection against failure of one element. Durability refers to both lifetime as well as resistance to damage and defines the ability to counteract external factors in the normal functioning of the system, i.e. before the occurrence of a critical situation [43, 44].

To ensure high availability, information, communication, and computational redundancy should be provided in environmental monitoring systems. *Information redundancy* is ensured by using a set of measuring sensors. In the event of damage to any of them, its indications can be reproduced by approximating the indications of the other sensors. *Communication redundancy* consists in multiplying communication channels connecting individual nodes. In the case of WSN networks, building a mesh network uses standard technology capabilities, which does not involve additional costs.

The monitoring and warning system components can be divided into remote, located directly in the monitored area, whose task is to collect and send information on the state of the environment and *central*, intended for their storage and processing. The architecture of the remote part of the system using communication redundancy is shown in Fig. 10. The most important remote components are measuring and routing nodes, integrated by means of aggregating nodes. Thanks to the abandonment of the simplest measuring nodes in construction, each measurement can reach the central node via several independent routes.

The basic computational elements of the system are common to the entire system and are in the central management node (CMN). Because all forecasts, decision-making process and informing entities is made by CMN, in order to guarantee the required level of availability, the hardware and software components of this node are mirrored by the resources of the redundant management node (RMN). In particular, the development of forecasts and long-term data storage is carried out simultaneously in several, usually two, independent management nodes. To ensure a sufficient level of resistance to damage, the measurement data is delivered to CMN and RMN independently, using separable communication channels. The functional architec-

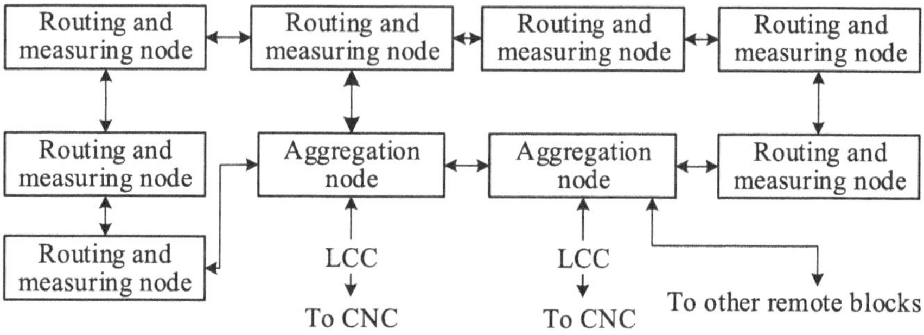

Fig. 10 Components of a cyber-physical monitoring and warning system

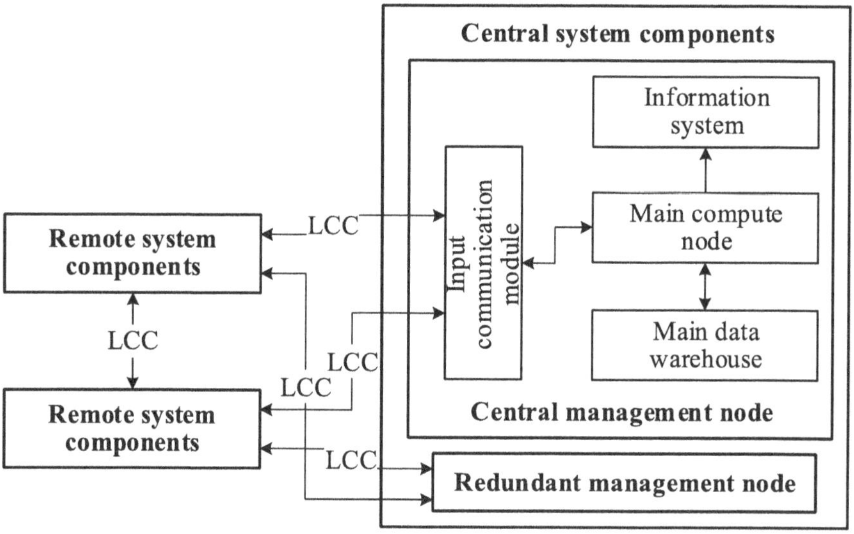

Fig. 11 Functional architecture of the central components of the monitoring system. Designations: LCC—long distance communication channel

ture of the central components of the monitoring system with communication and computational redundancy is shown in Fig. 11.

If the normal functioning of CMN is disturbed, its role is taken over by RMN, which temporarily obtains the status of a central node. If the correct operation of the original CMN is restored, it is initially connected to the system as a redundant node, and after the so-called *grace time*, reducing the possibility of re-disconnection, again becomes CMN. The procedure of attaching any redundant node begins with data synchronization, however, to limit the size of inter-node communication, it is carried out from the last checkpoint available in both nodes. Checkpoints are created periodically when monitored threats are minimal. The triggering factor can be the time or level of changes made to the archive. Communication relations between aggregating nodes as well as CMN and RMN are shown in Fig. 12.

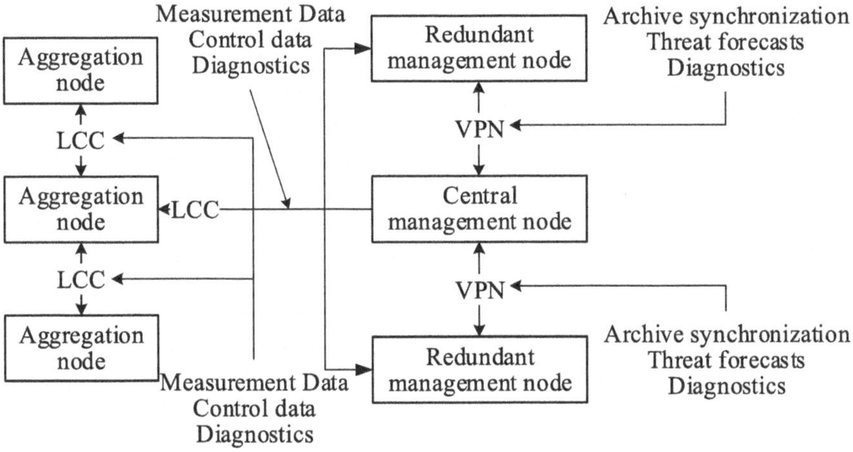

Fig. 12 Information interoperability of aggregation nodes and the central and redundant management node

Connection of CMN with redundant nodes is carried out via the Internet, in particular based on VPN channels combined using it. These channels, in addition to synchronization of archives, are used for diagnostics and sending forecasts of threats generated additionally in redundant nodes.

The level of cooperation of redundant nodes with CMN depends on the mode of operation of the system. Typically, the system operates in three basic modes: *standard*; *emergency call* and *disaster*. In *standard* mode, the role of the redundant node is limited to archiving of measurement data and periodic diagnosis of CMN status. In other modes, the RMN doubles the computations performed in CMN. If different forecasts are obtained based on the same algorithms and input data, the system switches to diagnostic mode, whose task is to determine the reason for the appearance of differences. A failed node is eliminated from the system. The main data warehouse stores uploaded data on the state of the environment and forecasts based on them. The above information is additionally protected in the RMN. As noted earlier, in *emergency* mode, when a threat becomes probable, the RMN expands its operation to forecast changes in the state of the environment [45, 46].

If during the functioning of the system, further probability occurs, it goes into *disaster* mode. In a situation where the prepared forecast indicates a high probability or inevitability of a threat, an information and warning system (IWS) is activated, which sends voice or text messages to all persons in the affected area. The IWS module is already activated when the monitoring system goes into alarm mode. However, in this case, messages are sent only to emergency services. Because of the costs, redundant management n usually do not have an information and warning module [45, 46, 50].

6 Resident Alert Subsystem

The regional environmental monitoring system described in the work has been prepared and is implemented in the city of Boguchwala, several thousand, located in the immediate vicinity of Rzeszow—a regional center of science, industry, and culture. The inhabitants of Boguchwala will be able to use the resources of the system and take an active part in its creation. The system provides for, among others, the use of interactive websites informing about the state of ecological security. Information on the state of the environment can be sent in the form of an SMS, email, Google message, map fragment for a given location. The above information can be sent to the resident's smartphone, tablet, or computer. A prerequisite to ensure the safety and comfort of life of residents is to maximally inform residents about potential threats. However, are the means of communication with residents used sufficient?

This problem should be included in the field of external communication in smart cities. Many publications summarizing the results of scientific work have been devoted to this topic. Among the scientific monographs describing this topic can be mentioned: [51–53]. Numerous papers in scientific journals and conferences are also devoted to this topic. Examples of such works can be [54–56]. The cited sources emphasize the key role of the Internet of Things in the information policy of smart cities. Smartphone, less often a tablet is a basic tool for communication between the resident and systems. In this context, the communication solutions used in the project implemented in Boguchwala should be considered in line with current trends. On the other hand, a few researchers note the reluctance of older people (hereinafter seniors) to use modern technologies, including smartphones. The generally applicable definition of old age is the age at which the senior will need help to perform everyday activities (to operate the smartphone). The age starting old age is 50. Researchers pay attention to a neglected fact: just having a smartphone by a senior does not mean using advanced features. Most often, seniors treat it like a regular cordless telephone. The above observations are contained both in the publications of foreign authors [57, 58] as well as Polish [59–61]. The cited studies indicate a systematic increase in the number of seniors using smartphones as a binding trend. For example, research from the UK society shows that in 2018 over 40% of senior citizens aged 55–64 had smartphones. For the age group 65+ this share falls to below 20%. American studies show the activity of senior citizens there. It is only in the age group 65+ that we observe a low level of smartphone owners. In studies published by the University of Information Technology and Management regarding Podkarpackie, in 2014 less than 30% of seniors aged 55–64 were smartphone owners. In the 65+ group this share dropped to around 10%. The research of these authors shows that only a dozen or so percent of seniors from the 55–64 age group used non-basic functions in the smartphone (voice communication, SMS, alarm clock). For the group over 65 this share was only 5%. Studies by all authors show a drastic decline in the use of more advanced services after the age of 50, even in the best-educated American society. A summary of the authors' research on the use of smartphones in the 50+ group on the entire adult population is presented in Fig. 13 [61].

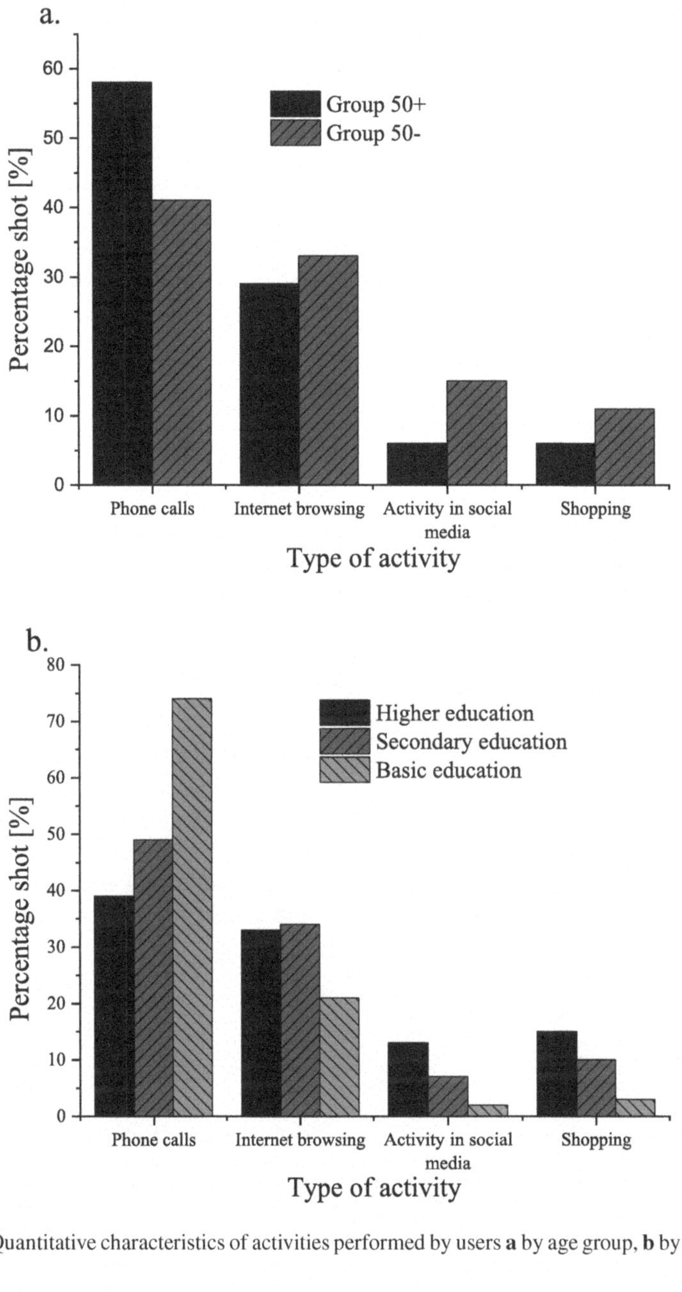

Fig. 13 Quantitative characteristics of activities performed by users **a** by age group, **b** by education

According to the authors, in order not to deepen the information exclusion of older people, one should take advantage of the experience of countries such as Japan, Taiwan, the USA and Spain. In these countries, bulky light panels are an obligatory external communication tool in smart cities. Such solutions have been described, among others in [62, 63]. Usually, light boards are used to inform seniors (and other city residents and guests) about: cultural and sporting events; planned interruptions in the supply of electricity, gas, water; public road accessibility and other security threats. Similar functions will be fulfilled by the table installed in Boguchwala. This is beneficial, if only because of the significant share of residents from the age group 50+ in their total number. These people would most likely be excluded from using the results of the designed system.

7 Summary

The system described in the work consists of dozens of dedicated stationary measuring nodes and one mobile drone-based one. With the system, as the needs arise, with the consent of the owners, several dozen measuring sensors are components of the local Internet of Things. Their role is to verify the correctness of measurements, and in particular to supplement them. The operation of the function of supplementing missing measurements is based on a specially prepared algorithm and is aimed at minimizing the uncertainty of measurement data, as well as their virtualization. Verification of the effectiveness of methods and measures to ensure information security, including the detection of attacks on accessibility, also plays an important role in research.

At present, the operation of the system is focused on air monitoring and the elements related to it are systematically developed first. Additional areas of monitoring functioning include hydrological, noise, electromagnetic threats as well as the safety of persons and property.

References

1. Katastrofy naturalne i cywilizacyjne, Wyższa Szkoła Oficerska Wojsk Lądowych, Wrocław (2006)
2. Woodcock, A., Davis, M.: Catastrophe Theory. Clarke, Irwin & Company Limited, Toronto (1978)
3. Davis, L.: Natural Disasters. Facts On File, New York (2009)
4. Nemchinov, I.V.: Catastrophic Events Caused by Cosmic Objects. Springer, Dordrecht (2008)
5. Brown, C.: Chaos and Catastrophe Theories. Sage Publications Inc., Thousand Oaks (1995)
6. Arnold, V.I.: Catastrophe Theory. Springer, Berlin (1986)
7. Morris, D., McGann, E.: Catastrophe. HarperCollins Publishers, Pymble (2009)
8. Lancaster, J.F.: Engineering Catastrophes Causes and Effects of Major Accidents. CRC Press, Boca Raton (2005)

9. Sagarin, R.D., Taylor, T.: Natural security how biological systems use information to adapt in an unpredictable world. Secur. Inf. **1**(14), 1–9 (2012)
10. Auker, MR, et al.: A statistical analysis of the global historical volcanic fatalities record. J. Appl. Volcanol. 1–24 (2013)
11. Desai, ChS, Zaman, M.: Advanced Geotechnical Engineering, Soil-Structure Interaction Using Computer and Material Models. CRC Press, Boca Raton (2014)
12. Randolph, M., Gourvenec, S.: Offshore Geotechnical Engineering. Spon Press, Abingdon (2011)
13. Ghafoori, N. (ed.): Challenges, Opportunities and Solutions in Structural Engineering and Construction. CRC Press, Leiden (2010)
14. Hori, M.: Introduction to Computational Earthquake Engineering. Imperial College Press, London (2006)
15. Day, R.W.: Geotechnical Earthquake Engineering Handbook. McGraw Hill, New York (2012)
16. Briaud, J.-L.: Introduction to Geotechnical Engineering: Unsaturated and Saturated Soils. Wiley, Hoboken (2013)
17. Grossi, P., Kunreuther, H.: Catastrophe Modeling: A New Approach to Managing Risk. Springer, New York (2005)
18. Risk, Earthquakes: Monitoring and Research. Nova Science Publishers Inc., New York (2009)
19. Lu, Y.Ch.: Singularity Theory and an Introduction to Catastrophe Theory. Springer, New York (1976)
20. Bac, S., Rojek, M.: Meteorologia i klimatologia. Panstwowe Wydawnictwa Naukowe, Warszawa (1981)
21. Kożuchowski, K.: Meteorologia i klimatologia. Wydawnictwa Naukowe PWN, Warszawa (2006)
22. Makki, Kia (ed.): Sensor and Ad Hoc Networks, Theoretical and Algorithmic Aspects. Springer, New York (2008)
23. Zheng, Jun (ed.): Wireless Sensor Network. Institute of Electrical and Electronics Engineers, Hoboken (2009)
24. Hajder, M., Loutskii, H., Strciwilk, W.: Informatyka. Wirtualna podróżw świat systemów i sieci komputerowych, Wydawnictwo Wyższej Szkoy Informatyki i Zarzdzania, Rzeszów (2002)
25. Grocki, R., Mokwa, M., Radczuk, L.: Organizacja i wdrażanie lokalnych systemów ostrzeżeń powodziowych. Biuro Koordynacji Projektu Banku światowego, Wrocław (2001)
26. Shidlovskiy, S.V.: Automated Control, Reconfigurable structure. Tomsk State University, Tomsk (2006)
27. Ha, A.: Wireless Sensor Network Designs. Wiley, Hoboken (2013)
28. Santi, P.: Topology Control in Wireless Ad Hoc and Sensor Networks. Wiley, Chichester (2005)
29. Freeman, R.L.: Fundamentals of Telecommunications. Wiley, Hoboken (2005)
30. Illyas, M., Mahgoub, I. (eds.): Handbook of Sensor Networks: Compact Wireless and Wired Sensing Systems. CRC Press, Boca Raton (2012)
31. Raychaudhuri, D., Gerla, M.: Emerging Wireless Technologies and the Future Mobile Internet. Cambridge University Press, Cambridge (2011)
32. Tannenbaum, A.S.: Strukturalna Organizacja Systemów Komputerowych. Helion, Gliwice (2006)
33. Hennessy, J.L., Patterson, D.A.: Computer Architecture a Quantitative Approach. Morgan Kaufmann, San Francisco (2002)
34. Perahia, E., Stacey, R.: Next Generation Wireless LANs: Throughput, Robustness, and Reliability in 802.11n. Cambridge University Press, Cambridge (2008)
35. May, E.: Wireless Communications & Networks. Prentice Hall, New Jersey (2004)
36. Mesarovic, M.D., Macko, D., Takahara, Y.: Theory of Hierarchical, Multilevel Systems. Academic Press, New York (1970)
37. Smith, N.J., Sage, A.P.: An Introduction to Hierarchical Systems Theory. Information and Control Sciences Center, SMU Institute of Technology, Dallas (2005)
38. Hajder, M., Dymora, P., Mazurek, M.: Projektowanie topologii transparentnych sieci optycznych. Konferencja Polski Internet Optyczny: technologie. usługi i aplikacje, pp. 183–195. Instytut Informatyki Politechniki Poznańskiej, Poznań (2002)

39. Wiersma, G.B.: Environmental Monitoring. CRC Press, Boca Raton (2004)
40. Fraden, J.: Handbook of Modern Sensors. Physics, Designs, and Applications. Springer Science, New York (2013)
41. Wang, B.: Coverage Control in Sensor Networks. Springer, London (2010)
42. Gbica, P., Starkel, L., Cebulak, E., Limanówka, D., Pyrc, R., Hajder, M., Kolbusz, J.: Ulewy i powodzie opadowe w województwie podkarpackim. Studium przebiegu, skutków i przeciwdziałania, Wydawnictwo Uniwersytetu Rzeszowskiego, Rzeszów (2019)
43. Freeman, Roger L.: Telecommunication System Engineering. Wiley, New York (2003)
44. Bagad, V.S., Dhotre, I.A.: Data Communication And Networking. Technical Publications, Pune (2006)
45. Hajder, M., Nycz, M.: Analiza systemu hydrologicznego i przeciwdziałanie zagrożeniom powodziowym Tarlaki. In: Hajder, M. (ed.) Innowacyjna gmina. Nowoczesne technologie na usługach samorządu terytorialnego, pp. 123–139. Urząd Gminy Leżajsk, Lżajsk (2011)
46. Florek, B., Hajder, M.: Teoria adaptacyjnych systemów monitoringu środowiska. In: Hajder, M. (ed.) Innowacyjna gmina. Nowoczesne technologie na usługach samorzdu terytorialnego, pp. 109–122. Urząd Gminy Leżajsk, Leżajsk (2011)
47. Hajder, M., Nycz, M., Hajder, P., Liput, M.: Security of cyber-physical environmental monitoring systems based on Internet of Things-basic challenges. In: Social and Technicals Aspects of Security vol. 2, Oficyna wydawnicza Politechniki Rzeszowskiej, Rzeszow (2019)
48. Hajder, M., Nycz, M., Hajder, P., Liput, M.: Security of cyber-physical environmental monitoring systems based on Internet of Things–reliability and survivability. In: Social and Technicals Aspects of Security vol. 2, Oficyna wydawnicza Politechniki Rzeszowskiej, Rzeszow (2019)
49. Hajder, M., Hajder, P., Nycz, M., Liput, M.: Analiza i synteza regionalnych systemów monitoringu środowiskowego. Oficyna Wydawnicza Politechniki Rzeszowskiej, Rzeszow (2020)
50. Hajder, M., Nycz, M.: Architektura systemu wykrywania zagro zeń powodziowych o zasigu lokalnym. In: Hajder, M. (ed.) Innowacyjna gmina. Nowoczesne technologie na usługach samorządu terytorialnego, pp. 83–96. Urzád Gminy Leżajsk, Leżajsk (2011)
51. Song, H., Srinivasan, R., Sookoor, T., Jeschke, S.: Smart Cities: Foundations, Principles and Applications. Wiley, New York (2018)
52. Cicirell, F., Guerrieri, A., Mastroianni, C.: The Internet of Things for Smart Urban Ecosystems. Springer, Cham (2019)
53. Barton, A., Manning, R.: Smart Cities: Technologies, Challenges and Future Prospects. Nova Science Pub Inc., Hauppauge (2019)
54. Min, W., Yu, L., He, S.: People logistics in smart cities. Commun. ACM **11**, 54–59 (2018)
55. Beckwith, R., Sherry, J.: Smart city: give the people what they want In: 1st ACM/EIGSCC Symposium on Smart Cities and Communities (SCC'18). New York (2018)
56. Yonezawa, T., Ito, T., Nakazawa, J., Tokuda, H.: SOXFire: a universal sensor network system for sharing social big sensor data in smart cities. In: Proceedings of the 2nd International Workshop on Smart (SmartCities'16). New York (2016)
57. Berenguer, A., Goncalves, J., Hosio, S., Ferreira, D., Anagnostopoulos, T., Kostakos, V.: Are smartphones ubiquitous?: an in-depth survey of smartphone adoption by seniors. IEEE Consum. Electron. Mag. **1**, 104–110 (2017)
58. Lu, Y., Chang, Y., Sung, T.: Domestication of smartphone on a senior community: a case study. In: 2017 International Conference on Applied System Innovation (ICASI). Sapporo (2017)
59. Niemczy, A.: Seniorzy wobec nowych technologii. Studia Ekonomiczne. Zeszyty Naukowe Uniwersytetu Ekonomicznego w Katowicach, **303**, 101–113 (2016)
60. Hajder, M., Kolbusz, J., Florek, B.: Społeczeństwo informacyjne Podkarpacia. Studium analityczne, Wydawnictwo W]'szej Szkoły Informatyki i Zarzdzania z siedzibw Rzeszowie, Rzeszsieciowa osb starszych-przypadek Podkarpacia In: Nycz, M. (ed.) Social and technical aspects of security, pp. 21–35. Oficyna Wydawnicza Politechniki Rzeszowskiej, Rzeszw (2019)
61. Liput, M., Mysakowec, A., Nycz, M.: Aktywność sieciowa osób starszych-przypadek Podkarpacia In: Nycz, M. (ed.) Social and Technical Aspects of Security, pp. 21–35. Oficyna Wydawnicza Politechniki Rzeszowskiej, Rzeszów (2019)

62. Wang , N., Mao, B.: The research on the problems of smart old-age care in the background of smart city construction. In: 2019 International Conference on Intelligent Transportation, Big Data & Smart City (ICITBS). Changsha (2019)
63. Vechione, M.: Smart mobility for seniors: challenges and solutions in El Paso, TX, and New York, NY. In: 2018 IEEE International Smart Cities Conference (ISC2). Kansas City (2018)

Risk Identification and Risk Assessment of Communication Networks in Smart Grid Cyber-Physical Systems

Amitkumar Vidyakant Jha, Abu Nasar Ghazali, Bhargav Appasani, and Dusmanta Kumar Mohanta

Abstract Electricity, which is the greatest invention of nineteenth century, is distributed to the geographically distributed loads through the help of the power grid, which acts as a backbone to the system. Power grid is a complex infrastructure that is intended to transmit the electricity from the generation stations to the consumers by encompassing several distribution stations. The conventional power grid was not designed to cope up with a dynamically changing environment. Thus, for uninterrupted and reliable transmission, generation, and distribution of the electricity, the conventional power grid is being upgraded to the smart-grid (SG). The success key to the SG is the integration of power network infrastructures providing capability of seamless interaction among its components. The cyber-physical system (CPS) is the correct attempt for integration and interaction of the various components of the SG. In the smart grid paradigm, the communication networks act as the cyber system, while the processing, sensing and controlling devices act as the physical system. Thus, the smart grid cyber-physical system (SGCPS) consists of different intelligent devices, which exchange the data over the communication networks for effective operation. Due to high level of integration among the various entities, SGCPS is more complex and thus, it is also susceptible to different cyber as well as physical vulnerabilities. Moreover, most of the smart grid applications have stringent requirements such as low latency and high reliability. Hence, the communication networks of the SGCPS are subjected to many challenges and risks. This chapter identifies different risks in designing the communication networks for various applications in the smart grid cyber-physical system and proposes the methodologies for risk assessment and risk mitigation. A systematic approach is presented to identify the risk factors pertaining to the design of communication networks for various SGCPS applications such as synchrophasor application, advanced metering application, and electric vehicular application. Further, risk assessment strategies for these SGCPS applications are formulated with detailed discussion. To elucidate the work, a case

---

A. V. Jha · A. N. Ghazali · B. Appasani (✉)
School of Electronics Engineering, KIIT Deemed University, Bhubaneswar 751024, India
e-mail: bhargav.appasanifet@kiit.ac.in

D. K. Mohanta
Department of Electrical and Electronics Engineering, Birla Institute of Technology, Mesra 835215, India

A. I. Awad et al. (eds.), *Security in Cyber-Physical Systems*, Studies in Systems, Decision and Control 339, https://doi.org/10.1007/978-3-030-67361-1_8

study in each of these applications of SGCPS have been presented in this chapter. Nevertheless, the practical power grid of Bihar, India has been considered as a case study for synchrophasor applications of the SGCPS.

Keywords Smart grid · Cyber-physical system · Smart grid cyber-physical system · Communication network · Risk assessment · Reliability · Case study of SGCPS · Synchrophasor · AMI · Electric vehicular

1 Introduction and Motivation

1.1 Introduction

The cyber-physical system (CPS) in an engineering marvel involving multi-disciplinary domains, disruptive technologies, embedded hardware, ubiquitous computations, complex designs, etc. [1]. The different organizations have defined cyber-physical system differently depending upon the characteristics and thus, no unique definition fits to all the applications. The term cyber-physical system was coined by U.S National Science Foundation (NSF) in 2006, which describes it as an integration of recent embedded and computing technologies (cyber component) with the physical world (physical components) [2]. The U.S vision of cyber-physical system is mainly focused on integration of embedded system to physical components. Unlike U.S NSF, the European version of cyber-physical system refers it to as interaction of cloud-based cyberspace components with human factors [3]. The European vision of cyber-physical system explains it as "smart everywhere vision" by conjunction of smart internet technologies and physical world. On the other side, China refers to the CPS in more general and broader sense as an amalgamation of the sensing, processing, monitoring and controlling capabilities with physical world through cyber-communication infrastructure [4]. In simple words, we can refer to a CPS as a smart integration of cyber units and physical units of a system to interact in real-time and having capability to provide seamless communication among several entities of the CPS.

A typical framework of the cyber-physical system where physical world interact to cyber world over communication infrastructure is as illustrated in Fig. 1. The physical world of the CPS consists of various devices, generally having small computing capabilities such as actuators, controllers, sensors, etc. These devices are operated and controlled by the cyber world (also referred to as cyber space) consisting of many disruptive technologies such as artificial intelligence, machine learning, data analytics, cloud computing, Internet of Things (IoT), etc. The cyber world and physical world of the CPS are integrated to establish a real-time and efficient interaction using the communication infrastructure or communication network. Thus, it can be seen that the communication infrastructure acts as a backbone to the cyber-physical system.

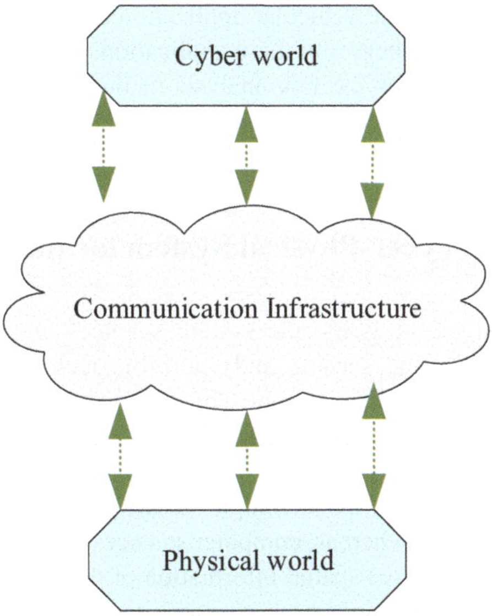

Fig. 1 A CPS framework

1.2 Motivation

The cyber-physical system is an integration of various cyber-physical components across many domains. The various components of the cyber-physical system interacts over a complex communication infrastructure. The components of the CPS including the communication infrastructure are having different characteristics and are vulnerable to several challenges. For e.g., either a component of the CPS may fail during its operation because of wear and tear or the whole communication infrastructure may fail because of natural disasters. The failure of one components may (with less intensity) or may not hamper the performance of the system. However, the failure of communication infrastructure may disrupt the whole system. Hence, different level of risk (correlated with impact or loss) are associated with different components of the cyber-physical system. Consequently, the level of risks associated with different components of the CPS are identified in this chapter. Moreover, this chapter also presents the risk assessment strategies for the cyber-physical system.

The rest of the chapter is conceived into seven sections which are summarized as follows:

Section 2 presents an overview of the cyber-physical system and its potential application in the smart grid for the power system, which are referred to as smart grid cyber-physical system (SGCPS) throughout the chapter. The different communication technologies and their use cases with challenges are described in Sect. 3 of the chapter. In Sect. 4, the different futuristic application of the cyber-physical system for the smart grid such as synchrophasor application, advanced metering application and electric vehicular application are discussed. The risk identification and assessment strategies of these application of the SGCPS are presented in Sect. 5. The case studies for the risk analysis of the SGCPS applications are presented in Sect. 6. Finally, Sect. 7 concludes the chapter with a way forward for further advancements.

2 cyber-Physical System for the Smart Grid

In recent years, the system sciences and engineering has seen rapid growth in modelling, sensing, and controlling technology, which are essential for synthesis and analysis of the system. The developments such as real-time computing, processing, visualizing, analytical designing, etc., in the field of computer sciences and engineering have also seen tremendous growth [4, 5]. The system sciences and engineering are the techniques to deal with the qualitative temporal information of the system, whereas, computer sciences and engineering are good tools to deal with the quantitative spatial information of the systems [6]. These two fields can be merged together using cyber-physical system to solve many challenges pertaining to next generation applications, which are complex in nature and high in dimension.

The cyber-physical system is disruptive technology, which is revolutionizing the modern world. From the very beginning, many approaches have been made to integrate the physical world with cyber system to address the real world problems. One such dominant approach is to use the smart grid technology. Earlier, the concept of smart grid was introduced in the power system, where the smart grid was envisaged to replace the traditional grid for improving the performance of the power system [7]. The last few decades have seen unprecedented technological advancements in the field of computing, processing, analyzing, and networking along with the development of highly efficient sensor, actuators and other sophisticated devices. Consequently, it has been possible to deploy the cyber-physical system based smart grid technology for many other next generation applications such as synchrophasor communication, advance metering infrastructure, electric vehicular applications, etc. [8].

2.1 Overview of the Smart Grid Power System

In this section, we first introduce an overview of the smart grid which envisages to modernize the traditional power system. The smart grid was envisaged with an objective to provide efficient and uninterrupted power flow to the end users or consumers.

In last few years, there have been unprecedented technological advancements in various micro electro-mechanical devices including the various components of the power grid. Further, the disruptive advancements in the fields of communication technologies and protocols have increased the efficacy of the power grid system in terms of real time monitoring and controlling of the grid. Today's grids are modernized using many such disruptive technologies and advanced power grid components embedded with computing capabilities. These modern power grid systems are widely referred to as the "smart grid" (SG) [9]. The smart grid is further improved as flexible grid to cater the demand of several next generation diverse applications such as smart monitoring and control of the grid, smart metering, inclusion of renewable energy sources into the grid, inclusion of energy generation capability of the consumers, etc. Thus, the smart grid is an engineering marvel, which envisages an uninterrupted power flow from generation station to the customers, and ensures efficient energy management over power infrastructure.

The overview of the smart grid power system is as shown in Fig. 2. The smart grid comprises of renewable energy system such as solar system, wind system, etc., which contributes in the overall energy generation capacity of the power system. Further, both residential as well as commercial consumers are the active consumers in the sense that they also contribute in the total energy generation capacity of the

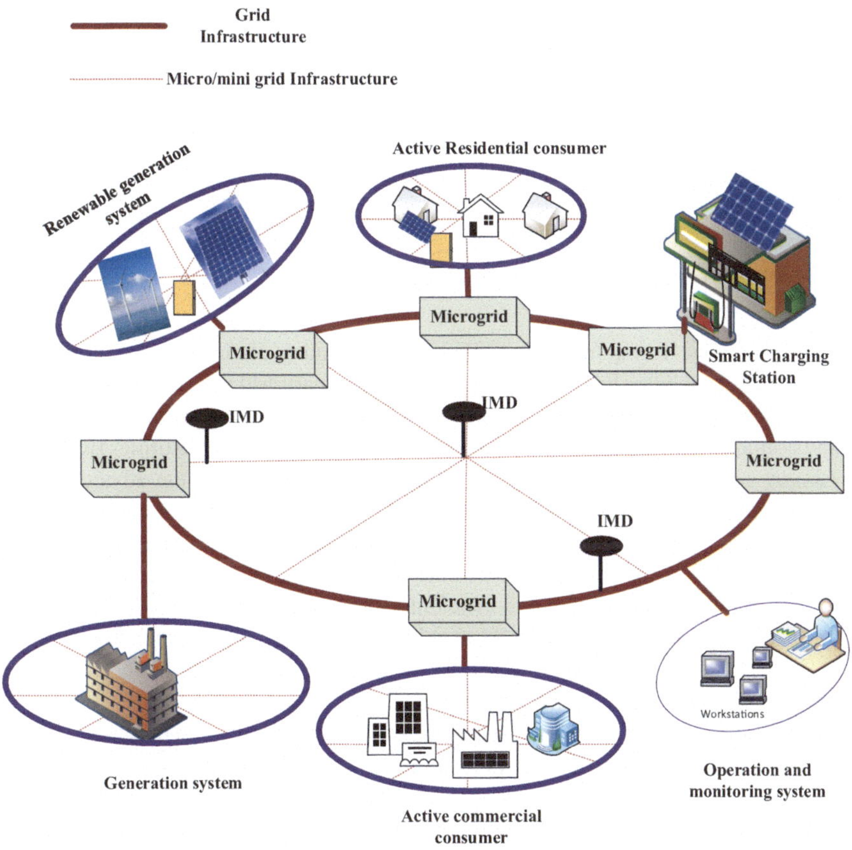

Fig. 2 An overview of the smart grid

power system. Hence, these consumers are referred to as an active consumer which not only acts as consumers but also acts as contributors in the power generation. The active consumer helps in reducing the load during peak consumption time, which helps in reducing the burden on the power grid. The smart grid system consists of several microgrids which are interconnected to ensure the smooth flow of the power. The renewable energy sources along with the traditional energy sources are responsible as generating sources in the SG. The energy flow is maintained efficiently from the source to the active consumers using the smart grid communication infrastructure. The smart grid technology includes diverse consumers such as residential consumers, industrial consumers, commercial consumers (electric vehicle), etc. The status of the smart grid system is continuously monitored and controlled by deploying several sensors and actuators, which are collectively referred as intelligent monitoring devices (IMD). The overall operations of the smart grid under dynamic environment is coordinated by the operation and monitoring systems. The smart grid envisages to heal itself in case of fault, however, the intervention of the operation and management system becomes the ultimate requirement in some of the cases to ensure reliable operation. Further, the operation and management system play vital role in distribution and billing system. Thus, in a nutshell, there exists a two-way coordinated communication between the various units, components, and systems which collectively forms the smart grid. Such smart grid system to ensure an efficient, reliable and uninterrupted operation of the power system is as shown in Fig. 2.

2.2 Smart Grid Cyber-Physical System

The cyber-physical system can potential combine many disruptive technologies to bring innovations into different applications. The smart grid cyber-physical system (SGCPS) is one such applications, which uses various disruptive technologies across many domains of the cyber-physical systems [10]. The SGCPS constitutes two parts: (a) *smart grid*; and (b) *cyber-physical system*. Thus, it includes the advantages of both the smart grid as well as cyber-physical system. The typical advantages of the smart grid includes grid monitoring and control, integration of renewable energy sources, efficient energy management, and operations, etc. The typical advantages of the cyber-physical system includes efficient integration of power infrastructure and cyber components. Thus, the SGCPS provides typical capabilities such as real-time monitoring and control of the grid, energy management, interaction, and operation of the smart grid power system efficiently.

According to NIST framework, the smart grid is thought of seven domains: *generation, distribution, transmission, operation, customer, market* and *service provider domains* [11]. The some of the key responsibilities of these smart grid domains are summarized in Table 1. Even though all these domains have different functionali-

Table 1 SGCPS communication technologies

Smart grid domains	Key responsibilities
Generation	It is responsible for generation of energy. It also includes renewable sources of energy generation along with the traditional non-renewable sources. It may also have storage capability
Distribution	Its responsibility includes the distribution of energy to the different consumers. It may also have storage and generation capabilities
Transmission	It is responsible to transport the energy from generation to the distribution domains. It may also have storage and generation capabilities
Operation	It is responsible for management and operation of the power system such as connectivity, maintenance, billing, etc.
Customer	These are primarily the consumers of the energy such as residential houses, commercial building, industries, offices, etc. It may also have storage and generation capabilities
Market	These are the platform which enables the different operators, providers, and managers to participate in the smart grid market for pricing and operation related utilities

ties, these domains are interrelated and must be integrated for effective operations from the SGCPS perspective. The integration of all these domains can be achieved by integrating the various entities (cyber entities as well as physical entities) of all the domains, which can be achieved using the burgeoning cyber-physical system approach.

Based on NIST framework of the smart grid, a smart grid cyber-physical system is as shown in Fig. 3. All domains of the smart grid integrated to interact with each other based on the cyber-physical system technology. The core of the cyber-physical system is the communication network or communication infrastructure which bridges the communication gap between the various domains of the smart grid. As shown in Fig. 3, a SGCPS basically comprises of generation, transmission, distribution and consumer systems, which are primarily responsible for power generation, transmission, distribution and consumption of the electricity respectively. These four components are the basic constituents of the physical infrastructure of the smart grid cyber-physical system. On the other side, the major constituents of the cyber system includes service provider, market and operation domains. These are primarily responsible to provide various features of the SGCPS such as real-time monitoring, control, billing, pricing, energy management, and operations, etc. All the domains of the smartgrid are integrated using the cyber-physical system communication infrastructure. The communication network's key responsibility is to provide seamless information regarding the operation and management of the smart grid system. Moreover, the various data pertaining to the operation and management of the smart grid are exchanged over the communication network among the various entities of both cyber as well as physical systems of the SGCPS. Hence, information flow between the different domains of the SGCPS through the communication

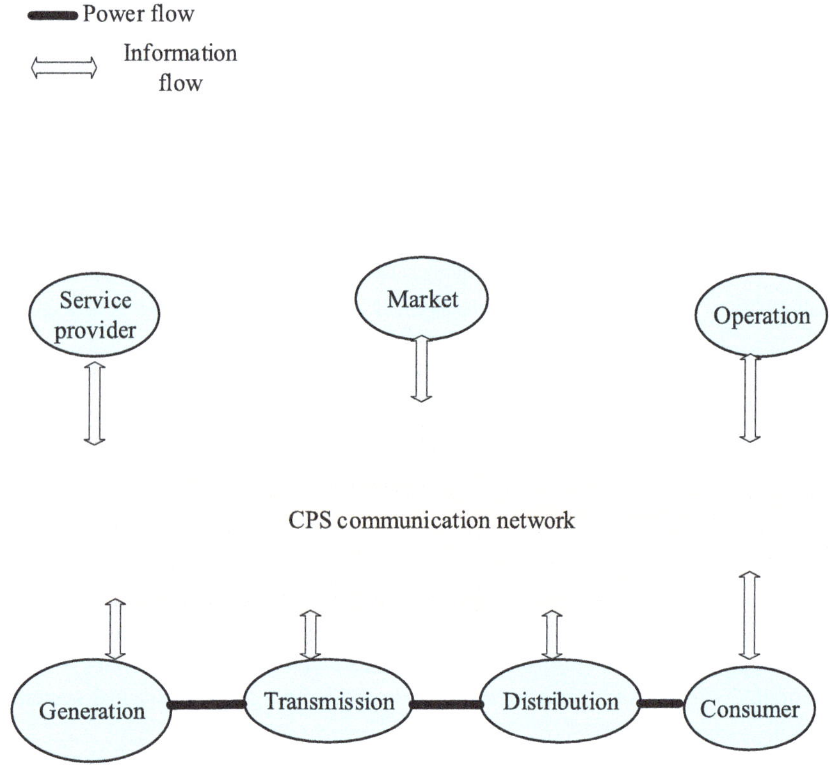

Fig. 3 The smart grid cyber-physical system

network. However, power flow is mainly from generation side to consumer side through power infrastructure of the SGCPS.

The smart grid cyber-physical system exhibits the following key characteristics:

- Integration of physical world with real world under dynamically varying scenario.
- Seamless and efficient communication of several application data such as synchrophasor communication data, energy management data, etc.
- Real-time processing, computing and analyzing the data for timely decision to ensure the reliable operation of the smart grid.
- Incorporation of new technologies from many industries to ensure secure and safe energy management of the grid.
- The self-configurable, self-adaptable and self-learning capabilities to ensure reliability of the smart grid CPS.
- Resiliency of the SGCPS communication network to provide uninterrupted power flow and energy management under catastrophic events.

3 Communication Networks for SGCPS

The data belonging to diverse applications of the smart grid cyber-physical system is communicated from geographically separated physical world to real world and vice versa over wide area communication network. These data are having requirements that are generally different than the data of other cyber-based applications. For example, data from the synchrophasor applications (discussed in Sect. 4) needs to have minimum delay and high reliability for availability estimation of the SGCPS. Failure in achieving this, may result into disastrous events to the power grid such as power outage or even blackout. On the other side, confidentiality estimation characteristics such as security, safety, etc., are of the highest priority for other industrial applications. Comparatively, the higher delay is permissible in the data pertaining to industrial applications as compared to that of synchrophasor applications. Thus, existing communication technologies are modified to suit the application-specific requirements of the SGCPS. Some of the major communication technologies for SGCPS applications are classified in Fig. 4. The key attributes and characteristics of these communication technologies are summarized in Table 2.

Due to the requirement of high level of integration among the various entities, SGCPS is complex in nature and high in dimension. Thus, it is also susceptible to different cyber as well as physical vulnerabilities. Also, as discussed earlier, the most of the smart grid applications have stringent requirements such as low latency and

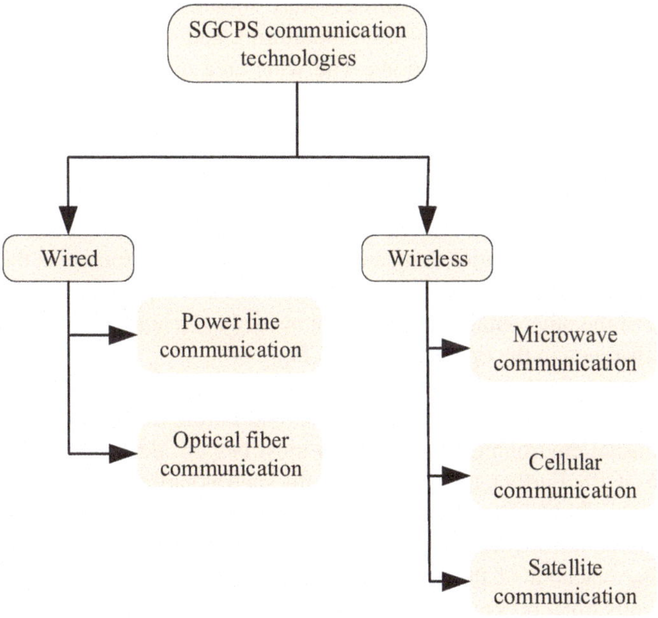

Fig. 4 Different communication technologies for SGCPS

Table 2 SGCPS communication technologies

Communication technologies	Key features
Power line communication	Uses existing transmission line for communication and provides the infrastructure for both: narrowband and broadband
Optical fiber communication	It provides wired communication infrastructure with higher data rate, improved security, and high reliability
Microwave communication	It enables low cost wireless data communication for SGCPS
Cellular communication	Provides high speed, less delay and efficient communication through cellular networks
Satellite communication	Uses satellite links to establish the communication spanning large geographical coverage

high reliability. These pose severe challenges to the communication network designer. Further, there exists some other challenges in the implementation of smart grid cyber-physical system such as reliability, resiliency, safety, security, data integrity, etc. The success of the SGCPS heavily relies on solving these diversified challenges, which is possible if and only if these challenges are identified properly. The challenges pertaining to SGCPS can be broadly categorized into four aspects.

- *Infrastructure challenges*: The conventional power system infrastructure does not support the requirements of the current generation. The components installed in these are quickly ageing and needed to be replaced for upgrading the traditional power system. The smart grid infrastructure should be upgraded using cyber-physical system to support real-time monitoring and control of the grid status, and other utility related operations. The wide area measurement system, wide area monitoring systems, protection and control of the SGCPS equipments are needed to be improved further to increase the reliability of the SGCPS.
- *Market/customer challenges*: Unlike the conventional power system, the customers in SGCPS are active customers, which do not merely consume the energy but also contribute the energy through the microgrids. Hence, efficient cyber-physical technologies that can incorporate active customers is the requirement of the SGCPS. Moreover, the proper power market policies and regulations are some of the other vital challenges that needed to be answered to bring liberty and transparency in the operation of SGCPS.
- *Environmental challenges*: The aging of the equipments and components of the SGCPS as a result of its normal operation, and failure rate of such components under unusual environmental conditions must be considered while designing a reliable SGCPS. Further, the natural calamity such as earthquakes, tsunami, hurricanes, storm, floods, tornados, and other geological events can cause immense damage to the infrastructure of the SGCPS and may even result in the disastrous events like blackout of the power system. Such blackout of the power system consequently hinders economic growth and development of the regions, or states. As an example, the hurricane Katrina in 2005 had resulted in power system blackout in the southeastern regions of the U.S with huge economical losses

[12]. Thus, the reliable operation of the SGCPS is the urgent need of the time to support uninterrupted power supply in the dynamically changing environment.

- *Innovative technological challenges*: In recent time, there have been tremendous technological development in the field of cyber-physical system. However, these technologies still do not full fill the complete requirement of safety and security. Thus, these technologies have been inappropriately used in many applications, particularly in SGCPS. Recently, new power electronic materials, development in the field of micro-electromechanical sensors, advance power system technologies, improved networking and communication technologies, and computation technologies, etc., have taken place in the design of modern smart grid power system. However, the concern of security and safety for these innovations are still at the frontier for the design engineers and researchers.

4 Applications of Smart Grid CPS

Numerous applications of smart grid has been found in the literatures with cyber-physical system as technology enablers. This chapter discusses few of the dominant applications such as synchrophasor, advanced metering and electrical vehicular applications of the SGCPS.

4.1 SGCPS for Synchrophasor Applications

The existing power grid, which is used for several diverse applications is a complex network. These complex networks, like other communication networks, are also susceptible to faults due to many disturbances such as failure of components, extreme weather events, malicious attacks, etc. Sometime, extreme weather events such as earthquake, tsunami, rainfall, flood, and other natural disasters can have a huge impact on the operation of such power system. Nevertheless, the fault due to components failure may also cause a huge impact on the operation of power system. If such faults are not detected in time for proper action, then it may lead to power blackout resulting in huge economic losses [13]. For e.g., according to the report of Electricity Consumers Resource Council, the 2003 power blackout has caused about 10 billion US dollar losses to U.S and Canada [14]. The subsequent finding suggested that this kind of catastrophic events can be prevented if the there would be an early warning system for due time action and control [15]. As a lesson from this and similar other blackouts across the world, it has necessitated the power system scientists and engineers to design the wide area synchrophasor measurement system (WASMS) for early warning system [16]. Moreover, an exemplary approach for the risk assessment of the WASMS is presented by Appasani and et al. [17].

The synchrophasor application is a WASMS which is responsible for monitoring the health of the smart grid cyber-physical system. In synchrophasor application,

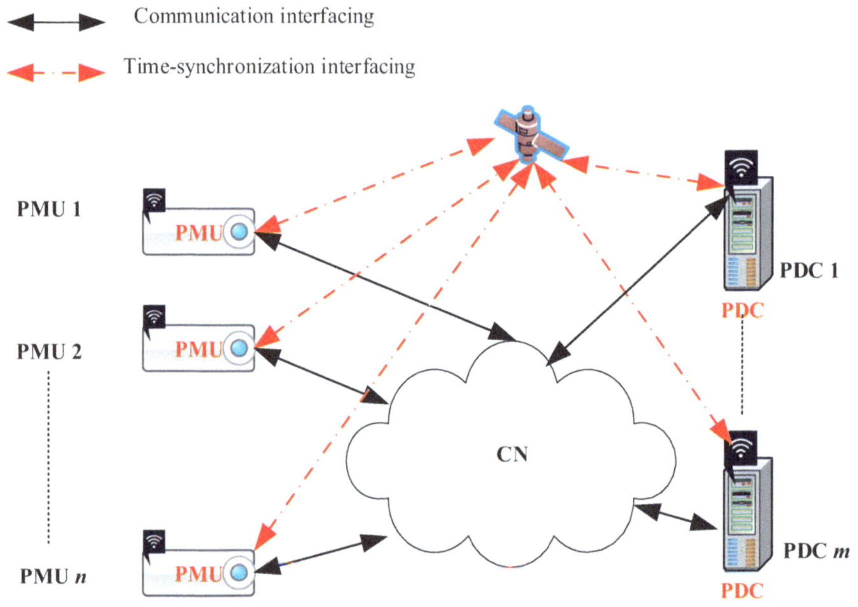

Fig. 5 Overview of SGCPS for synchrophasor application

there are large number of phasor measurement units (PMUs) which are used to estimate time-synchronized phasor measurements (magnitude and angle) of voltage signal or current signal of the power grid. The real-time synchronization of the grid phasor measurements are provided through global positioning system (GPS). Such real-time synchronized phasor measurement of grid voltage or current is known as the synchrophasor measurement, and the corresponding data is known as synchrophasor data. The PMUs are installed as a node on different geographically separated electrical buses to monitor the health of the overall grid of the power system. All the PMUs send the real-time synchrophasor data to the remotely located control center which is also referred to as a phasor data concentrator (PDC). The communication network over which the PMUs and PDC communicate to monitor and control the health of the power system is called synchrophasor communication network (SPCN) [18]. Hence, the prime constituents of the synchrophasor application of the SGCPS are PMU, PDC and synchrophasor communication network as shown in Fig. 5.

4.2 *SGCPS for Advanced Metering Applications*

The advanced metering infrastructure (AMI) plays vital role in modernizing the today's power grid. The AMI can be used for a wide variety of applications of which advanced metering application is one of the important applications. The advanced metering is one of the applications that can be effectively deployed using smart grid cyber-physical system. Advanced metering applications of the SGCPS includes many responsibilities such as communication of data pertaining to user's power consumption, electricity pricing, tariff regulations, distribution line losses, etc. It envisages to revolutionize the modern power system as many advantages are associated with it. These are as follows: enhancement in safety and security, improved customer experience, reduction in electricity tariffs, reduction in electricity theft, reduction in energy wastage, improvement in operational efficiency, capacity savings, and many more. Through AMI (with the help of website or app), customers can access the real-time data pertaining to the electricity use and can regularize their use. This will not only save the money for individual customers but also save the electricity consumption during peak time, thus easier for operator from the view of load scheduling [19].

Advanced metering applications is based on advanced metering infrastructure, which has smart meter (SM) as one of the prime constituents. The conventional electro-mechanical meter used in existing grid has been obsolete. Thus, it will be replaced by smart meter for the deployment of cyber-physical technology in the smart grid power system. The smart meter is one of the key components of the AMI, which provides an interface for cyber and physical components of the SGCPS [20]. The smart meter data is collected from the geographically separated customer's utility, and communicated to operator's utility. These data from the smart meter may be used to for various applications such as energy thefts, outage management, load forecast, power scheduling, etc. [21].

Figure 6 shows a typical architecture of the advanced metering (AM) application of the SGCPS. As shown in Fig. 6, the advanced metering application consists of smart meter which is used to connect the customers to the utility operators. The smart meter data is communicated between customers and utility operators over communication networks of the SGCPS. The generic architecture of the advanced metering application of the SGCPS consists of following major components:

- *Smart meter (SM)*: to exchange data from customers to utility operators and vice versa.

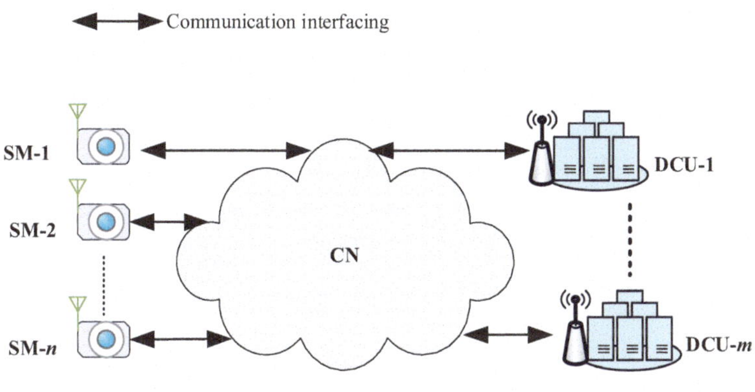

Fig. 6 Overview of SGCPS for advanced metering application

- *Data concentrator unit (DCU)*: to aggregate data from several smart meters for interpretations and actions.
- *Communication network (CN)*: to interconnect various networks and components of the SGCPS-based advanced metering application.

Further, the key components of the AMI such as smart meter and DCU are interconnected to the physical components of the SGCPS such as transformer, substation, etc. using the communication technology. The several communication technologies exists as discussed in Sect. 4, however, for advanced metering application the best suitable communication technologies would be ZigBee and PLC [22]. On one side, a large number of smart meters can be interconnected to a module known as ZigBee-PLC bridge using ZigBee technology. This provides wireless interface between smart meter and ZigBee-PLC bridge. On the other side, PLC can be used to provide wired interface to ZigBee-PLC bridge and DCU as illustrated in Fig. 7.

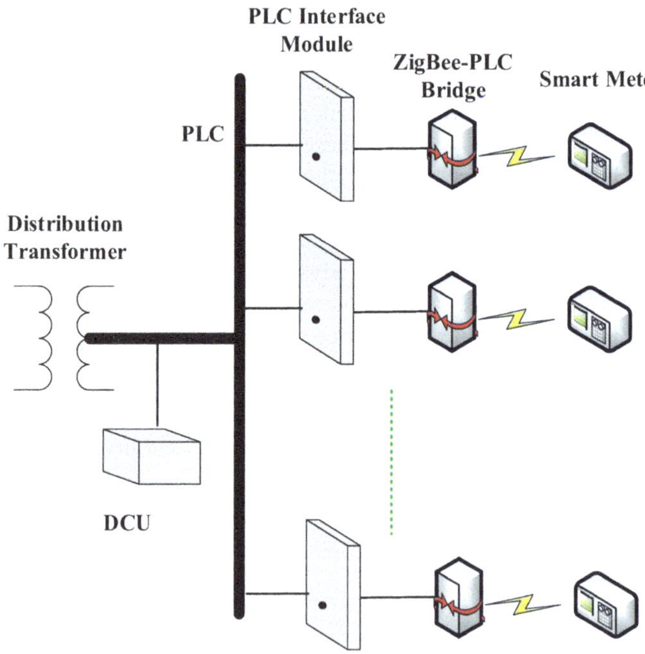

Fig. 7 Overview of AMI

4.3 SGCPS for Electrical Vehicular Applications

The concern of the people regarding the clean energy, limited source of energy, environmental pollution, etc., has demanded the paradigm shift in traditional vehicular system. This has initiated the birth of electric vehicular (EV) system where electricity acts as the prime source of energy. The electric vehicular system is a continuously evolving disruptive technology that combines many industrial sectors like automobile, power grid, information technology, etc.

The paradigm of electric vehicular system suffers from many challenging issues. The core of the electric vehicle is electric charging station (ECS). The charging station must be embedded with recent technologies for optimum charging of the electric vehicle and to support the real-time market requirements [23]. The electric vehicle is connected to the ECS for charging of the battery. The electric charging station is basically an advanced power station which is connected to the micro grid of the power grid system. The EV system poses many challenges to the power grid system. Some of the challenges faced in electric vehicle applications to the power grid system are: it can impact the power quality of the power grid system, it can overload the distribution system and its equipments, it can affect the capability of generation stations [24]. A large number of research has been done in the literatures to address these challenges for realizing electric vehicular application at large scale. It has been found that the smart grid cyber-physical system approach can be used to address all these major challenges faced in the electric vehicular application.

The electrical vehicular application is one of the rapidly evolving areas, which can harness the potentials of the smart grid cyber-physical system. It is also an optimum approach to save the depleting energy resources. The electric vehicular system is an opportunistic next generation application with many hurdles in its way [25]. The SGCPS has potential to improve the efficacy and performance of the electric vehicular system. The dominant challenges faced to the power grid system in realizing electric vehicular application can be answered using SGCPS as discussed below in Table 3.

From the perspective of smart grid, the basic conceptual architectural framework of an electric vehicular application using smart grid cyber-physical systems is as shown in Fig. 8. As shown in Fig. 8, the EV application of SGCPS consists of following key components as,

- *Charging station*: to charge the electric vehicle on plug-and-play principle.
- *Microgrid substation*: to supply the energy to the electric charging station.
- *Electro-vehicular grid monitoring and control system*: to monitor and control the status of the electric vehicular system.

The electro-vehicular grid monitoring and control system (EVGMCS) acts as a communication infrastructure that connects various cyber and physical components of the SGCPS for electric vehicular applications. The EVGMCS is responsible for rendering various services to electric vehicular application, which are summarized below:

Table 3 Major challenges in EV system and approach using SGCPS

Major challenges faced in EV application	Approach to the challenges using SGCPS
Impact on the power quality of the power grid system	Monitor the power quality using suitable methods of SGCPS. For e.g., similar to synchrophasor application which is used in WAMS
Overloading of distribution system	Use real-time monitoring system at the distribution site and enable communication capability to each distribution system. For e.g., an approach similar to AMI can be employed using SGCPS
Impact on generation capability of power grid system	Identify and integrate different energy contributors such as from renewable energy using SGCPS. For e.g., integration of renewable energy (from the consumers as well) is one of fundamental features of smart grid CPS

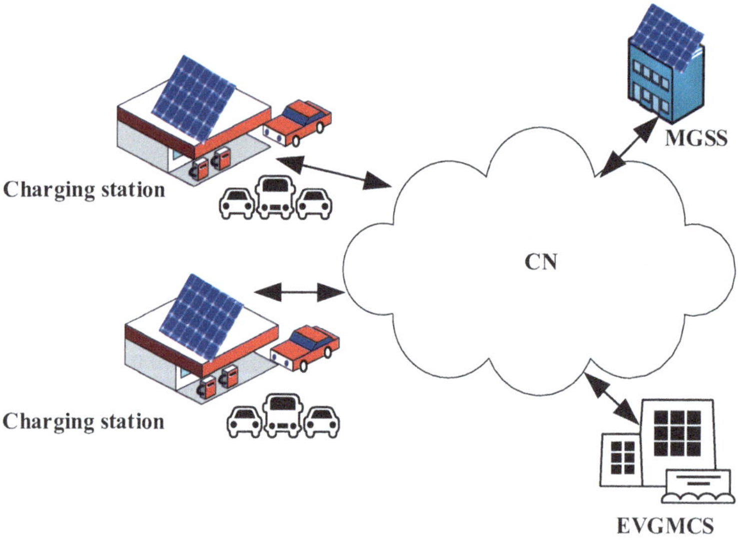

Fig. 8 Overview of SGCPS for electric vehicular application

- Power quality monitoring
- Demand forecast
- Locally generated renewable energy integration to the ECS
- Load shifting and management of ECS
- Voltage and load frequency regulation of charging station.

5 Risk Identification and Risk Assessment of SGCPS

Risk assessment is the process of analyzing the system vulnerability in presence of threat and challenges. The risk assessment is the successor of risk identification [26]. The communication network for the smart grid acts as a backbone and it is based on several networking technologies such as Ethernet, WiMax, cellular systems, microwave, radio and optical fiber, etc. The data between different components of the smartgrid cyber-physical systems are exchanged over the Internet infrastructure. Internet being an open to all, poses several security challenges to the SGCPS. Several IP based technologies to trace back the source of threats can be found in [27, 28]. The different attacks and their impact on the smart grid synchrophasor system is discussed [29, 30]. Moreover, these issues in focus to advanced metering application of the SGCPS is more comprehensively discussed in literature [31–35]. The state-of-art review on risk analysis is presented by Sun et al. [36]. The most comprehensive challenges in the domain of electric vehicular applications has been discussed by Dutta et al. [41]. The blockchain-based mechanism has been shown to be more effective in case of security management in electric vehicular applications [42]. Moreover, some of the seminal works in the direction of vulnerability studies for different applications are reported in Table 4.

The risk assessment models and approaches for different components of the power system are presented in [46]. The similar methodologies can be applied to analyze the risks and its evaluation in the context of smart grid cyber-physical system. As discussed in previous sections, the smart grid cyber-physical system comprises of large number of sensors, actuators, networking device, power equipments, etc., which are interconnected using cyber technology to provide seamless interaction over communication networks. Thus, communication network of the SGCPS acts as a backbone to the system, since whole communication depends upon it. Similarly, the several other components of the SGCPS play different role in the performance of the SGCPS. The failure of each components have different impact on the overall performance of the SGCPS. The impact of different components of the SGCPS can

Table 4 Vulnerability studies in literatures for different SGCPS applications

SGCPS applications	Seminal work	Domains of studies	References
Synchrophasor	Vulnerability assessment	GPS signaling and data	[29, 30]
AMI	Vulnerability assessment Risk assessment standards	Components of CPS, communication aspects, and privacy	[31–35, 37–40]
EV application	Challenges including security and privacy, methodology for risk analysis	Security and privacy concern for connected vehicle, smart energy management, Cloud computing for EV system	[41–45]

be measured in terms of risk which in turn evaluates the losses due to failure of different components [47]. Thus, the risks identification and assessment methodologies for different components of the SGCPS for different applications are discussed in this section.

5.1 Synchrophasor Application

The synchrophasor application of the smart grid cyber-physical system comprises of key components such as PMU, PDC and other sensors. Since, PMU is located on different electrical buses, thus, if PMU fails, then the corresponding bus cannot be observable. On the other hand, if PDC fails, then the entire system become unobservable as all PMUs send the data to PDC for monitoring and controlling the smart grid. Hence, we can comment that different levels of risks are associated with different components or equipments of the SGCPS. Thus, it becomes extremely important to analyze the level of risks associated with different components of SGCPS and their impacts.

5.1.1 Risk Application

The synchrophasor application of smart grid cyber-physical system consists of mainly PMUs, PDCs, and communication networks (i.e., synchrophasor communication network). Without loss of generality, we assume that the cyber components are the constituents of the communication network of SGCPS for synchrophasor application. Thus, the following risks are identified for synchrophasor applications of SGCPS:

- Failure of the PMUs
- Failure of the PDC
- Failure of synchrophasor communication network

The risk associated with the different constituents has different level of impact on the performance of SGCPS for synchrophasor application. For e.g., the risk associated with a PMU failure has less impact on the SGCPS as compared to the risk failure associated with the communication network failure. The later has potential to disrupt the whole SGCPS, whereas, the former may affect only the part of SGCPS. The level of risks associated and its potential impact on the synchrophasor applications in SGCPS is summarized in Table 5. Here, L_i indicates the level of risks of the corresponding constituents such that $L_1 < L_2 < L_3$. Obviously, L_1, L_2 and L_3 are categorized as low, medium and high level of risks.

Table 5 Key components, associated risks, and its impact in synchrophasor application

Constituents of SGCPS	Risk level	Impact
PMU	L_1	On observability of particular electrical bus
PDC	L_2	On the synchrophasor application of SGCPS
Synchrophasor communication network	L_3	On the whole SGCPS

5.1.2 Risk Assessment of SGCPS for Synchrophasor Application

In this section, the metrics to measure the risks and the strategies to reduce or mitigate them are discussed. Let us consider an event 'i'. We can define the severity and the risk of the event 'i' as follows.

Severity: This is defined as the measure of the impact of the risk associated with a particular kind of risks. In other words, it is the ratio of the total number of events/components that are dependent on ith to the total number of events/components in the system.

$$(\text{Severity})_i = \frac{\text{Total number ofevents dependent on } i\text{th event}}{\text{Total number of events in the system}} \tag{1}$$

In particular, let us consider an event/component 'i' is present in the system such that its failure leads into the failure of 'k' out of 'N' events/components of the system. Then, the severity of the event/component 'i' can be given by following equation.

$$S_i = \frac{k}{N} \tag{2}$$

Risk: This measures the impact of severity in terms of failure probability. In other words, it measures the consequences on the system in terms loss as a result of probable failure of individual events/components in the system.

The risk associated with ith event/component of the system can be described using following equation.

$$\text{Risk}_i = (\text{Failure Probability})_i \times (\text{Severity})_i \tag{3}$$

In particular, if p_i be the probability of failure of the ith event/component of the system, then the risk α_i associated with this event/component can be given by the following equation.

$$\alpha_i = p_i \times S_i \tag{4}$$

On substituting the severity from Eq. (1), we have

$$\therefore \alpha_i = p_i \times \frac{k}{N} \tag{5}$$

W.L.O.G, the failure probability p_i includes both the software failures probability as well as hardware failure probability.

Consider the synchrophasor communication application of SGCPS for wide area measurement system. The wide area measurement system consists of several high speed sensors known as phasor measurement units (PMUs) that continuously monitor or observe the grid and send the data to the PDC at the control center. Based on the data fed from different PMUs, a proper control command is initiated by the control center to maintain desire performance level of the smart grid cyber-physical system. It is not necessary to install the PMU on each electrical bus to observe the status of the entire grid. This is because, a PMU installed on a bus can observe the status of more than one bus simultaneously [22].

Risk assessment for PMU: Let us consider the synchrophasor application of SGCPS consists of n number of PMUs and N number of electrical buses such that $n \leq N$. Consider a PMU is represented by PMU_{xi}, where x, $i$denotes the PMU number and corresponding bus location respectively such that $x \in \{1, 2, \ldots, n\}$ and $i \in \{1, 2, \ldots, N\}$. If a PMU-$x$ located on a bus i (denoted as PMU_{xi}) can observe the status of k_x out of N number of buses simultaneously in the grid. Then, the severity (S_{xi}) of the PMU can be given by,

$$S_{xi} = \frac{k_x}{N}; \forall x \in \{1, 2, \ldots, n\} \tag{6}$$

If p_x be the failure probability of the PMU_{xi}, then the risk of the xth PMU (α_x) can be given as below.

$$\alpha_x = p_x \times \frac{k_x}{N}; \forall x \in \{1, 2, \ldots, n\} \tag{7}$$

Therefore, the overall risk (α) including all PMUs can be estimated using equation given below.

$$\alpha = \sum_{x=1}^{n} \left(p_x \times \frac{k_x}{N} \right) \tag{8}$$

$$\therefore \alpha = p_1 \times \frac{k_1}{N} + p_2 \times \frac{k_2}{N} + \cdots + p_n \times \frac{k_n}{N} \tag{9}$$

If $k_x = k; \forall x \in \{1, 2, \ldots, n\}$, then Eq. (9) becomes,

$$\alpha = \frac{k}{N} \sum_{x=1}^{n} p_x \tag{10}$$

Observation: From the Eq. (10), it can be observed that the overall risk associated with the PMUs completely depends upon their failure probability. Further, since all PMUs are mutually independent, hence overall risk can be minimized by minimizing sum of the probability of failure of all PMUs.

Risk assessment for PDC: Let us consider the synchrophasor application of SGCPS consists of m number of PDCs and N number of electrical buses such that $m \leq N$. Consider a PDC is represented by PDC_{yi}, where y, i denotes the PDC number and corresponding bus location respectively such that $y \in \{1, 2, \ldots, m\}$ and $i \in \{1, 2, \ldots, N\}$. Let, a PDC-$y$ located on a bus i (denoted as PDC_{yi}) acts as a server for ∂_y number of PMUs such that $\partial_y \leq n$. Hence, the severity (S_{yt}) of the PDC can be given by,

$$S_{yi} = \frac{\partial_y}{n}; \forall y \in \{1, 2, \ldots, m\} \tag{11}$$

If p_y be the failure probability of the PDC_{yi}, then the risk of the yth PDC (β_y) can be given as below.

$$\beta_y = p_y \times S_{yi}; \forall y \in \{1, 2, \ldots, m\}$$

$$\beta_y = p_y \times \frac{\partial_y}{n} \tag{12}$$

Therefore, the overall risk (β) including all PDCs can be estimated using equation given below.

$$\beta = \sum_{y=1}^{m} \left(p_y \times \frac{\partial_y}{n} \right)$$

$$\therefore \beta = p_1 \times \frac{\partial_1}{n} + p_2 \times \frac{\partial_2}{n} + \cdots + p_m \times \frac{\partial_m}{n} \tag{13}$$

If $\partial_y = \partial; \forall y \in \{1, 2, \ldots, m\}$, then Eq. (13) becomes,

$$\beta = \frac{\partial}{n} \sum_{y=1}^{m} p_y \tag{14}$$

Observation: *From* the Eq. (14), it can be observed that the overall risk associated with the PDCs completely depends upon their failure probability. Further, since all PDCs are mutually independent, hence overall risk can be minimized by minimizing sum of the probability of failure of all PDCs.

Risk assessment for communication network: Further, the communication networks acts as a backbone to the SGCPS through which different cyber and physical components of the SGCPS interacts. The severity of the communication network depends on the number of PMUs connected to the network and is given by S_{CN}. Depending on the architecture, different metrics are used to indicate the severity. One popular metric is the number of buses that become unobservable, when a network fails, given by S_{CN} If p_z denotes the failure probability of the communication network, then the risk of the communication network (γ) completely depends upon its failure probability as given by Eq. (15).

$$\gamma = p_z S_{CN} = p_z \tag{15}$$

5.2 *Advanced Metering Application*

In this section, we consider an advanced metering application of the SGPCS. We identify the level of risks associated with different key components and their impact on the performance of the SGCPS for advanced metering application. Moreover, the risk assessment methodologies for advanced metering application of the SGCPS is also presented in this section.

5.2.1 Risk Identification of SGCPS for Advanced Metering Application

As seen in Sect. 4, the advanced metering application of SGCPS is based on advanced metering infrastructure. The advanced metering infrastructure consists of following key elements: *smart meter* (*SM*), *data concentrator unit* (*DCU*), and *communication networks*. Without loss of generality, we assume that the cyber components are the constituents of the communication network of SGCPS for electric vehicular application. Thus, the following risks are identified for advanced metering applications of SGCPS:

- Failure of smart meter
- Failure of a DCU
- Failure of communication network

The risk associated with the different constituents has different level of impact on the performance of SGCPS for advanced metering application. For e.g., if a smart meter fails, it leads into failure of the utility related operation of corresponding consumer only. On the other side, if a data concentrator unit fails, it may affect the operation of all smart meters that communicate the data to it. Obviously, the failure of communication network may lead into failure of entire AMI. This is because the communication network acts as a bridge to cyber and physical components of the SGCPS. The level of risks associated with various components are identified and its

Table 6 Key components, associated risks, and its impact in advanced metering application

Constituents of SGCPS	Risk level	Impact
Smart meter	L_1	On billing and utility related capability of the corresponding consumer of the SGCPS
Data concentrator unit	L_2	On billing and utility related capability of all the smart meters that communicated the data to this DCU
Communication network	L_3	On the entire AMI of SGCPS

potential impact on the advanced metering applications in SGCPS is summarized in Table 6. Here, L_i indicates the level of risks such that $L_1 < L_2 < L_3$. Obviously, L_1, L_2 and L_3 are categorized as low, medium and high level of risks.

5.2.2 Risk Assessment of SGCPS for Advanced Metering Application

Risk assessment for smart meter: Let us consider an advanced metering application of the SGCPS having n number of smart meters. Consider a smart meter is represented by SM_x, where x denotes the identification number of the corresponding smart meter such that $x \in \{1, 2, \ldots, n\}$. W.L.O.G, we assume that each smart meter is connected to a distinct consumer. Thus, it can be observe that we have n number of consumers that are served using AMI system. The failure of one smart meter affects only one consumer out of n consumers in the system. Hence, each smart meter will have equal severity index (S_x) which can be given by following equation.

$$S_x = \frac{1}{n}; \forall x \in \{1, 2, \ldots, n\} \tag{17}$$

If p_x be the failure probability of the smart meter SM_x, then the risk of the xth smart meter (α_x) can be given as below.

$$\alpha_x = p_x \times \frac{1}{n}; \forall x \in \{1, 2, \ldots, n\} \tag{18}$$

Therefore, the overall risk (α) including all smart meters can be estimated using equation given below.

$$\alpha = \sum_{x=1}^{n} \left(p_x \times \frac{1}{n} \right) \tag{19}$$

$$\therefore \alpha = p_1 \times \frac{1}{n} + p_2 \times \frac{1}{n} + \cdots + p_n \times \frac{1}{n}$$

$$\therefore \alpha = \frac{1}{n}\sum_{x=1}^{n} p_x \tag{20}$$

Observation: From the Eq. (20), it can be observed that the overall risk associated with the smart meters completely depends upon their failure probability. Further, since all smart meters are mutually independent, hence overall risk can be minimized by minimizing their total failure probability.

Risk assessment for DCU: We consider an advanced metering system having $m \leq n$ number of data concentrator. Let, a DCU-y (denoted as DCU_y) acts as a server for ∂_y number of smart meters such that $\partial_y \leq n$ Hence, the severity (S_y) of the DCU_y can be given by,

$$S_y = \frac{\partial_y}{n}; \forall y \in \{1, 2, \ldots, m\} \tag{21}$$

If p_y be the failure probability of the DCU_y, then the risk of the yth DCU (βy) can be given as below.

$$\beta_y = p_y \times S_y; \forall y \in \{1, 2, \ldots, m\}$$

$$\therefore \beta_y = p_y \times \frac{\partial_y}{n} \tag{22}$$

Therefore, the overall risk (β) including all DCUs can be estimated using equation given below.

$$\beta = \sum_{y=1}^{m}\left(p_y \times \frac{\partial_y}{n}\right)$$

$$\therefore \beta = p_1 \times \frac{\partial_1}{n} + p_2 \times \frac{\partial_2}{n} + \cdots + p_m \times \frac{\partial_m}{n} \tag{23}$$

If $\partial_y = \partial; \forall y \in \{1, 2, \ldots, m\}$, then Eq. (23) becomes,

$$\beta = \frac{\partial}{n}\sum_{y=1}^{m} p_y \tag{24}$$

Observation: From the Eq. (24), it can be observed that the overall risk associated with the DCUs completely depends upon their failure probability. Since all PDCs are mutually independent, hence, overall risk can be minimized by minimizing their total failure probability.

Risk assessment for communication network: Further, the communication networks of the AMI application acts as a backbone to the SGCPS. Hence, the severity of the communication network is always equal to 1, i.e., $S_{CN} = 1$. If p_z denotes the failure probability of the communication network, then the risk of the communication network (γ) completely depends upon its failure probability as given by Eq. (25).

$$\gamma = p_z S_{CN} = p_z \tag{25}$$

5.3 *Electric Vehicular Application*

Let us consider now an electric vehicular application of the smart grid cyber-physical system. Now, the risk pertaining to the SGCPS for electric vehicular application is identified in this section. To understand the significance of each components present in the system, the risk assessment strategies is also elaborated in this section which can be used to evaluate the overall performance of the system.

5.3.1 Risk Identification of SGCPS for Electric Vehicular Application

As seen in Sect. 4 earlier, the electric vehicular application of SGCPS consists of following key components.

- Electric charging station (ECS)
- Microgrid substation (MGSS)
- Electro-vehicular grid monitoring and control system (EVGMCS)
- Communication network (CN)

Further, the electro-vehicular grid monitoring and control system consists of several cyber-physical components that builds the communication infrastructure of EV application of the SGCPS. Moreover, it is responsible for monitoring and controlling of the overall status of SGCPS for electrical vehicular application in real-time. The smart meter, data concentrator unit are some of the vital components of the EVGMCS. All the components of the E2V system are interconnected and communicate using the communication network. Without loss of generality, we assume that the cyber components are the constituents of the communication network of SGCPS for electric vehicular application. Thus, the following risks are identified for electric vehicular applications of SGCPS:

- Failure of electric charging station
- Failure of microgrid substation
- Failure of smart meter
- Failure of a DCU
- Failure of EVGMCS communication network

Table 7 Key components, associated risks, and its impact in EV application

Constituents of SGCPS	Risk level	Impact
Smart meter	L_1	On billing and utility related capability of the corresponding electric vehicle or meter of charging station
DCU	L_2	On billing and utility related capability of the entire electric vehicular charging station
Charging station	L_3	On the corresponding charging station performance capability
Microgrid substation	L_4	On the performance of all the charging station that are connected through this microgrid substation
EVGMCS	L_5	On the observability and controlling capabilities of the entire electric vehicle application of SGCPS

The risk associated with the different constituents has different level of impact on the performance of SGCPS for electric vehicular application. For e.g., if either microgrid substation fails it leads into failure of more than one electric charging station. On the other side, if an electric charging station fails, it does not hamper much to the other charging station. Similarly, failure of the EVGMCS may lead into the failure of entire system, which depends on this as a communication infrastructural backbone. The level of risks associated with various components are identified and its potential impact on the electric vehicular applications in SGCPS is summarized in Table 7. Here, L_i indicates the level of risks such that $L_1 < L_2 < L_3 < L_4$ L_5.

5.3.2 Risk Assessment of SGCPS for Electric Vehicular Application

Risk assessment for charging station: Let us consider an electric vehicular application of the SGCPS having n number of electric charging station. Each charging station are equipped with smart meters and DCU for which risk can be analyzed in the similar manner as seen in AMI application. Now consider a charging station is represented by CS_x, where x denotes the identification number of the corresponding charging station such that $x \in \{1, 2, \ldots, n\}$. Let each charging station work independently. Thus, failure of one charging station does not affect the other charging station. Hence, each charging station will have equal severity index (CS_x) which can be given by following equation.

$$S_x = \frac{1}{n}; \forall x \in \{1, 2, \ldots, n\} \tag{27}$$

If p_x be the failure probability of the charging station CS_x, then the risk of the x charging station (α_x) can be given as below.

$$\alpha_x = p_x \times \frac{1}{n}; \forall x \in \{1, 2, \ldots, n\} \tag{28}$$

Therefore, the overall risk (α) including all charging station can be estimated using equation given below.

$$\alpha = \sum_{x=1}^{n}\left(p_x \times \frac{1}{n}\right) \tag{29}$$

$$\therefore \alpha = p_1 \times \frac{1}{n} + p_2 \times \frac{1}{n} + \cdots + p_n \times \frac{1}{n}$$

$$\therefore \alpha = \frac{1}{n}\sum_{x=1}^{n} p_x \tag{30}$$

Observation: From the Eq. (30), it can be observed that the overall risk associated with the charging station completely depends upon their failure probability. Since, all charging station are mutually independent, hence overall risk can be minimized by minimizing their total failure probability.

Risk assessment for microgrid substation: We consider an electric vehicular application of SGCPS having $m \leq n$ number of microgrid substation (MGSS). Let, a MGSS-y (denoted as $MGSS_y$) acts as a coordinator for ∂_y number of charging stations such that $\partial_y \leq n$. Hence, the severity (S_y) of the $MGSS_y$ can be given by,

$$S_y = \frac{\partial_y}{n};\ \forall y \in \{1, 2, \ldots, m\} \tag{31}$$

If p_y be the failure probability of the $MGSS_y$, then the risk of the yth MGSS (β_y) can be given as below.

$$\beta_y = p_y \times S_y;\ \forall y \in \{1, 2, \ldots, m\}$$

$$\therefore \beta_y = p_y \times \frac{\partial_y}{n} \tag{32}$$

Therefore, the overall risk (β) including all MGSSs can be estimated using equation given below.

$$\beta = \sum_{y=1}^{m}\left(p_y \times \frac{\partial_y}{n}\right)$$

$$\therefore \beta = p_1 \times \frac{\partial_1}{n} + p_2 \times \frac{\partial_2}{n} + \cdots + p_m \times \frac{\partial_m}{n} \tag{33}$$

If $\partial_y = \partial;\ \forall y \in \{1, 2, \ldots, m\}$, then Eq. (33) becomes,

$$\beta = \frac{\partial}{n} \sum_{y=1}^{m} p_y \tag{34}$$

Observation: From the Eq. (34), it can be observed that the overall risk associated with the MGSSs completely depends upon their failure probability. Since all MGSSs are mutually independent, hence, overall risk can be minimized by minimizing their total failure probability.

Risk assessment for communication network: Further, the communication networks of the electric vehicular application acts as a backbone to the SGCPS. Hence, the severity of the communication network is always equal to 1, i.e., $S_{CN} = 1$. If p_z denotes the failure probability of the communication network, then the risk of the communication network (γ) completely depends upon its failure probability as given by Eq. (35).

$$\gamma = p_z S_{CN} = p_z \tag{35}$$

Observation: From above equation, it is observed that the risk of the communication network completely depends upon its failure probability. For the failure probability equal to 1 (i.e., $p_z = 1$), the risk associated with the communication network is equal to 1 (i.e., $\gamma = 1$). This indicates that the failure of the communication network leads into the failure of the whole SGCPS for electric vehicular application.

6 Case Studies

In this section, three case studies are presented to estimate the risk associated with the failure of the communication networks for different applications in a SGCPS. The first case study deals with the synchrophasor application of SGCPS. Second case study deals with the advanced metering (AM) application of SGCPS. Finally, third case study deals with the electric vehicular (EV) application of SGCPS.

6.1 Synchrophasor Applications of SGCPS

We consider the case study of synchrophasor application for the power grid of Bihar, India. The single line diagram (SLD) for this grid is as shown in Fig. 9. The complete information regarding the geometric location of the buses, PMU and PDC in the grid is given in [48].

The buses in red color are locations where the PMUs are placed in the grid. For the sake of estimating the risk associated with the communication networks it is assumed that the communication network between each PMU and the PDC is not shared with

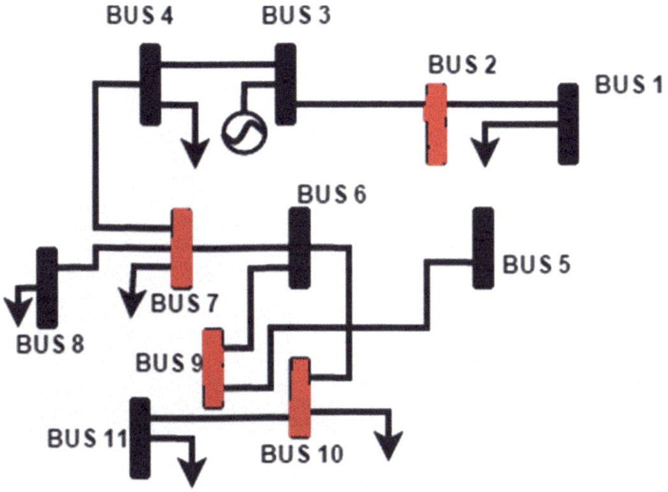

Fig. 9 SLD for Bihar's power grid

any other PMU. This assumption is valid because, the PMUs are generally separated by large geographical distances. The severity of failure for the networks for PMU's at bus 2, 7, 9 and 10 are 2, 2, 1and 1 respectively [48]. The severity is the number of buses that become unobservable due to the failure of communication networks. The overall risk associated with the communication networks for this system is the summation of the individual risks. Failure of communication system may be due to hardware failures or packet losses. To get an overall estimate of the risk, we perform a Monte-Carlo simulation of 100,000 runs and plot the histogram for the overall risk, which is illustrated in Fig. 10.

Fig. 10 Distribution of risk for synchrophasor application

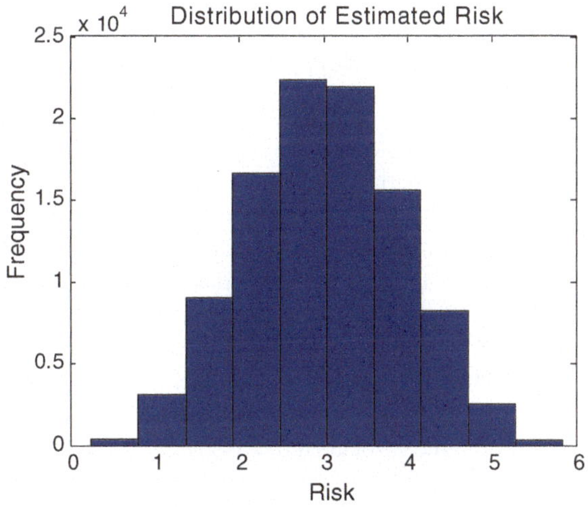

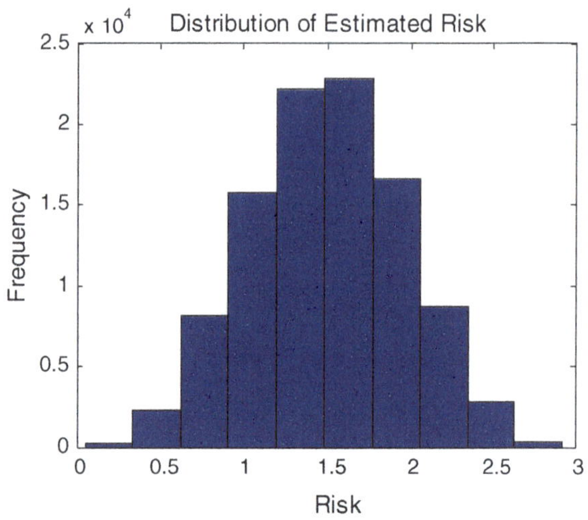

Fig. 11 Distribution of risk with reduced probability of failure for synchrophasor application

The distribution of risk for this system can be approximated by a normal distribution having a mean of 3 and variance of 0.84. Thus, the estimated risk for this system is 3. By improving the redundancy of hardware and reducing the packet loss, the risk associated can be minimized. Let us assume that because of the measures, the maximum probability of failure has come down to 0.5. The distribution of risk for the system with reduced probability of failure is shown in Fig. 11.

The distribution of risk for this system with reduced probability of failure can be approximated by a normal distribution having a mean of 1.5 and variance of 0.21. Thus, the estimated risk for the system with reduced failure probability is 1.5. Thus, improving the availability of the communication network reduces the overall risk associated with the network.

6.2 Advanced Metering Application of SGCPS

To estimate the risk associated with the AMI application in SGCPS a simple case study of 100 smart meters connected to a single DCU is considered as shown in the Fig. 12. This can be considered as a fundamental unit for AMI application in SGCPS. We have considered a local area network (LAN) on which 100 smart meters are connected to exchange their data to the DCU, which is also connected on the same LAN. In general, many such LANs can be used for large scale applications. Each LAN is interconnected to form a wide area network. This is simply represented by connecting the LAN under consideration to other network. If the communication network between an SM and the DCU fails, information pertaining to a single unit gets lost and the associated risk is given by Eq. (25).

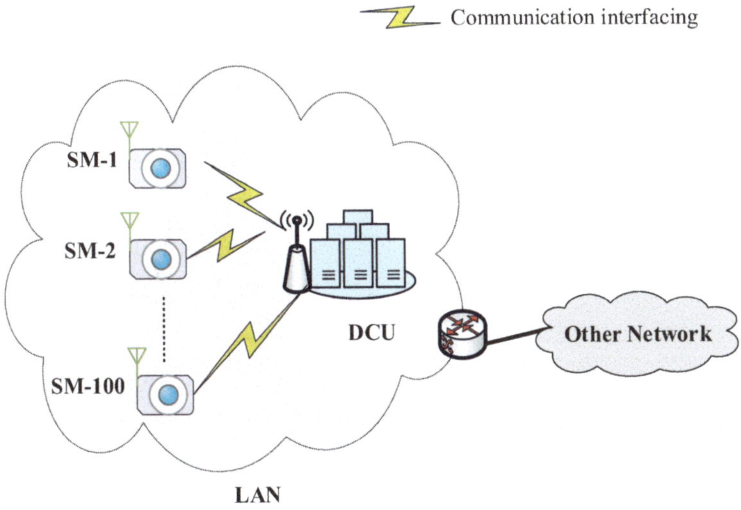

Fig. 12 Risk assessment model for advanced metering application of SGCPS

To get an overall estimate of the risk, we perform a Monte-Carlo simulation of 100,000 runs and plot the histogram for the overall risk, which is illustrated in Fig. 13.

The distribution of risk for this system can be approximated by a normal distribution having a mean of 50 and variance of 8.4. Thus, the estimated risk for this system is 50. More realistic estimates can be obtained by taking appropriate distribution functions for the probability of failures instead of taking a uniform distribution between 0 and 1. This, choice depends on the actual network configuration and the nature of network traffic. By improving the redundancy of hardware and reducing

Fig. 13 Distribution of risk for advanced metering application

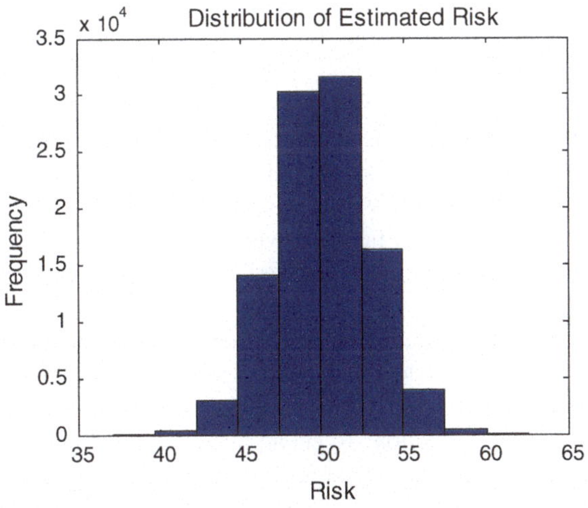

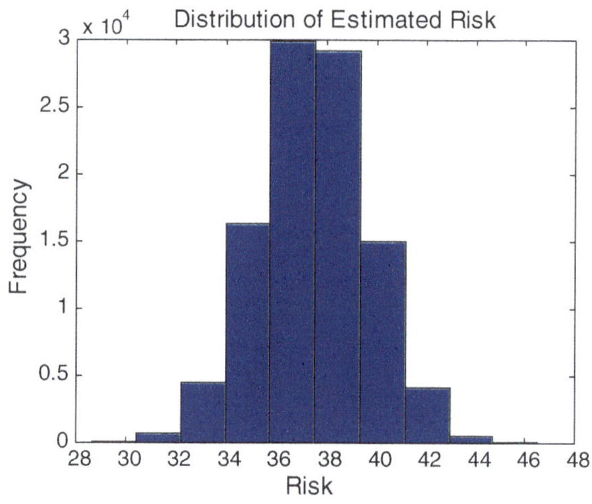

Fig. 14 Distribution of risk with reduced probability of failure for AM application

the packet loss, the risk associated can be minimized. Let us assume that because of the measures, the maximum probability of failure has come down to 0.75. The distribution of risk for the system with reduced probability of failure is shown in Fig. 14.

The distribution of risk for this system with reduced probability of failure can be approximated by a normal distribution having a mean of 37.5 and variance of 4.7. The estimated risk for this system is 37.5. Thus, improving the availability of the communication network reduces the overall risk associated with the network.

6.3 Electric Vehicular Application of SGCPS

The analysis equations for risk assessment of communication networks for electric vehicular application of SGCPS has been thoroughly presented in the previous sections. Now, we consider the risk assessment model for electric vehicular application of the SGCPS as shown in Fig. 15. We consider a case study where there are 10 electronic charging stations (ECS). These ECS are connected to each other and to the other units of EV system such as MGSS, EVGMCS using a local area network. The large scale application involves many such interconnected LANs. This has been incorporated by the fact that a LAN is connected to the other network as shown in Fig. 15.

This fundamental risk assessment model unit can be scaled in size and in quantity to get the estimate of risk for any other network. When the communication network between an ECS and MGSS fails partly or in full, it results in loss of information of one charging station. To get an overall estimate of the risk associated with communication network, we perform a Monte-Carlo simulation of 100,000 runs and plot the histogram for the overall risk, which is as illustrated in Fig. 16.

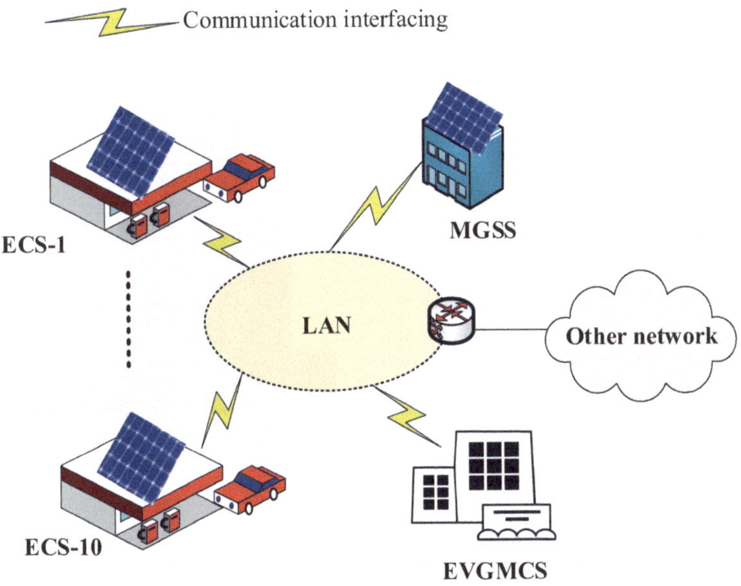

Fig. 15 Risk assessment model for electric vehicular application of SGCPS

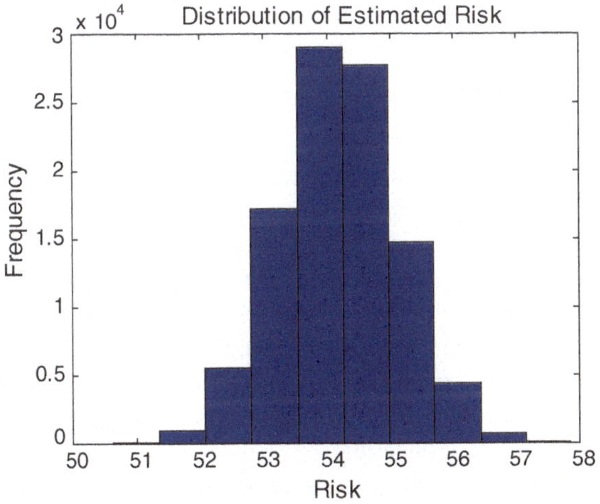

Fig. 16 Distribution of risk for EV application

The distribution of risk for this system can be approximated by a normal distribution having a mean of 54.2 and variance of 0.834. Thus, the estimated risk for this system is 54. By improving the redundancy of hardware and reducing the packet loss, the risk associated can be minimized. Let us assume that because of the measures, the maximum probability of failure has come down to 0.6. The distribution of risk for the system with reduced probability of failure is shown in Fig. 17.

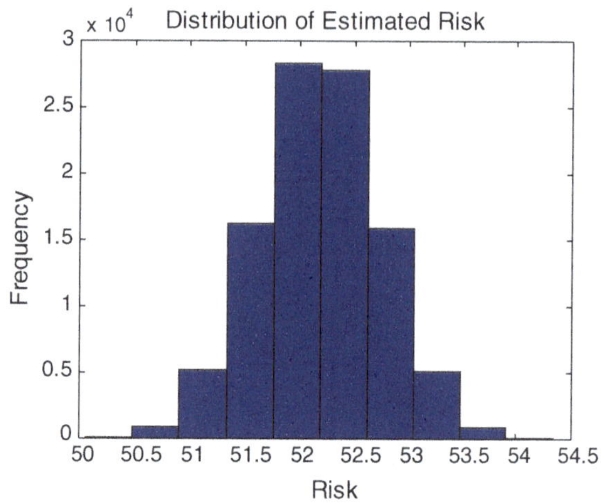

Fig. 17 Distribution of risk with reduced probability of failure for EV application

The distribution of risk for this system with reduced probability of failure can be approximated by a normal distribution having a mean of 52.2 and variance of 0.3. Hence, the estimated risk for this system is 52. Thus, improving the availability of the communication network reduces the overall risk associated with the network for electric vehicular application of the smart grid cyber-physical system.

7 Conclusion

Risk assessment of communication networks for different applications in a SGCPS is an important contribution to the power system design. In this chapter, a detailed discussion on the identification and assessment of the risk for three important applications, namely, the synchrophasor application, the advanced metering application and the electric vehicular application has been presented. Using suitable case studies, risk was estimated for these applications. The estimation was carried out using Monte-Carlo simulation assuming the probability of failure to be uniform. However, more realistic results can be obtained by choosing an appropriate distribution for these failure probabilities. The risk estimation by reducing the failure probability is also shown, which emphasize on designing the robust and reliable communication network. Thus, it has been identified that improving the availability of communication networks by reducing the hardware failures and packet losses can lead to a significant reduction in the risk.

References

1. Serpanos, D.: The cyber-physical systems revolution. Computer **51**(3), 70–73 (2018)
2. Edward, A.L., Sanjit, A.S.: Introduction to Embedded Systems: A Cyber-Physical Systems Approach, 1st edn. (2011)
3. Gunes, V., Peter, S., Givargis, T., Vahid, F.: A survey on concepts applications challenges in cyber-physical systems. KSII Trans. Internet Inf. Syst. (2015)
4. Rajkumar, R., Lee, I., Sha, L., Stankovic J.: Cyber-physical systems: The next computing revolution. In: Proceedings of 47th Design Automation Conference, pp. 731–736 (2010)
5. Framework for Cyber-Physical Systems. NIST Special Publication. https://s3.amazonaws.com/nist-sgcps/cpspwg/files/pwgglobal/CPS_PWG_Framework_for_Cyber_Physical_Systems_Release_1_0Final.pdf
6. Baheti, R., Gill, H., Samad, T., Annaswamy, A.M.: Cyber-physical systems. In: The Impact of Control Technology, pp. 161–166 (2011)
7. Kezunovic, M., Mccalley, J.D., Overbye, T.J.: Smart grids and beyond: Achieving the full potential of electricity systems. Proc. IEEE **100**(no. Special Centennial Issue), 1329–1341 (2012)
8. Leitão, P., Karnouskos, S., Ribeiro, L., Lee, J., Strasser, T., Colombo, A.W.: Smart agents in industrial cyber-physical systems. Proc. IEEE **104**(5), 1086–1101 (2016)
9. IEEE vision for smart grid controls: 2030 and beyond reference model. In: IEEE Vision for Smart Grid Control: 2030 and Beyond Reference Model, pp. 1–10, 12 September 2013
10. Yu, X., Xue, Y.: Smart grids: a cyber-physical systems perspective. Proc. IEEE **104**(5), 1058–1070 (2016)
11. NIST framework and roadmap for smart grid interoperability standards, release 3.0. NIST Special Publication 1108r3, https://www.govinfo.gov/content/pkg/GOVPUB-C13-fa3e5c3146fc29b2b9fd2fb8a1bc62f3/pdf/GOVPUB-C13-fa3e5c3146fc29b2b9fd2fb8a1bc62f3.sp.1108r3.pdf
12. Kwasinski, A., Weaver, W.W., Chapman, P.L., Krein, P.T.: Telecommunications power plant damage assessment for Hurricane Katrina—Site Survey and follow-up results. IEEE Syst. J. **3**(3), 277–287 (2009)
13. Pourbeik, P., Kundur, P.S., Taylor, C.W.: The anatomy of a power grid blackout—root causes and dynamics of recent major blackouts. IEEE Power Energy Mag. **4**(5), 22–29 (2006)
14. The Economic Impacts of the August 2003 Blackout, Electricity Consumers Resource Council (2004)
15. Phadke, A.G.: The wide world of wide-area measurement. IEEE Power Energy Mag. **6**(5): 52–65 (2008)
16. Phadke, A.G.: Synchronized phasor measurements in power systems. IEEE Comput. Appl. Power **6**(2), 10–15
17. Appasani, B., Mohanta, D.K.: Uncertainty analysis and risk assessment for effective decision-making using wide-area synchrophasor measurement system. In: Aleem, S.A., et al. (eds.) Decision Making Applications in Modern Power Systems, 1st edn., pp. 63–88. Academic Press (2020)
18. Phadke, A.G., Thorp, J.S.: Synchronized phasor measurements and their applications. Springer, New York (2008)
19. Smart Metering and Infrastructure Program Business Case, BC Hydro, Burnaby, BC, Canada, 2011. http://www.bchydro.com/etc/medialib/internet/documents/smi/smi_business_case.Par.0001.File.smi_business_case.pdf
20. Liu, X., Zhu, P., Zhang, Y., Chean K.: Cyber collaborative intrusion detection mechanism against false data injection attack in advanced metering infrastructure. IEEE Trans. Smart Grid **6**(5), 2435–2443 (2015)
21. Luan, W., Peng, J., Maras, M., Lo, J., Harapnuk, B.: Smart meter data analytics for distribution network connectivity verification. IEEE Transactions on Smart Grid **6**(4), 1964–1971 (2015)

22. Li, P., Tsai, C., Chen, C., Chen, P.: High connectivity, low power, low cost advanced metering infrastructure. In: 2018 3rd International Conference on Intelligent Green Building and Smart Grid (IGBSG), Yilan, pp. 1–3 (2018)
23. Jin, C., Tang, J., Ghosh, P.: Optimizing electric vehicle charging with energy storage in the electricity market. IEEE Trans. Smart Grid **4**(1), 311–320 (2013)
24. Jewell, N., Naber, J., McIntyre, M., Turner, M.: A power monitoring and control system to minimize electricity demand costs associated with Electric Vehicle charging stations. In: 2012 IEEE International Electric Vehicle Conference, Greenville, SC, pp. 1–5 (2012)
25. Liu, C., Chau, K.T., Wu, D., Gao, S.: Opportunities and challenges of vehicle-to-home, vehicle-to-vehicle, and vehicle-to-grid technologies. Proc. IEEE **101**(11), 2409–2427 (2013)
26. Wang Ren, L., Song, J., Hu, H., He, Q., Fang, S.: The state of the art of risk assessment and management for information systems. In: 2013 9th International Conference on Information Assurance and Security (IAS), Gammarth, pp. 66–71 (2013). https://doi.org/10.1109/isias.2013.6947735
27. Cronin, J.: Automated IP tracking system and Method. U.S. Patent Application No. 09/781, 362, 12 February 2001
28. Robert, S.: CenterTrack: an IP overlay network for tracking DoS floods. In: Proceedings of the USENIX Security Symposium, Denver, CO, USA, vol. 21, 14–17 October 2000
29. Fan, Y., Zhang, Z., Trinkle, M., Dimitrovski, A.D., Song, J.B., Li, H.: A cross-layer defense mechanism against GPS spoofing attacks on PMUs in smart grids. IEEE Trans. Smart Grid **6**, 2659–2668 (2015)
30. Jiang, X., Zhang, J., Harding, B.J., Makela, J.J., Domınguez-Garcıa, A.D.: Spoofing GPS receiver clock offset of phasor measurement units. IEEE Trans. Power Syst. **28**, 3253–3262 (2013)
31. Rana, S., Zhu, H., Lee, C.W., Nicol, D.M., Shin, I.: The not-so-smart grid: preliminary work on identifying vulnerabilities in ANSI C12.22. In: Proceedings of the IEEE Globecom Workshops, Anaheim, CA, USA, 3–7 December 2012
32. Ye, F., Qian, Y., Hu, R.Q.: A Security protocol for advanced metering infrastructure in smart grid. In: Proceedings of the IEEE Global Communications Conference, Austin, TX, USA, 8–12 December 2014
33. Yan, Y., Qian, Y., Sharif, H.: A secure and reliable in-network collaborative communication scheme for advanced metering infrastructure in smart grid. In: Proceedings of the IEEE Wireless Communications and Networking Conference, Cancun, Quintana Too, Maxco, 28–31 March 2011
34. Security Profile for Advanced Metering Infrastructure, AMI-SEC Task Force (UCAlug). http://osgug.ucaiug.org/utilisec/amisec/Shared%20Documents/AMI%20Security%20Profile%20(ASAP-SG)/AMI%20Security%20Profile%20-%20v1_0.pdf. Accessed 24 June 2010
35. Kenchington, H.S.: Deputy Assistant Secretary, Office of Electricity Delivery and Energy Reliability, Department of Energy, Smart Grid Cybersecurity Lessons Learned From More Than 11 Million Smart Meters Deployed. TCIPG Seminar, 2013. http://tcipg.org/sites/tcipg.org/files/slides/2013_03-01_Kenchington-TCIPG-FINAL-Revised.pdf. Accessed 16 June 2014
36. Sun, C.-C., Liu, C.-C., Xie, J.: Cyber-physical system security of a power grid: state-of-the-art. Electronics **5**, 40 (2016)
37. Liu, J., Xiao, Y., Li, S., Liang, W., Chen, C.L.P.: Cyber security and privacy issues in smart grids. IEEE Commun. Surv. Tutor. **14**, 981–997 (2012) [Google Scholar] [CrossRef]
38. Liu, Y., Hu, S., Ho, T.Y.: Vulnerability assessment and defense technology for smart home cybersecurity considering pricing cyberattacks. In: Proceedings of the 2014 IEEE/ACM ICCAD, San Jose, CA, USA, 3–6 November 2014
39. McLaughlin, S., Podkuiko, D., McDaniel, P.: Energy theft in the advanced metering infrastructure. In: Critical Information Infrastructures Security, pp. pp. 176–187. Springer, Berlin Heidelberg CY: Berlin/Heidelberg, Germany (2010) [Google Scholar]
40. IEEE802.15.4, IEEE Standard 802, part 15.4: Wireless Medium Access Control (MAC) and PHY Specifications for low rate Wireless Personal Area Networks (WPANs). http://ieeexplore.ieee.org/stamp/stamp.jsp?tp=&arnumber=4152704&isnumber=4152703. Accessed 2 August 2020

41. Datta, S.K., et al.: Vehiclesas connected resources: opportunities and challenges for the future. IEEE Vehic. Tech. Mag. **12**(2), 26–35 (2017)
42. Liu, H., Zhang, Y., Yang, T.: Blockchain-enabled security in electric vehicles cloud and edge computing. IEEE Netw. **32**(3), 78–83 (2018). https://doi.org/10.1109/MNET.2018.1700344
43. Zhou, Z., et al.: Software defined machine-to-machine communication for smart energy management. IEEE Commun. Mag. **55**(10), 52–60 (2017)
44. Hou, X., et al.: Vehicular fog computing: a viewpoint of vehicles as the infrastructures. IEEE Trans. Vehic. Tech. **65**(6), 3860–3873 (2016)
45. Zhang, et al.: Securing vehicle-to-grid communications in the smart grid. IEEE Wirel. Commun. **20**(6), 66–73 (2013)
46. Li, W.: Risk Assessment of Power Systems: Models, Methods, and Applications, pp. 313–350. Wiley, NJ (2014)
47. Appasani, B., Mohanta, D.K.: Co-optimal placement of PMUs and their communication infrastructure for minimization of propagation delay in the WAMS. IEEE Trans. Ind. Inform. **14**(5), 2120–2132 (2018)
48. Appasani, B., Mohanta, D.K.: Optimal placement of synchrophasor sensors for risk hedging in a smart grid. IEEE Sensors J. **17**(23), 7857–7865 (2017)

An Overview of Cybersecurity for Natural Gas Networks: Attacks, Attack Assessment, and Attack Detection

Zisheng Wang, Bining Zhao, and Rick S. Blum

Abstract Cyber technology is used frequently today to control and monitor many important physical systems. Such Cyber-Physical Systems (CPSs) are of great importance as they provide fundamental services to society. The focus here is on one particular Cyber-Physical System (CPS) focusing on the delivery of natural gas. Natural gas pipeline system operations rely heavily on industrial control systems (ICSs) and Supervisory Control and Data Acquisition (SCADA) systems. While the ICSs and SCADA systems introduce many advantages, they also introduce more vulnerabilities by providing opportunities for malicious cyber-attackers. However, academic investigations on cyber-attacks on gas systems have appeared only sparingly. Therefore, we intend to inform the community about this important topic that needs further study. We introduce the typical structure of natural gas systems and the natural gas markets. We also introduce the partial differential equations (PDEs) that describe the dynamics of the gas flows through pipelines. We focus on cyber-physical-attacks on natural gas systems that can be classified into three categories: man-in-the-middle attacks (MiMA), spoofing attacks, and topology attacks. We provide an overview of cyber-physical-attack detection approaches for gas networks. We provide a detailed discussion on the models, theories, and representative detection approaches for natural gas networks. Using numerical examples, we analyze the damage caused by three particular cyber-physical-attacks on natural gas systems. We also illustrate the performance of a representative attack detection approach.

Keywords Cybersecurity · Natural gas networks · Attack detection · Attack assessment

Z. Wang (✉) · B. Zhao · R. S. Blum
Lehigh University, Bethlehem, USA
e-mail: zs.wang.prc@gmail.com

B. Zhao
e-mail: bbz5089@psu.edu

R. S. Blum
e-mail: rblum@lehigh.edu

A. I. Awad et al. (eds.), *Security in Cyber-Physical Systems*, Studies in Systems, Decision and Control 339, https://doi.org/10.1007/978-3-030-67361-1_9

Abbreviations

ICS	Industrial Control System
SCADA	Supervisory Control and Data Acquisition
PDF	Partial Differential Equation
MiMA	Man-in-the-Middle Attacks
LDC	Local Distribution Company

1 Introduction and Background

Cyber technology is used frequently today to control and monitor many important physical systems. Such Cyber-Physical Systems (CPSs) are of great importance as they provide fundamental services to society. In this chapter, we demonstrate the security of one particular Cyber-Physical System (CPS) focusing on the delivery of natural gas. The operation and control of natural gas systems are increasingly dependent on cyber-infrastructure, typically in the form of industrial control systems (ICSs) and Supervisory Control and Data Acquisition (SCADA) systems. The increased usage of cyber-infrastructure introduces more vulnerabilities into natural gas systems by providing opportunities for malicious cyber-attackers. However, the deployment of cybersecurity measures in the industry isn't keeping pace with the growth of digitalization in oil and gas operations [1], which may make the natural gas system even more vulnerable to cyber-attacks.

Cybersecurity, including cybersecurity in energy systems, has been described as a very dangerous concern by the U.S. Department of Homeland Security (DHS) [2]. In a workshop hosted by the Department of Energy Office of Science (DOE/SC) [3], it was reported that security, reliability, and resilience mechanisms are often not supported over the complete data life cycle in networks, and traditional security mechanisms, such as firewalls, cannot meet current network performance requirements [4]. Therefore, the Advanced Scientific Computing Research (ASCR) program of DOE recognizes that the growing number of sensors in the field should be protected, mechanisms to identify data corruption and its source should be designed, and advanced cyber security mechanisms that protect both data and infrastructure should be investigated [4]. The Transportation Security Administration (TSA) also provides gas pipeline security guidelines for industry practice [5]. Currently, the natural gas industry uses a number of standards and protocols to improve the resilience and security of crucial components in natural gas systems [6]. These standards include but are not limited to the NIST cybersecurity framework [7], API Standard 1164 [8], and NISTIR 7628 [9]. However, the cyber-security environment which currently exists in the natural gas industry raises concerns. Roughly, 67% of oil and gas companies believe that the risk level to industrial control systems has significantly increased because of cyber-threats, and their operations have had security compromises that resulted in the loss of confidential information or operational technology disruption [1].

Damages caused by attacks and failures on pipeline networks can be extremely severe. For example, transmission pipeline explosions in San Bruno, California in September, 2000, and in Allentown, Pennsylvania in February 2011 caused eight and five casualties respectively [10]. Two pipeline explosion accidents killed six and two people in January, 2002 and in April, 2004 in China [11]. A leak at the Aliso Canyon natural gas storage facility was discovered in late 2015 and operations were not upgraded to normal status until late 2017. The leak affected more than 70% of the local electricity generating capacity [12]. There are other examples listed in [13].

On the other hand, interdependencies between the natural gas and electric power systems may make the cyber-security of the natural gas systems even more crucial. Natural gas-fired generation units serve as the main connections between the natural gas and electric power systems. Their generation capacity accounts for 44% of total installed generation capacity in the United States in 2017 [14], and over 60% of generation capacity additions in 2018 [15]. Therefore, a dependable fuel-supply for natural gas-fired units is crucial for the reliable and secure operation of electric power systems employing a significant number of natural gas-fired generating units. Recent studies have attempted to investigate the interdependencies in the natural gas and electric power systems to enhance the security, reliability, and economy of the electricity systems. Some studies focus on the coordinated operation of the two systems [16–20], while other studies investigate the coordinated planning of the two systems [12, 21–24].

However, in addition to the coordination issues between the two energy systems that are investigated in these studies, attacks on the natural gas systems can be another threat to electric power systems [25–27]. In [25], the authors investigated how the failures of natural gas pipelines, power transmission lines, and/or connections between gas systems and natural gas-fired units can impact the operations in an interdependent gas-electric system. The authors in [25] propose a robust optimization model to protect the most vulnerable components in the system. Their work focuses on physical component failure without specifying the source of the attacks. We claim that such component failure in an energy system can be caused by either physical- or cyber- attacks [28]. In [26], the authors propose a false-data-injection attack targeting the information on natural gas availability to natural gas-fired units. As stated in [26], because of the gas-supply contracts held by the natural gas-fired units and the lack of coordination between the natural gas and electric power systems, false natural gas availability information is hard for the electricity system operators to detect. These types of cyber-attacks can increase the operational cost of the electricity systems and even cause unserved energy in extreme cases. To mitigate the impact of false-data-injection attacks, the authors propose a screening methodology that allocates a limited budget for best protecting critical fuel supply information. The authors in [26] also propose strategies to sign firm supply contracts to reduce natural gas supply uncertainties through a tri-level optimization problem. Due to the interdependency between between natural gas and electricity systems, the existing literature shows that attacks that result in either natural system component failure or gas-supply curtailments for gas-fired units are likely to affect the operation of electric power systems. Therefore, the study of attacks and associated mitigation strategies in natural gas systems is important for not only the security of natural gas systems but also the reliability of electric power systems.

Unfortunately, the theory of cyber-attacks on natural gas systems remains under investigated. The security risks of a wireless sensor network employed at a gas compression station is investigated in [29] and the results show that attacks on sensor availability and on sensor data can lead to compressor station failures. In [30], the resilience of natural gas pipeline systems is studied under two types of cyber-physical attacks: a pressure measurement integrity attack and a cyber-attack that propagates from the electric power systems to the natural gas system. Vulnerabilities and communication security threats in the SCADA system of an oil pipeline are described in [31] along with possible solutions for securing communications. A multi-objective optimization model is proposed to minimize the loss and recovery time from pipeline failures in [32]. While [32] does not focus on cyber-attacks, pipeline failures can be caused by cyber-attacks. In general, the existing literature lacks mathematical models, quantitative analysis, detection approaches and mitigation approaches for cyber-attacks on natural gas systems. It should be noted that this is not the case for electrical networks, where simple linear models are typically employed. For gas networks, the accepted model incorporates the Weymouth equation which is quadratic. As all the existing work on detection, mitigation, and quantitative analysis are heavily based on the system model, this makes all the electrical work unusable for gas networks. Even nonlinear electrical models, which are seldom used, are far different from the Weymouth equation making it impossible to use the model-based electrical cyber attack detection approaches in gas networks.

Therefore, in this chapter we want to highlight the importance of cyber-security studies for natural gas systems, provide an overview of some cyber-attacks that can occur in natural gas systems, and evaluate the damage that these cyber-attacks can cause in the natural gas systems through numerical examples. We hope this chapter can help to attract further investigation into this field. We need to understand the impacts of cyber-attacks on natural gas systems, the detection and protection approaches, and many related topics.

The remainder of this chapter is organized as follows. Sections 2 and 3 describe the physical structure of an entire natural gas system, including an overview of the natural gas system market. Section 4 describes models for the gas flow dynamics in the pipelines. Section 5 describes an accepted natural gas steady-state operation model. Section 6 introduces some possible types of cyber-attacks that can be launched on natural gas transmission systems. Section 7 overviews some cyber physical attack detection approaches on natural gas networks. Section 8 provides a statistical-based cyber-attack detection approach for natural gas networks. Section 9 presents some numerical examples showing the damage provided by cyber-attacks introducing incorrect system state measurements. Section 10 provides conclusions and future study recommendations. In every section, we provide references and discussions on the state-of-the-art work that relates to the the topics discussed in that section.

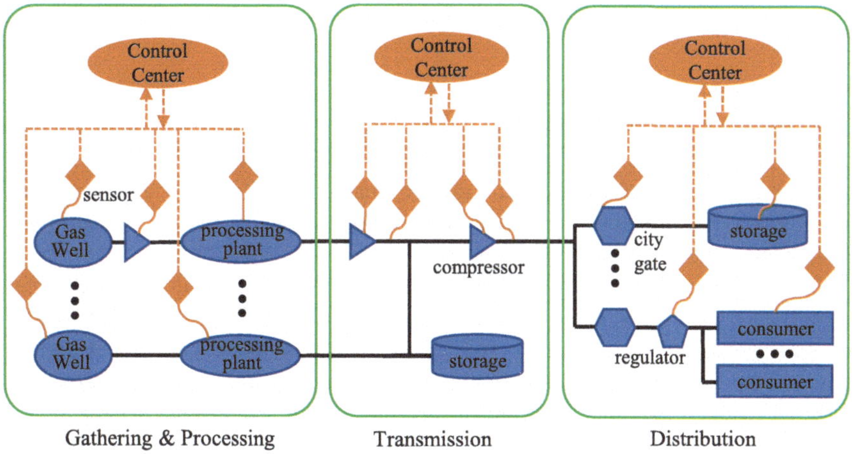

Fig. 1 Illustration of integrated gas-cyber systems

2 Physical Structure of Natural Gas System

The natural gas system consists of three sections: a gathering and processing section, a transmission section, and a delivery section [33]. In the gathering and processing section, raw natural gas extracted from multiple gas wells will be collected, compressed and sent to a processing plant. After being cleaned by processing plants, clean and dry gas will be compressed and sent into long-distance transmission gas pipelines to be transported in the transmission section. Due to frictional losses in these long-distance pipelines, compression stations are installed along the transmission section to increase gas pressure. The transmission section ends once the gas reaches city gates, and the system after a city gate is referred to as a distribution system. Natural gas will be metered at city gates, then the gas will be either delivered to consumers by local distribution companies (LDCs), or stored in gas storage facilities for future use. Some large consumers like large capacity gas-fired power plants are connected to transmission pipelines directly [33]. In the distribution system, gas pressure will be reduced by pressure regulators to certain levels so that the gas can be delivered to consumers.

Compression stations and pressure regulators are controlled and adjusted to maintain the appropriate gas pressure level and meet the demand of consumers. The states of the gas system are monitored by different entities in different sections of the system. These states include composition, temperature, gas pressure, gas flow rate, as well as the status of gates, compressors, and regulators. The operation of a natural gas network is controlled by a cyber-system. Sensors that are installed at certain points of the gas system to collect measurements, and then the measurement data will be transmitted back to a central control center called a SCADA system. Control signals will then be sent from the central control center to compression stations or regulators to adjust their ratios. The integrated gas-cyber systems are shown in Fig. 1.

3 Natural Gas Market Overview

The natural gas market in the United States does not have a central controller like the independent system operator (ISO) in the electricity market. Natural gas can be traded in several ways from a producer to end-users. Some producers are able to sell their gas directly to local distribution companies (LDCs) or to large industrial consumers; other producers sell their gas to marketers, who are responsible to sell gas to LDCs or to consumers directly [34].

There are three types of markets in the natural gas industry: the commodity gas market, the transmission capacity market, and the financial market. On-system consumers (consumers who connect to pipelines directly), marketers, and other participants contact pipeline operators to nominate capacity, specifying the demand of gas, the point of injection, and the point of receipt in the transmission capacity nomination process. Pipeline operators schedule deliveries with a priority order which is determined by the types of contracts the participants hold. Pipeline operators are federally regulated and do not participate in the buying, selling, or processing of natural gas, instead, they only provide transportation service, in which they transport gas on behalf of buyers and sellers of the gas [34]. In the commodity market, market participants trade natural gas on a spot basis every day at specified trading points [34]. There is also a monthly spot market that occurs in the "bid week", which is the last five business days in a month [34]. In addition to these spot markets, many consumers purchase gas through longer-term contracts. The financial market involves derivatives and sophisticated financial instruments in which the buyer and seller never take physical delivery of the natural gas, thus these aspects are out of the scope of our discussion.

There are three types of contracts in the natural gas market, interruptible contracts, baseload contracts, and firm contracts. Interruptible contracts have the lowest priority to use transmission capacity, while firm contracts holders have the highest priority. Neither the buyers nor the sellers with interruptible contracts and baseload contracts are obligated to deliver or receive the exact volume specified, however, both parties in a firm contract are legally obligated to either receive or deliver the amount of gas specified in the contract [35].

4 Gas Pipeline Dynamics

An accurate model that describes gas flow through pipelines typically employs a set of complex partial differential equations (PDEs).

- The pipeline system is an isothermal system, which means the gas in pipelines has constant temperature over space and time.
- The gas in a pipeline has slow transients which do not excite waves or shocks.

Under these assumptions, the gas flow in a gas network can be described by the simplified Euler equations in one dimension

$$\partial_t \rho + \partial_x \phi = 0, \tag{1}$$

$$\partial_t \phi + a^2 \partial_x \rho = -\frac{\lambda}{2D} \frac{\phi |\phi|}{\rho}, \tag{2}$$

where ρ stands for gas density, ϕ stands for gas mass flux, t represents time, $x \in [0, L]$ is the pipeline length, λ represents the friction factor of the pipeline, D is the pipeline diameter, and a is the speed of sound in a gas pipeline. Given the assumptions, the speed of sound in a gas pipeline, a, is a constant. The relationship between gas pressure p and density ρ is $p = a^2 \rho$, and the relationship between gas mass flux ϕ and gas flow φ is $\varphi = A\phi$, where A is the cross-sectional area of the pipeline.

The model can be further simplified in a steady state gas system, whose pipeline boundary conditions in terms of pressure and flow change slowly, so that the rates of changes of gas density ρ and mass flux ϕ with respective to time are zero. Then (1) and (2) can be simplified to

$$\frac{d\phi}{dx} = 0, \tag{3}$$

$$a^2 \frac{d\rho}{dx} = -\frac{\lambda}{2D} \frac{\phi |\phi|}{\rho}. \tag{4}$$

Equation (3) implies that the gas mass flux is constant along a gas pipeline. Equation (4) indicates that the gas density change along a gas pipeline is determined by the gas mass flux. Integrating equation (4) along x yields

$$\underline{\rho}^2 - \overline{\rho}^2 = \frac{\lambda L}{D a^2} \phi |\phi|, \tag{5}$$

where $\underline{\rho}$ and $\overline{\rho}$ stand for the gas density at $x = 0$ and $x = L$, respectively. Instead of using gas density ρ and flux ϕ, Eq. (5) can also be represented by using gas pressure p and gas flow ψ as

$$\underline{p}^2 - \overline{p}^2 = \frac{a^2 \lambda L}{D A^2} \psi |\psi|. \tag{6}$$

For the sake of simplicity, we define $\beta = \frac{a^2 \lambda L}{D A^2}$. Note that β is a constant for a particular pipeline.

5 Gas System Steady-State Operation Model

As discussed in Sect. 4, the steady-state physics of natural gas flow in gas pipelines is modeled using Eq. (6), which is also referred as the Weymouth equation

$$\underline{p}_{(i,j)}^2 - \overline{p}_{(i,j)}^2 = \beta_{(i,j)}\psi_{(i,j)}|\psi_{(i,j)}|, \forall (i, j) \in \Lambda, \tag{7}$$

where (i, j) represents a gas pipeline with starting and ending nodes i and j, and Λ is the set of pipelines in the system. Gas flow balance at each gas node i requires

$$s_i - d_i + d_i^{\text{shed}} = \sum_{j \in \underline{\Lambda_i}} \psi_{(j,i)} - \sum_{j \in \overline{\Lambda_i}} \psi_{(i,j)}, \ \forall i \in N, \tag{8}$$

where s_i, d_i, and d_i^{shed} denote the gas supply, gas demand, and unserved gas demand at node i, respectively. In (8), $\underline{\Lambda_i}$ and $\overline{\Lambda_i}$ represent the sets of pipelines with node i as its starting node and ending node, respectively.

Gas pipelines that are equipped with compression stations at the starting/ending nodes are called active pipelines, while pipelines without compression stations are referred to as passive pipelines. The relationship between the pipeline ending pressure and the nodal pressure is

$$\underline{p}_{(i,j)} = \underline{\alpha}_{(i,j)} p_i, \ \forall (i, j) \in \Lambda, \tag{9}$$

$$\overline{\alpha}_{(i,j)} \overline{p}_{(i,j)} = p_j, \ \forall (i, j) \in \Lambda, \tag{10}$$

where $\underline{\alpha}_{(i,j)}$ is the compression boost ratio of the compressor installed at the starting node of a pipeline, while $\overline{\alpha}_{(i,j)}$ is the compression boost ratio of the compressor installed at the ending node of pipeline. We have $\underline{\alpha}_{(i,j)} = \overline{\alpha}_{(i,j)} = 1$ when (i, j) is the passive pipeline.

Finally, gas nodal pressures, pipeline ending pressures, gas supplies and compression boost ratios are bounded by the following equations due to physical operation limits

$$p_{(i,j)}^{\min} \leq \underline{p}_{(i,j)} \leq p_{(i,j)}^{\max}, \ \forall (i, j) \in \Lambda \tag{11}$$

$$p_{(i,j)}^{\min} \leq \overline{p}_{(i,j)} \leq p_{(i,j)}^{\max}, \ \forall (i, j) \in \Lambda \tag{12}$$

$$p_i^{\min} \leq p_i \leq p_i^{\max}, \ \forall i \in N \tag{13}$$

$$s_i^{\min} \leq s_i \leq s_i^{\max}, \ \forall i \in N \tag{14}$$

$$\underline{\alpha}_{(i,j)}^{\min} \leq \underline{\alpha}_{(i,j)} \leq \underline{\alpha}_{(i,j)}^{\max}, \ \forall (i, j) \in \underline{\Lambda}^{A} \tag{15}$$

$$\overline{\alpha}_{(i,j)}^{\min} \leq \overline{\alpha}_{(i,j)} \leq \overline{\alpha}_{(i,j)}^{\max}, \ \forall (i, j) \in \overline{\Lambda}^{A} \tag{16}$$

The upper bounds in Eqs. (11)–(16), confining the gas system operation model, are stricter than the designed limitations of a pipelines in order to keep enough margin and avoid internal explosion.

It is convenient to write the gas system operation model (7)–(16) in a vector and matrix form for detection and mitigation applications. Let N_s, N_d and E denote the number of supplier nodes, demand nodes, and pipelines, respectively. Let p_i denotes the gas pressure at node i. Let us number the pipelines in the set Λ using the index

$l = 1, 2, \ldots, E$. Let ψ_l and β_l denote the gas flow and pipeline constant for the lth pipeline. When indexing nodes, we start with the supplier nodes followed by the demand nodes. Define

$$\boldsymbol{p} = (p_2, \ldots, p_N)^T, \tag{17}$$
$$\boldsymbol{\psi} = (\psi_1, \psi_2, \ldots, \psi_E)^T, \tag{18}$$

and

$$\boldsymbol{q} = (-s_2, \ldots, -s_{N_s}, d_1, d_2, \ldots, d_{N_d})^T. \tag{19}$$

Note that the pressure of the first node is excluded in (17) since the first node is chosen as the reference node and its pressure is assumed to be known. Similarly, the gas supply of the first node is excluded in (19). The natural gas network is modeled as a directed graph. We also define $\overline{\boldsymbol{\alpha}} = (\overline{\alpha}_1, \overline{\alpha}_2, \ldots, \overline{\alpha}_E)^T$, $\underline{\boldsymbol{\alpha}} = (\underline{\alpha}_1, \underline{\alpha}_2, \ldots, \underline{\alpha}_E)^T$ and $\boldsymbol{\beta} = (\beta_1, \beta_2, \ldots, \beta_E)^T$. If the flow in pipeline l is directed from node i to node j where $(i, j) \in \Lambda$ then the (l, n) entry of the $L \times N$ incidence matrix $\widetilde{\boldsymbol{S}}$ is defined as

$$\left[\widetilde{\boldsymbol{S}}\right]_{l,n} = \begin{cases} 1, & n = i \\ -1, & n = j \\ 0, & \text{otherwise} \end{cases}, \tag{20}$$

with $l = 1, 2, \ldots, E$, $n = 1, \ldots, N$ and $0 \le i, j \le N$. Isolating the first column corresponding to the reference node, the matrix $\tilde{\boldsymbol{S}}$ can be partitioned as $\tilde{\boldsymbol{S}} = (\boldsymbol{s},\ \boldsymbol{S})$. Let $\boldsymbol{\omega} = (\boldsymbol{p}^T,\ \boldsymbol{\psi}^T)^T$ denote the state of the natural gas network. Therefore, the gas system operation model (7)–(10) can be represented in vector and matrix form as

$$\boldsymbol{S}^T\boldsymbol{\psi} = \boldsymbol{q}, \tag{21}$$
$$\boldsymbol{B}(\boldsymbol{p} \odot \boldsymbol{p}) = \boldsymbol{\beta} \odot \boldsymbol{\psi} \odot |\boldsymbol{\psi}| + \boldsymbol{b}p_0^2, \tag{22}$$
$$\boldsymbol{B} \triangleq \text{diag}\{\underline{\boldsymbol{\alpha}}\}[\boldsymbol{S}]_+ - \text{diag}\{\overline{\boldsymbol{\alpha}}\}[-\boldsymbol{S}]_+, \tag{23}$$

and

$$\boldsymbol{b} \triangleq \text{diag}\{\underline{\boldsymbol{\alpha}}\}[\boldsymbol{s}]_+ - \text{diag}\{\overline{\boldsymbol{\alpha}}\}[-\boldsymbol{s}]_+, \tag{24}$$

where $\odot$ denotes the element-wise product and $[x]_+ = \max(0, x)$ is an element wise operator.

6 Categorization of Cyber-Physical-Attacks on Natural Gas Systems

In this section, we focus on categorizing cyber-physical-attacks on natural gas systems. First we acknowledge that attackers who threaten the cyber-security of natural gas systems can be classified into the two groups,

1. **Insiders**: Based on the investigation in [1], the top cyber-security threat is launched by careless or malicious insiders. Such individuals are extremely dangerous. Insiders have full knowledge and control of the networks. They can manipulate the data such that attacks can not be detected by an intrusion detector deployed in the natural gas system. There are studies that investigate insider attacks in detail [36] but their focus is broader than just natural gas networks.
2. **Outsiders**: These attackers also threaten natural gas network operations, although these attacks are generally more difficult to launch. These attacks might target the natural gas network information system or they may manipulate sensor data or communications. These attacks cause inappropriate operation or physical failure of natural gas systems. Several natural gas companies recently reported attempts to disrupt critical operation systems and communication systems [37].

As mentioned in Sect. 1, there are few studies of cyber-attacks on natural gas systems. However, different types of cyber-attacks on electric power systems have been investigated in depth in recent years [30, 38–53], so we also provide references for electric power system examples in this section. Although the operational model of a natural gas system is very different from that of an electric power system, the types of cyber-attacks that can occur in the electric power system are similar to those which can occur in the natural gas systems [54]. In this chapter, we focus on cyber-physical-attacks which can be grouped into the three following categories.

1. **Man-in-the-middle attack (MiMA)**: A MiMA is launched by a cyber-attacker intercepting a communication packet and changing its contents [30, 50, 52]. Examples are replay attacks [44], which replace real-time data with previously acquired data, or false data injection attacks, in which an attacker compromises measurements sent from sensors in order to introduce errors into the state estimation algorithm [49]. Data contamination caused by such attacks may lead to financial misconduct and loss of profit in energy markets [39, 46] as well as inappropriate system control or degraded performance. Due to the interdependencies between the natural gas and electricity systems, such attacks on natural gas information can lead to increased operational costs in the electricity system and even electricity load shedding in extreme cases. The attacks in this category can cooperate with physical attacks, such as intentional sabotage of pipeline systems and alter corresponding measurements at the same time, so that the physical damage could be covered by manipulated data [42]. Although recently developed meters

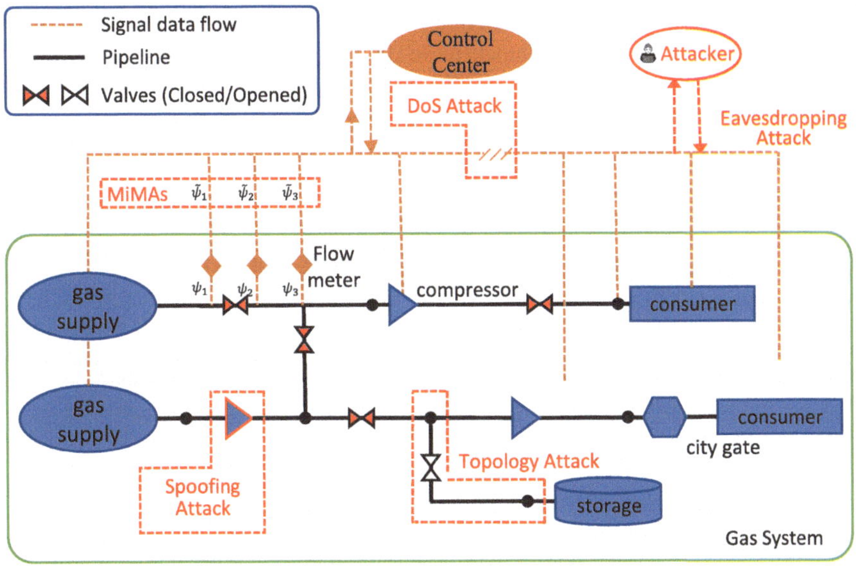

Fig. 2 The location of cyber attacks on a natural gas distribution network

may limit data manipulation, these systems are still vulnerable, especially if some nodes do not use secured meters [43, 55]. In Fig. 2, the flow measurements are manipulated by MiMAs so that the operator has an incorrect knowledge of the gas supply.

2. **Spoofing attacks**: A spoofing attack directly modifies the data measured by a sensor [41, 52]. GPS spoofing attacks are one example where the attacker generates a fake GPS signal. GPS spoofing attacks can also impact phasor measurement units (PMU) in the electrical grid [56] since some PMUs use GPS to time stamp data. Some other spoofing attacks are discussed in [44, 57]. In Fig. 2, the pressure measured at the compressor is spoofed. The spoofed compressor measurement will indicate a smaller gas pressure than the operator desires. In order to balance the gas supply, the operator will increases the gas pressure causing the compressor to operate in a dangerous range defined in (15) and (16).
3. **Topology attack**: In a topology attack, the operator believes the topology is different from the actual one due to malicious modification of a data base or corrupted communications [53, 58]. Topology attacks can be launched in combination with MiMAs or physical attacks, and have been studied in electricity systems [48, 51]. In Fig. 2, the operator is led to believe that one pipeline valve is open when it is actually closed. Given the incorrect topology, the operator will employ inappropriate control.

Note that any attacks falling under any of these categories can be either an insider or outsider attack. Denial of Service (DoS) attacks are a special case of spoofing attacks. For readers interested in DoS attacks, see [59, 60]. Here we only consider attacks which attempt to disrupt system operation directly. We note that eavesdropping attacks might be used to launch more directed attacks but they still must be employed with some other mechanisms to cause damage to system operation. For eavesdropping attacks, see [61, 62]. DoS attacks and eavesdropping attacks are shown in Fig. 2 for illustration.

7 An Overview of Cyber-Physical-Attack Detection on Natural Gas Networks

In this section, we give an overview of cyber attack detection approaches for natural gas networks. We also provide references for detection approaches of general SCADA systems. Based on the techniques used, we categorize the detection approaches as: data-based approaches [63–70], model-based approaches [71–76], and combined approaches [53, 77].

7.1 Data-Based Approaches

The data-based approaches make decisions based on hidden characteristics learned from data by using machine learning and pattern recognition algorithms. These methods do not employ operational models of natural gas networks, like (7)–(16). Some of these methods also require little understanding of the system operation.

In [64, 66, 68, 78] the authors evaluate the performance of general machine learning algorithms for detecting MiMAs. The authors of [64] show the K-means algorithm achieves the best performance among clustering algorithms for some specific attacks under some specific conditions of a gas compressor control system. The authors in [66] show that the random forest algorithm achieves the best performance among naive bayes, random forests, nearest-neighbor-like algorithms, and support vector machine algorithms for some different specific attacks and some different specific operating conditions. In [63], a neural network based algorithm, which detects if the command or pressure data are manipulated by MiMAs, is proposed. This algorithm achieves at least 84.9% accuracy for detecting MiMAs for the cases tested. The authors of [67] propose a long short term memory learning network based a softmax classifier to detect manipulation of data. The proposed network performs better than the Gaussian mixture model, principle component analysis, support vector data description model, and isolation forest model.

However, since natural gas networks are critical infrastructure, it is difficult to halt the system to acquire enough labeled data which covers many attacks and operation conditions. Generally the only available data for learning is acquired without attacks during normal operation. Therefore, one class classification approaches, which only employ unattacked training data, are used for natural gas networks in [69, 70]. One class classification only requires learning data under normal operation. However, in [67], the authors show that one class classification has a lower performance than a neural network based classifier. In fact, employing the statistic of only one of several possible hypotheses is generally known to reduce performance [79] based on mathematical proofs. Moreover, the performance of data-based approaches is not guaranteed for a natural gas network with operating conditions or an attack which is different from the that used to obtain the training dataset which is important, since it is typically not possible to obtain data for all attacks and operating conditions.

7.2 Model-Based Approaches

Model-based approaches identify the presence of cyber attacks by checking if the measurements match a model of the network. Based on a linear dynamic system model, the finite moving average test [73], the sequential probability ratio test of Wald [71], and the residue-based test [72] have been studied to detect MiMAs under some linearization assumptions. To employ the approaches in [71–73, 80], we can use techniques proposed in [24] to linearize (7)-(16) under some assumptions that limit the size of signal changes.

Since liquid flowing in a pipeline follows the same mathematical equations as gas in a pipeline, the model-based algorithm [74] developed for water distribution systems is also of interest. The authors of [74] propose a detection algorithm which checks if sensor measurements match the reference values generated by a water hydraulics model. Moreover, natural gas leak detection algorithms [76] are also useful since a leak may be the outcome of a cyber attack. The model-based approaches will achieve reliable performance when the model accurately describes the natural gas network systems.

7.3 Combined Approaches

Utilizing the model of a natural gas network with the help of massive data can achieve better performance than using only the data-based approach or the model-based approach. The authors of [81] propose a leakage detector by combining the gas network model and a neural network. The authors aim to apply the approach to natural gas networks but the linear model they employ is not frequently used in practice for gas networks [82, 83]. In [53], the authors propose an algorithm to detect topology attacks on a natural gas network by modifying a classical statistical approach using noisy sensor measurements and the model in (21)–(24) while also learning unknown model parameters. Given a sufficient amount of data the algorithm can achieve good performance. In [58], the authors extend the work in [53] by providing a new closed-form expression for the asymptotic performance of the algorithm that is accurate for the typical amount of data required to achieve acceptable performance.

In the following section, we provide a more detailed discussion of the approach in [53] which is based on the most popular model for gas networks [82].

8 Models and Theory of Cyber-Physical Attacks and Illustrative Detection Algorithms

In this section, we describe representative approaches for the detection of topology attacks, MiMAs, and spoofing attacks on natural gas networks. We first introduce a steady-state model for the sensor measurements.

8.1 Sensor Measurement Model

Recall the model developed in (21)–(24) for the gas network. We define three variables that describe if measurements from the flow sensor on the lth pipeline, the pressure sensor on the nth node, or the injection/withdraw sensor on the nth node will be employed as

$$\delta_{\psi,l} = \begin{cases} 1, & \text{if the } l\text{th pipeline employs a flow sensor} \\ 0, & \text{otherwise} \end{cases} \tag{25}$$

$$\delta_{p,n} = \begin{cases} 1, & \text{if the } n\text{th node employs a pressure sensor} \\ 0, & \text{otherwise} \end{cases} \tag{26}$$

and

$$\delta_{q,n} = \begin{cases} 1, & \text{if the } n\text{th node employs an injection sensor} \\ 0, & \text{otherwise} \end{cases} \tag{27}$$

where $l = 1, \dots, E$ and $n = 2, \dots, N$. Define $\boldsymbol{\delta} = (\delta_{p,2}, \dots, \delta_{p,N}, \delta_{q,2}, \dots, \delta_{q,N}, \delta_{\psi,1}, \dots, \delta_{\psi,E})^T$.

At time $k = 1, 2, \dots, K$, we can (depending on (25)–(26)) observe noisy versions of the deterministic gas pressure vector $\boldsymbol{p}$, the injection/withdraw vector $\boldsymbol{q}$, and the gas flow vector $\boldsymbol{\psi}$ from (17)–(19) so that

$$\widetilde{\boldsymbol{p}}_k = \boldsymbol{p} + \boldsymbol{n}_{p,k}, \tag{28}$$

$$\widetilde{\boldsymbol{q}}_k = \boldsymbol{q} + \boldsymbol{n}_{q,k}, \tag{29}$$

and

$$\tilde{\boldsymbol{\psi}}_k = \boldsymbol{\psi} + \boldsymbol{n}_{\psi,k}, \tag{30}$$

where $\boldsymbol{n}_{p,k}$, $\boldsymbol{n}_{q,k}$, and $\boldsymbol{n}_{\psi,k}$ denote zero-mean noise variables with standard deviation vectors $\boldsymbol{\sigma}_p$, $\boldsymbol{\sigma}_q$ and $\boldsymbol{\sigma}_\psi$, respectively. Since $\boldsymbol{q} = \boldsymbol{S}^T\boldsymbol{\psi}$, the pressure vector $\boldsymbol{p}$ and the flow vector $\boldsymbol{\psi}$ are sufficient to determine any of the components in $(\boldsymbol{p}, \boldsymbol{\psi}, \boldsymbol{q})$. Define the noise vector as $\boldsymbol{n}_k = (\boldsymbol{n}_{p,k}^T, \boldsymbol{n}_{q,k}^T, \boldsymbol{n}_{\psi,k}^T)^T$. Let $\boldsymbol{y}_k = (\tilde{\boldsymbol{p}}_k^T, \tilde{\boldsymbol{q}}_k^T, \tilde{\boldsymbol{\psi}}_k^T)^T$ denote the potential sensor measurements at time k determined by the state $\boldsymbol{\omega} = (\boldsymbol{p}^T\ \boldsymbol{\psi}^T)^T$. Define the measurement matrix as

$$\boldsymbol{H} = \begin{pmatrix} \boldsymbol{I}_{N-1} & \boldsymbol{0} \\ \boldsymbol{0} & \boldsymbol{S}^T \\ \boldsymbol{0} & \boldsymbol{I}_E \end{pmatrix}, \tag{31}$$

where $\boldsymbol{I}_{N-1}$ and $\boldsymbol{I}_E$ denote $(N-1)\times(N-1)$ and $E \times E$ identity matrices, respectively. Then, the compact model of the potential sensor measurements (28)–(30) is

$$\boldsymbol{y}_k = \boldsymbol{H}\boldsymbol{\omega} + \boldsymbol{n}_k. \tag{32}$$

Let $\boldsymbol{y} = \left[\boldsymbol{y}_1^T \ \dots \ \boldsymbol{y}_K^T\right]^T$ denote the full vector of the potential sensor measurements. Let $\tilde{\boldsymbol{y}}$ denote a vector of all elements of $\boldsymbol{y}$ which have a sensor measuring them. We make the following assumption throughout the chapter.

Assumption Let $\boldsymbol{n}_1, \boldsymbol{n}_2, \dots, \boldsymbol{n}_K$ denote an independently distributed sequence of zero mean Gaussian random variables.

Let $p(x|\mu, \sigma^2)$ denote a Gaussian probability density function (pdf) with the mean μ and the variance σ^2. Denote the pdf of the noisy sensor measurements $\tilde{\boldsymbol{y}}$ as $g(\tilde{\boldsymbol{y}}|\boldsymbol{\theta})$ which is parameterized by $\boldsymbol{\theta} = (\text{vec}\,(\boldsymbol{S})^T, \boldsymbol{\omega}^T, \boldsymbol{e}^T)^T$ where vec $(\boldsymbol{S})$ denotes the vector of the stacked columns of $\boldsymbol{S}$. We have

$$g(\tilde{\boldsymbol{y}}|\boldsymbol{\theta}) = \prod_{k=1}^{K} \left\{ \prod_{\substack{n=1 \\ \boldsymbol{\delta}_{q,n} \neq 0}}^{N-1} p[\tilde{\boldsymbol{q}}_{n,k}|\boldsymbol{q}_n + \boldsymbol{e}_{q,n}, \boldsymbol{\sigma}_{p,n}^2] \prod_{\substack{l=1 \\ \boldsymbol{\delta}_{\psi,l} \neq 0}}^{E} p[\tilde{\boldsymbol{\psi}}_{l,k}|\boldsymbol{\psi}_l \right.$$
$$\left. + \boldsymbol{e}_{\psi,l}, \boldsymbol{\sigma}_{\psi,l}^2] \prod_{\substack{n=1 \\ \boldsymbol{\delta}_{p,n} \neq 0}}^{N-1} p[\tilde{\boldsymbol{p}}_{n,k}|\boldsymbol{p}_n + \boldsymbol{e}_{p,n}, \boldsymbol{\sigma}_{p,n}^2] \right\}. \tag{33}$$

Define the weight matrix $\boldsymbol{W} = \operatorname{diag}\left(\boldsymbol{\sigma}_p^T, \boldsymbol{\sigma}_q^T, \boldsymbol{\sigma}_\psi^T\right)$. Taking the natural logarithm of (33), we have

$$\ln g(\widetilde{\boldsymbol{y}}|\boldsymbol{\theta}) = C - \sum_{k=1}^{K} \left\| \boldsymbol{W} \operatorname{diag}(\boldsymbol{\delta}) \left(\boldsymbol{y}_k - \boldsymbol{H}\boldsymbol{\omega}\right) \right\|^2, \tag{34}$$

where $C = -K(2N + E - 1)\ln(2\pi)/2 - K\ln[\operatorname{trace}(\boldsymbol{W})]$.

8.2 Topology Attack

We describe an algorithm to verify the topology which can also detect topology attacks. The following assumptions are made.

Assumption The total set of nodes which could be employed is known to the operator.

Similar to (20), let $\boldsymbol{S}_0$ describe the topology that the operator believes to be present. To test if the topology matches what the operator believes to be true, we will test if the observed data comes from one of two different possible sets of pdfs. The first set includes all pdfs $g(\boldsymbol{y}|\boldsymbol{\theta})$ with $\boldsymbol{\theta}$ such that $\boldsymbol{S} = \boldsymbol{S}_0$. The other set includes all other possible pdfs. Due to the fact that the gas pressures and the flows have to follow (22), then $\boldsymbol{\theta}$ must satisfy $[\boldsymbol{B} \ \ -\operatorname{diag}(\boldsymbol{\beta})](\boldsymbol{\omega} \odot \boldsymbol{\omega}) - \boldsymbol{b}p_0^2 = \boldsymbol{0}$. Let $\boldsymbol{J} = [\boldsymbol{B} \ \ -\operatorname{diag}(\boldsymbol{\beta})]$ denote a matrix corresponding to a general $\boldsymbol{S}$ as per (23). Let $\boldsymbol{J}_0$ correspond to $\boldsymbol{J}$ when $\boldsymbol{S} = \boldsymbol{S}_0$ in (23). Define the parameter sets Θ_{H_0} and Θ_{H_1} as

$$\Theta_{H_0} = \{\boldsymbol{\theta} | \boldsymbol{S} = \boldsymbol{S}_0,\ \boldsymbol{J}_0(\boldsymbol{\omega} \odot \boldsymbol{\omega}) - \boldsymbol{b}p_0^2 = \boldsymbol{0}\}, \tag{35}$$

and

$$\Theta_{H_1} = \{\boldsymbol{\theta} | \boldsymbol{S} \neq \boldsymbol{S}_0,\ \boldsymbol{J}(\boldsymbol{\omega} \odot \boldsymbol{\omega}) - \boldsymbol{b}p_0^2 = \boldsymbol{0}\}. \tag{36}$$

Verifying a topology requires solving the hypothesis testing problem

$$H_0 : \boldsymbol{\theta} \in \Theta_{H_0} \text{ versus } H_1 : \boldsymbol{\theta} \in \Theta_{H_1}, \tag{37}$$

involving the family of pdfs $g(\boldsymbol{y}|\boldsymbol{\theta})$ parameterized by $\boldsymbol{\theta} = (\operatorname{vec}(\boldsymbol{S})^T, \boldsymbol{\omega}^T)^T$. The generalized likelihood ratio test (GLRT) is a classical statistics-based approach to solve the hypothesis testing problem in (37). The GLRT makes a decision for H_1 if a test statistic is larger than a threshold which is described by

$$\frac{\sup_{\boldsymbol{\theta} \in \Theta_{H_1}} g(\widetilde{\boldsymbol{y}}|\boldsymbol{\theta})}{\sup_{\boldsymbol{\theta} \in \Theta_{H_0}} g(\widetilde{\boldsymbol{y}}|\boldsymbol{\theta})} \underset{H_0}{\overset{H_1}{\gtrless}} \rho, \tag{38}$$

The threshold ρ in (38) is chosen to fix the probability that (38) decides for H_1 given H_0 is true. This probability is called the false alarm probability, denoted by P_{fa}. The probability that (38) decides for H_1, when H_1 is true, is called the detection probability, denoted by P_d. The test in (38) is equivalent to

$$\sup_{\boldsymbol{\theta}\in\Theta_{H_1}} \ln g(\widetilde{\mathbf{y}}|\boldsymbol{\theta}) - \sup_{\boldsymbol{\theta}\in\Theta_{H_0}} \ln g(\widetilde{\mathbf{y}}|\boldsymbol{\theta}) \underset{H_0}{\overset{H_1}{\gtrless}} \ln \rho. \tag{39}$$

where the log likelihood function $\ln g(\mathbf{y}|\boldsymbol{\theta})$ was defined in (34). The test (39) involves non-linear mixed integer and non-convex continuous optimization problems. In [53], an efficient and high performance relaxation is described for the test in (39). The following theorem describes a suitable sensor placement for (38) to provide good performance.

Theorem 1 *Given the observed data comes under H_1, then a necessary and sufficient condition that the optimization in the numerator of* (38) *gives the correct $\boldsymbol{\theta}$ as $K \to \infty$ is that the sensor placement satisfies*

$$rank\,(diag\,(\boldsymbol{\delta})\,\boldsymbol{H}) \geq N - 1. \tag{40}$$

Proof The proof of Theorem 1 is given in Sect. 11.

In the sequel,[1] we assume the sensor placement $\boldsymbol{\delta}$ satisfies Theorem 1.

9 Numerical Examples

We examine the impact of topology and MiMA/spoofing attacks targeting the transmission section of an example natural gas system. This section presents the example system investigated and describes numerical results demonstrating attack impact. Moreover, we evaluate the performance of the representative tests from the previous section.

9.1 Example Natural Gas System

The topology of the example gas transmission system we consider is shown in Fig. 3. Node 1 represents the natural gas gathering and processing sections which provide natural gas to the network. The maximum gas supply of node 1 is 600 kg/s. Node 6 connects to a city gate supplying consumers (distribution network not shown). The gas consumer that connects to the transmission section directly (at node 4) is assumed

[1] The condition (40) will also imply that the optimization in the denominator of (38) gives the correct $\boldsymbol{\theta}$ as $K \to \infty$ when the observation comes under H_0.

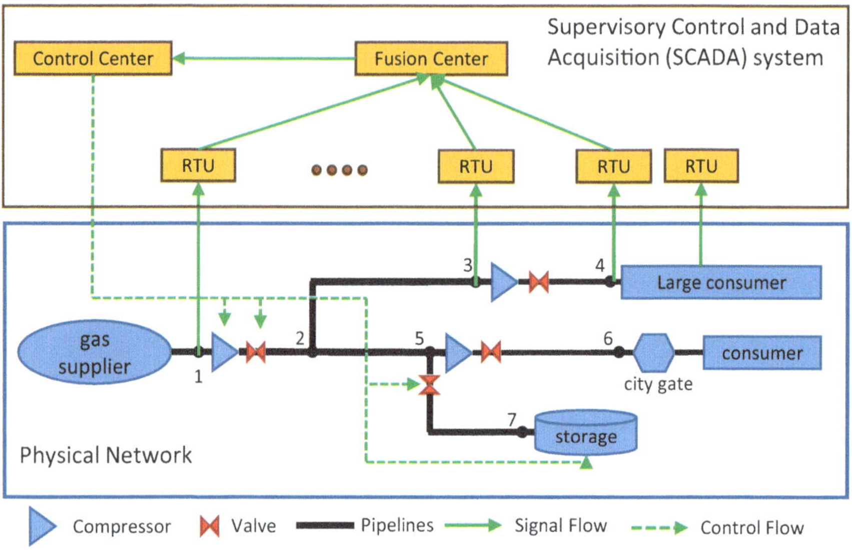

Fig. 3 The 7-node natural gas transmission system with an SCADA system

to be a large industrial consumer, possibly a large capacity natural gas-fueled power plant. In this example, we assume that the supply priority of the demand at node 4 is lower than that of the consumers after the city gate (at node 6), which is consistent with typical deployments.

We define the range of any physical variable as an interval over which the variable must lie for safety reasons as per (11)–(16). The operator will reject any values of variables not in this range, limiting the amount an attacker can modify variables. We assume any attacks trying to manipulate the operator's information will not cause any variable to become out of (11)–(16). There are three compression stations in the system. The ranges of the compression boost ratio for all three compression stations are between 1 and 2. Pressure ranges for both nodes and pipelines are between 2 MPa and 3 MPa. The pressure of node 1 is fixed at the pressure lower bound, which is 2 MPa. We assume pressure and flow measurements are available from all nodes and pipelines, respectively. For simplicity, we assume the measurements are noise-free. The operator at the control center determines the operating compression boost ratios and gas supply by solving the gas system operation model in (7)–(16) after obtaining the received sensor pressure and flow measurements.

Two operational days, the low-demand day and the high-demand day, are considered in this example. The demand data during the two days is shown in Fig. 4. The gas storage is in storing mode during low-demand days, and in releasing mode in high-demand days. The gas storage has physical limits resulting in storing/releasing rate limits. During low-demand days, the system will typically store natural gas subject to storing limits and the operational feasibility of the natural gas pipelines. Ranges for both the storing and releasing rates are 0 to 80 kg/s.

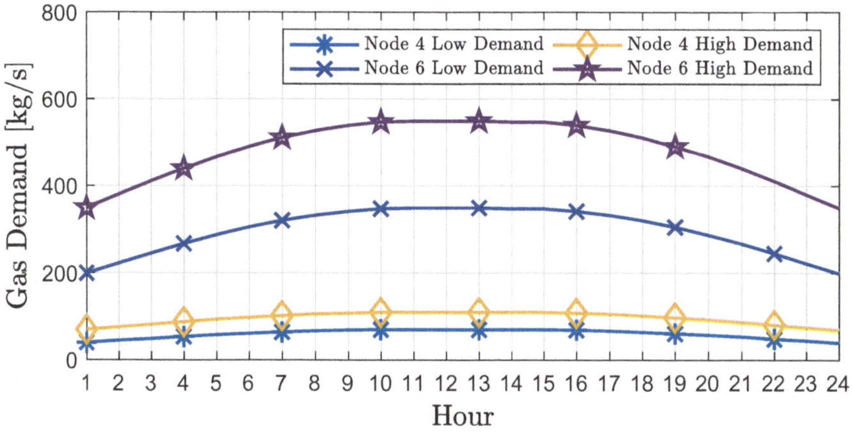

Fig. 4 Demand data in the low-demand day and the high demand day. Nodes not in the figure have no gas demand

In this example, the gas system is operated to achieve the objective of successfully serving all gas demands requested, subject to all variables in the suggested ranges, while trying to replenish the storage as much as possible during low-demand days. During high-demand days, the gas storage is only released if the gas supplier located at node 1 cannot serve all the demands, either due to the gas supply limit of node 1 (from 0 to 600 kg/s) or pipeline congestion. Otherwise all gas demand requested can be served. The gas demand requested at node 6 is served preferentially over the demand requested at node 4. We use two metrics to evaluate the impact of cyber attacks. The first one is the demand curtailment, i.e., unserved natural gas demands of consumers. The second one is the rate of storing natural gas in the storage connected to node 7. The natural gas storage will serve as a natural gas supplier during peak-load days and store gas during low-demand days. Hence, a lower rate of storing during low-demand days may lead to demand curtailment during peak-load days. Thus, we refer to the demand curtailment as an immediate impact, and the decreased rate of storing as a long-term impact.

9.2 MiMA/Spoofing Attacks: False Compression Boost Ratio Attack

The first cyber-attack considered is a specific type of MiMA/spoofing attack targeting the control signal of the compression boost ratio called the false compression boost ratio attack. This attack alters the values of the compression boost ratios of all three compression stations and sets the compression boost ratios to a value that is lower than desired by the operator based on the gas system operation model in (7)–(16). Let

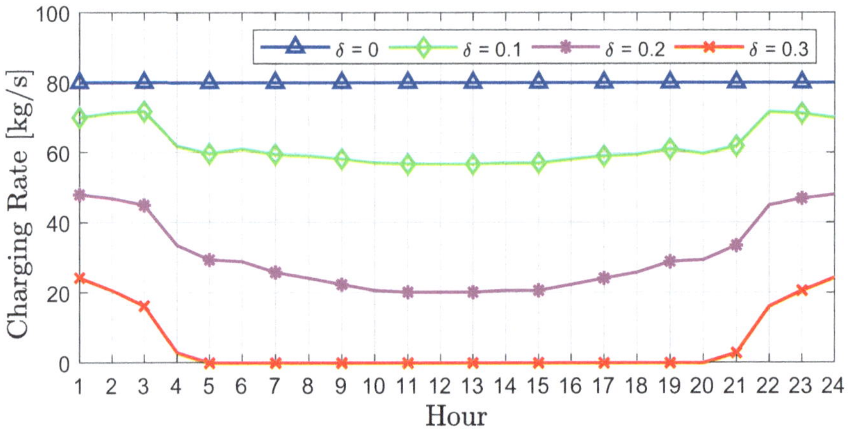

Fig. 5 Gas storage charging rate under either normal operation or a false compression boost ratio attack in a low-demand case

δ denote the value of the percentage change in the α above its lower bound of unity. Thus $\delta = (\alpha - \hat{\alpha})/(\alpha - 1)$ where $\hat{\alpha}$ is the attacked ratio, α is the desired value, and δ is an attack parameter that is determined by the attacker.

Figure 5 shows the gas storing rate during normal operation on a low-demand day for δ set to $0.1, 0.2$, and 0.3. For normal operation, we have $\delta = 0$. Fig. 6 shows the demand curtailment at node 4 in both the normal operation case and for false compression boost ratio attacks with δ set to $0.1, 0.2$, and 0.3. When $\delta = 0.3$, the consumer at node 4 suffers curtailment. Note that the storing rate decreases when the attacker decreases the compression boost ratios. Only when $\delta = 0.3$, do we have a non-zero demand curtailment. However, the false compression boost ratio attack can still be harmful when $\delta = 0.1$ or 0.2, since it decreases the rate of storing compared to normal operation. This will cause demand curtailment during high-demand days.

Figure 7 shows the storage releasing rates during normal operation and for three attacked cases where δ is set to either 0.3, 0.35, or 0.4 for a high-demand day. Storage releasing rates are increased with higher δ values. A higher δ value means that the true compression boost ratio is further decreased below the unattacked values, such that a reduced amount of gas can be delivered from the supplier at node 1 to the end users, and the releasing rate of natural gas storage will increase to serve the demand requested to mitigate the gas supply reduction. In that sense, the gas storage during high-demand days can sometimes mitigate the impact of the false compression boost ratio attacks. However, Fig. 8 shows that demand curtailment occurs at node 4 when the storage reaches its maximum releasing rate, as in hours 9 to 16 of the $\delta = 0.35$ case and in hours 8 to 17 of the $\delta = 0.4$ case. For these cases, the stored gas cannot sufficiently cover the deficit of gas supply.

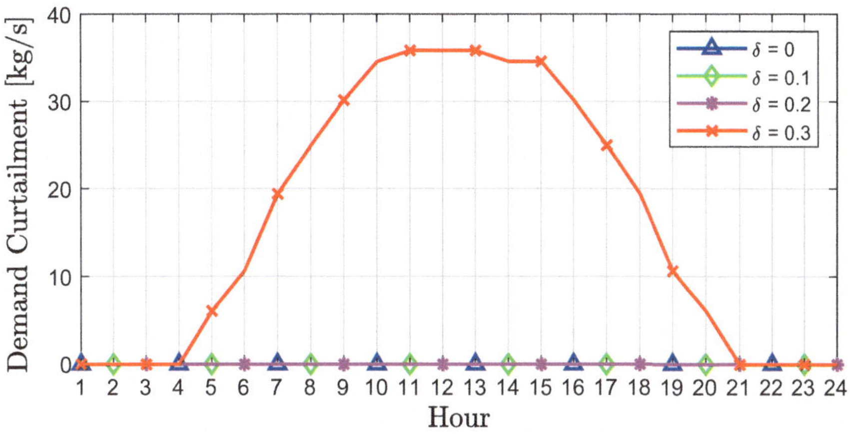

Fig. 6 Demand curtailment at node 4 under either normal operation or a false compression boost ratio attack in a low-demand case

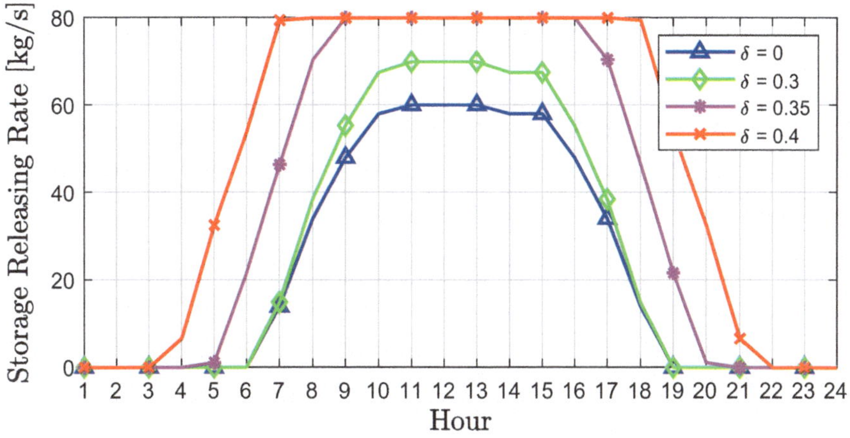

Fig. 7 Gas storage releasing rate in either normal operation or under false compression boost ratio attacks in the high-demand case

9.3 *MiMA/Spoofing Attacks: False Pressure and Flow Values*

The second cyber-physical-attack considered is another type of MIMA/spoofing attacks, which manipulates the pressure and gas flow measurements. The operator may employ inappropriate control commands, which may include inappropriate settings for compression boost ratios, gas storage storing/releasing rates, and gas supply rates, based on these false pressure and flow measurements. We call this type of attack a false measurement attack.

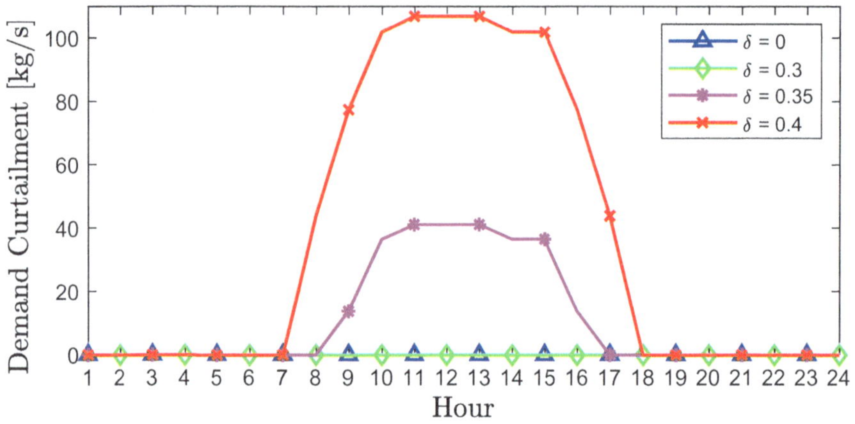

Fig. 8 Demand curtailment at node 4 in either normal operation or under false compression boost ratio attacks in the high-demand case

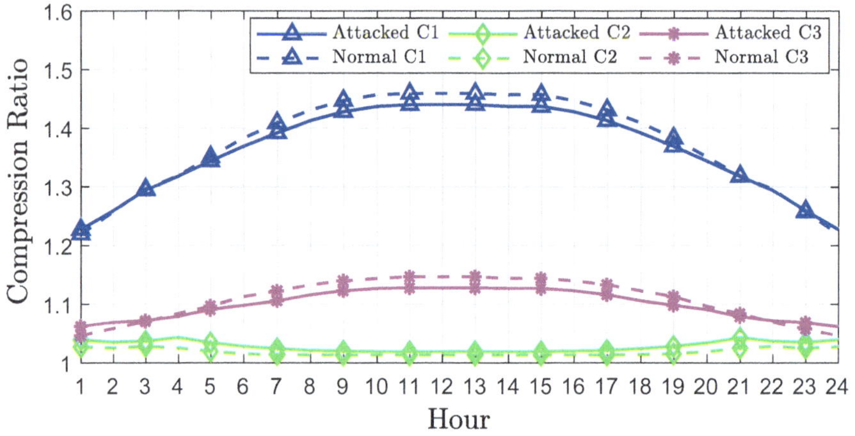

Fig. 9 Compression boost ratio in either normal operation or under wrong measurement attacks in low-demand case

As a particular example, consider the case where the attacks effectively increase the pressure and flow measurements received by the operator from all nodes and pipelines by a factor of 1.1 times the actual values during a low-demand day.

Figure 9 shows the compression boost ratios computed by (7)–(16) given both the false measurements and the correct ones. Figure 10 shows the demand curtailment at node 4 due to this attack. The storing rate of the natural gas storage in this particular example is the same as that without attacks, since the operator believes that the gas network is still able to provide enough gas to consumers. Thus the increased flow and pressure measurements cause immediate damage but have no impact on the amount of gas stored.

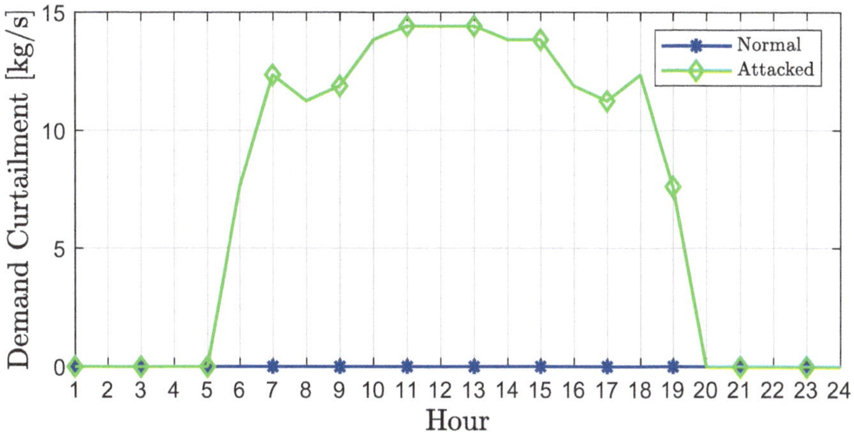

Fig. 10 Demand curtailment at node 4 in either normal operation or under wrong measurement attacks in low-demand case

9.4 Topology Attacks

The third cyber-attack considered is the topology attack. In a topology attack, the operator believes a topology is present which is different from the actual one present. The operator will employ control commands (including computing compression boost ratios, gas storage storing/releasing rates, and gas supply rates) that are obtained by solving the gas system operation model using the incorrect topology. We consider the case where the actual network is the one in Fig. 3 with the connection between

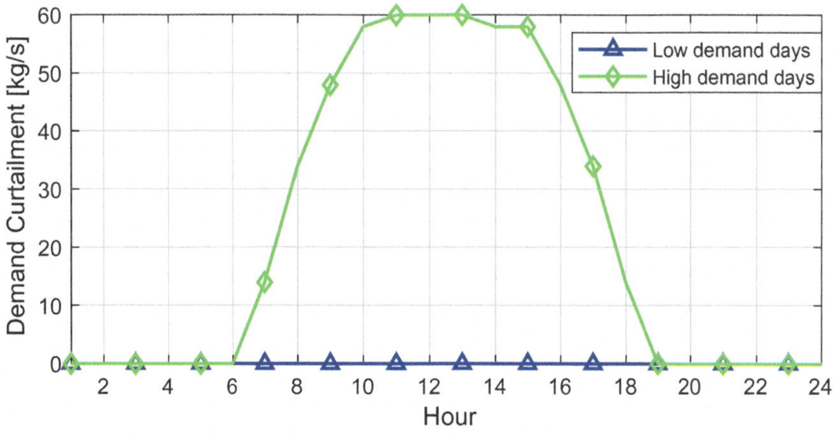

Fig. 11 Demand curtailment at node 4 in either normal operation or under topology attacks in low-demand case

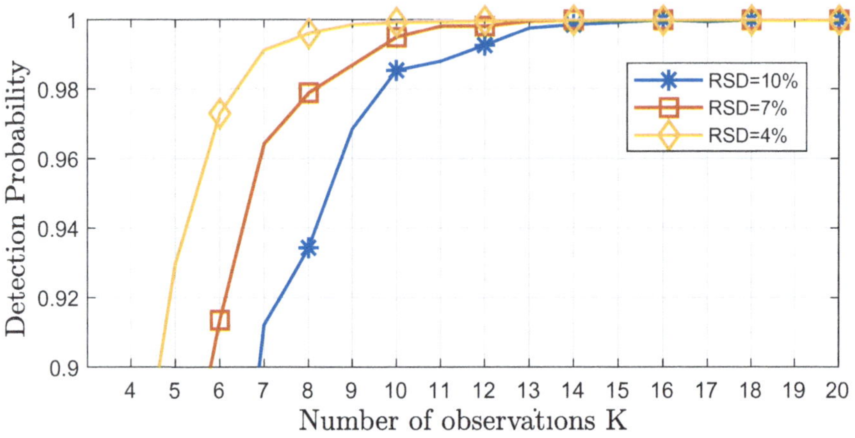

Fig. 12 Simulated detection probability of detecting topology attacks

node 5 and node 7 disconnected. The operator believes the topology given in Fig. 3 is present.

Figure 11 plots the demand curtailment at node 4 under the topology attack for the low demand day and high demand day, respectively. Figure 11 shows that there is no demand curtailment during the low-demand day. However, since the operator believes the connection between node 5 and node 7 is disconnected, the operator won't request the natural gas storage to store gas with a rate of 80 kg/s. This may cause demand curtailments during a later high-demand day. Figure 11 also shows that the topology attack causes the demand curtailment during the high-demand day. Therefore, the topology attacks cause the long-term impact during the low-demand day and short-term impact during the high-demand day.

We evaluate the performance of the test (39) to detect the topology attack during the low-demand day. Different from the previous section, in the rest of this chapter, we assume the measurements are noisy (we add noise to the noise-free values computed by (7)–(16)). The measure of the noise for each sensor is described using the relative standard deviation (RSD) $\boldsymbol{\tau} = \sigma/\mu \times 100\%$ in which μ and σ denote the mean and the standard deviation of the sensor measurements. We assume the different sensor observations have the same RSD. Figure 12 plots the detection probability is P_d versus the number of observations K when the false alarm probability $P_{\text{fa}} = 0.01$. Figure 12 shows that the test in (39) has a excellent performance ($P_d > 0.99$) when a reasonable number of observations are available (>13). The required number of observations to achieve $P_d > 99\%$ decreases as the RSD decreases.

10 Conclusion

In this chapter, we introduce the typical structure of natural gas systems and the functions of natural gas markets. We introduce a set of representative equations which accurately describes the gas flow through pipelines. We focus on three types of cyber-physical-attacks on natural gas system: MiMAs, spoofing attakcs, and topology attacks. We provide a comprehensive overview of cyber-physical-attack detection approaches for natural gas systems. The sensor measurement model, theories, and typical detection approaches are discussed in details. We describe a sufficient and necessary condition for sensor placement to ensure a good performance for detecting attacks. Numerical results show that the cyber-physical-attacks can cause both long-term and short-term impacts. We also show that given a reasonable number of observations the proposed cyber-physical-attack detection algorithm has a good performance in terms of the detection probability with a fixed false alarm probability.

11 Proof of Theorem 1

The maximum likelihood estimate used in (39) is equivalent to solving

$$\operatorname{diag}(\boldsymbol{\delta})\,\boldsymbol{H}\boldsymbol{\omega} = \frac{1}{K}\sum_{k=1}^{K} \boldsymbol{y}_k - \boldsymbol{n}_k, \tag{41}$$

and

$$\boldsymbol{J}\boldsymbol{\omega} \odot \boldsymbol{\omega} = \boldsymbol{b}p_0^2. \tag{42}$$

Define a vector $\boldsymbol{u} \triangleq \boldsymbol{\omega} \odot \boldsymbol{\omega}$. Then, (43) is equivalent to

$$\boldsymbol{J}\boldsymbol{u} = \boldsymbol{b}p_0^2, \tag{43}$$

and

$$\boldsymbol{u} - \boldsymbol{\omega} \odot \boldsymbol{\omega} = \boldsymbol{0}. \tag{44}$$

Define a function $\boldsymbol{f}(\boldsymbol{u}, \boldsymbol{\omega}) \triangleq \boldsymbol{u} - \boldsymbol{\omega} \odot \boldsymbol{\omega}$. Let $\boldsymbol{F}(\boldsymbol{u}, \boldsymbol{\omega}) = \left[\boldsymbol{I}_{N+E-1}\ 2\operatorname{diag}(\boldsymbol{\omega})\right]$ denote the gradient of $\boldsymbol{f}(\boldsymbol{u}, \boldsymbol{\omega})$ with respect to $(\boldsymbol{u}^T, \boldsymbol{\omega}^T)^T$. Note that $\boldsymbol{f} : \mathbb{R}^{2(N+E-1)\times 1} \rightarrow \mathbb{R}^{N+E-1}$ is differentiable and $\operatorname{rank}(\boldsymbol{F}(\boldsymbol{u}, \boldsymbol{\omega})) = N + E - 1$, which will be employed in the following theorem

Theorem 2 (Theorem 5.2 in [84]) *Assume* $\boldsymbol{f} : \mathbb{R}^{2(N+E-1)} \rightarrow \mathbb{R}^{N+E-1}$ *is a differentiable function and* $\boldsymbol{F}(\boldsymbol{u}, \boldsymbol{\omega})$ *has rank* $N + E - 1$ *whenever* $\boldsymbol{f}(\boldsymbol{u}, \boldsymbol{\omega}) = \boldsymbol{0}$. *Let*

$$\Theta = \left\{ \left(\boldsymbol{u}^T, \boldsymbol{\omega}^T \right)^T \mid \boldsymbol{f}(\boldsymbol{u}, \boldsymbol{\omega}) = \boldsymbol{0} \right\}. \tag{45}$$

Then, there exist an open set $U \subset \Theta$, *an open set* $W \subset \mathbb{R}^{N+E-1}$, *and an one-to-one differentiable function* $\boldsymbol{r} : W \to U$.

Based on Theorem 2, given that $\left(\boldsymbol{u}^T, \boldsymbol{\omega}^T\right)^T \in W \subset \Theta$, we have $\left(\boldsymbol{u}^T, \boldsymbol{\omega}^T\right)^T = \boldsymbol{r}(z)$. Substituting $\left(\boldsymbol{u}^T, \boldsymbol{\omega}^T\right)^T = \boldsymbol{r}(z)$ into (41) and (43), we have a linear equation with respect to $\boldsymbol{r}(z)$ given by

$$\underbrace{\begin{bmatrix} \boldsymbol{J} & \boldsymbol{0} \\ \boldsymbol{0} & \mathrm{diag}\,(\boldsymbol{\delta})\,\boldsymbol{H} \end{bmatrix}}_{\boldsymbol{D}} \boldsymbol{r}(z) = \underbrace{\begin{bmatrix} \boldsymbol{b} p_0^2 \\ \frac{1}{K}\sum_{k=1}^{K} \boldsymbol{y}_k - \boldsymbol{n}_k \end{bmatrix}}_{\boldsymbol{d}}, \tag{46}$$

where $\boldsymbol{D}$ and $\boldsymbol{v}$ are defined by (46). The quadratic term (44) is implied by $\boldsymbol{r}(z)$. The solutions of (46) in terms of z satisfy $\min_z \|\boldsymbol{D}\boldsymbol{r}(z) - \boldsymbol{v}\|_2^2 = 0$. When the Hessian matrix of $\|\boldsymbol{D}\boldsymbol{r}(z) - \boldsymbol{v}\|_2^2$ is positive definite for all z, $\|\boldsymbol{D}\boldsymbol{r}(z) - \boldsymbol{v}\|_2^2$ is convex in terms of z. Then, $\min_z \|\boldsymbol{D}\boldsymbol{r}(z) - \boldsymbol{v}\|_2^2 = 0$ has a unique solution.

Let $\boldsymbol{R}(z) \in \mathbb{R}^{2(N+E-1)\times(N+E-1)}$ denote the gradient of $\boldsymbol{r}(z)$ with respect to z. The Hessian matrix of $\|\boldsymbol{D}\boldsymbol{r}(z) - \boldsymbol{v}\|_2^2$ is

$$\nabla_z^2 \|\boldsymbol{D}\boldsymbol{r}(z) - \boldsymbol{v}\|_2^2 = 2\boldsymbol{R}^T(z)\boldsymbol{D}^T\boldsymbol{D}\boldsymbol{R}(z). \tag{47}$$

The matrix of the right hand side of (47) is symmetric which guarantees it is positive definite when it has full rank [85]. This guarantees a unique solution to $\min_z \|\boldsymbol{D}\boldsymbol{r}(z) - \boldsymbol{v}\|_2^2 = 0$. Using the chain rule and $\boldsymbol{f}(\boldsymbol{r}(z)) = \boldsymbol{0}$, we have, $\frac{\partial}{\partial z}\boldsymbol{f}(\boldsymbol{r}(z)) = \boldsymbol{F}(\boldsymbol{u}, \boldsymbol{\omega}_0)\,\boldsymbol{R}(z) = \boldsymbol{0}$. indicating that $\boldsymbol{R}(z)$ is a basis for the nullspace of $\boldsymbol{F}(\boldsymbol{u}, \boldsymbol{\omega}_0)$. Since $\boldsymbol{F}(\boldsymbol{u}, \boldsymbol{\omega}_0)$ has rank $N + E - 1$ and $2(N + E - 1)$ columns, we have the rank of $\boldsymbol{R}(z)$ is $N + E - 1$. The properties of the matrix on the right hand side of (47) implies

$$\begin{aligned} \mathrm{rank}(\boldsymbol{R}^T(z)\boldsymbol{D}^T\boldsymbol{D}\boldsymbol{R}(z)) &\le \min\{\mathrm{rank}\,(\boldsymbol{D}), \boldsymbol{R}(z)\} \\ &= \min\{\mathrm{rank}\,(\boldsymbol{D}), N + E - 1\}. \end{aligned} \tag{48}$$

Using (48) and noting that $\boldsymbol{R}(z)$ has $N + E - 1$ rows, for $2\boldsymbol{R}^T(z)\boldsymbol{D}^T\boldsymbol{D}\boldsymbol{R}(z)$ to be full rank we need $\mathrm{rank}\,(\boldsymbol{D}) \ge N + E - 1$. Based on the definition of $\boldsymbol{D}$ in (46), we have

$$\mathrm{rank}\,(\boldsymbol{D}) = \mathrm{rank}\,(\boldsymbol{J}) + \mathrm{rank}\,(\mathrm{diag}\,(\boldsymbol{\xi})\,\boldsymbol{H}_0) \tag{49}$$

$$= E - 1 + \mathrm{rank}\,(\mathrm{diag}\,(\boldsymbol{\xi})\,\boldsymbol{H}_0) \tag{50}$$

$$\ge N + E - 1, \tag{51}$$

which indicates

$$\mathrm{rank}\,(\mathrm{diag}\,(\boldsymbol{\xi})\,\boldsymbol{H}) \ge N - 1. \tag{52}$$

Hence when (52) is satisfied, (41)–(44) yields a unique solution. This finishes the proof of Theorem 1

Acknowledgements The work is supported by the Department of Energy under grant number DEOE0000779, the U.S. Army Research Laboratory and the U.S. Army Research Office under grant number W911NF-17-1-0331, the National Science Foundation under Grant ECCS-1744129, and a grant from the Commonwealth of Pennsylvania, Department of Community and Economic Development, through the Pennsylvania Infrastructure Technology Alliance (PITA).

References

1. Ponemon Institute LLC: The state of cybersecurity in the oil and gas industry: United States (2017). https://assets.new.siemens.com/siemens/assets/api/uuid:4ec3d46c-234e-4f48-9bc7-aef5889dcaba/version:1581364148/ponemoncyberreadinessinoilgasfinal.pdf
2. US Department of Homeland Security: U.S. Department of Homeland Security cybersecurity strategy (2018)
3. Carlson, R., Eggert, L., Papadopoulos, C., Rao, N., Tierney, B., Touch, J., Towsley, D., Zhang, L.: Network research problems and challenges for doe scientists workshop. Technical report (2016). https://www.orau.gov/networkresearch2016/
4. US Department of Energy: DOE network 2025: network research problems and challenges for DOE scientists. Workshop Report. Technical report (2016). https://www.osti.gov/biblio/1367529
5. Transportation Security Administration: Pipeline security guidelines. Technical report (2018)
6. Interstate Natural Gas Association of America: Is America's natural gas pipeline network prepared for cyber-attacks? (2018)
7. National Institute of Standards and Technology: Cybersecurity framework. Technical report (2018)
8. American Petroleum Institute: API standard 1164: pipeline SCADA security. Technical report (2009)
9. National Institute of Standards and Technology (2014) Guidelines for smart grid cybersecurity. Technical report (2014)
10. Siler-Evans, K., Hanson, A., Sunday, C., Leonard, N., Tumminello, M.: Analysis of pipeline accidents in the United States from 1968 to 2009. Int. J. Crit. Infrastruct. Prot. **7**, 257–269 (2014)
11. Tong, S., Lo, S., Zhang, P., Chen, B.: Jet fire consequence evaluation on the natural gas transported by pipelines. Procedia Eng. **52**, 349–354 (2013)
12. Zhao, B., Conejo, A.J., Sioshansi, R.: Using electrical energy storage to mitigate natural gas-supply shortages. IEEE Trans. Power Syst. **33**, 7076–7086 (2018)
13. Bajpai, S., Gupta, J.: Securing oil and gas infrastructure. J. Petr. Sci. Eng. **55**(1–2), 174–186 (2007)
14. US Energy Information Administration: Electric power annual 2017. Technical report (2018). https://www.eia.gov/electricity/annual/pdf/epa.pdf
15. Dubin, K.: More than 60% of electric generating capacity installed in 2018 was fueled by natural gas. Technical report, U.S. Energy Information Administration (2019). https://www.eia.gov/todayinenergy/detail.php?id=38632&src=email
16. Zhao, B., Zlotnik, A., Conejo, A.J., Sioshansi, R., Rudkevich, A.M.: Shadow price-based co-ordination of natural gas and electric power systems. IEEE Trans. Power Syst. **34**(3), 1942–1954 (2019)
17. Shao, C., Wang, X., Shahidehpour, M., Wang, X., Wang, B.: An MILP-based optimal power flow in multicarrier energy systems. IEEE Trans. Sustain. Energy **8**, 239–248 (2017)

18. Wu, F., Nagarajan, H., Zlotnik, A., Sioshansi, R., Rudkevich, A.M.: Adaptive convex relaxations for gas pipeline network optimization. In: 2017 American Control Conference, pp 4710–4716 (2017)
19. Zhao, B., Conejo, A.J., Sioshansi, R.: Unit commitment under gas-supply uncertainty and gas-price variability. IEEE Trans. Power Syst. **32**, 2394–2405 (2017)
20. Zhao, B., et al.: Electricity-gas systems: operations and expansion planning under uncertainty. Ph.D. thesis, The Ohio State University (2018)
21. He, C., Dai, C., Wu, L., Liu, T.: (a) Robust network hardening strategy for enhancing resilience of integrated electricity and natural gas distribution systems against natural disasters. IEEE Trans. Power Syst. **33**(5), 5787–5798 (2018)
22. He, C., Wu, L., Liu, T., Bie, Z.: (b) Robust co-optimization planning of interdependent electricity and natural gas systems with a joint $N - 1$ and probabilistic reliability criterion. IEEE Trans. Power Syst. **33**, 2140–2154 (2018)
23. Ding, T., Hu, Y., Bie, Z.: Multi-stage stochastic programming with nonanticipativity constraints for expansion of combined power and natural gas systems. IEEE Trans. Power Syst. **33**, 317–328 (2018)
24. Zhao, B., Conejo, A.J., Sioshansi, R.: Coordinated expansion planning of natural gas and electric power systems. IEEE Trans. Power Syst. **33**, 3064–3075 (2018)
25. Wang, C., Wei, W., Wang, J., Liu, F., Qiu, F., Correa-Posada, C.M., Mei, S.: Robust defense strategy for gas-electric systems against malicious attacks. IEEE Trans. Power Syst. **32**(4), 2953–2965 (2016)
26. Zhao, B., Lamadrid, A., Blum, R., Shalinee, K.: A three-level defender-attacker-operator problem against cyber-attacks in electric-gas systems. Electr. Power Syst. 1–20 (2020)
27. Tao, L., Mircea, E., Mohammad, S.: Interdependency of natural gas network and power system security. IEEE Trans. Power Syst. **23**(4), 1817–1824 (2008)
28. Yang, Y., Littler, T., Sezer, S., McLaughlin, K., Wang, H.: Impact of cyber-security issues on smart grid. In: 2011 2nd IEEE PES International Conference and Exhibition on Innovative Smart Grid Technologies, pp. 1–7. IEEE (2011)
29. Pricop, E., Mihalache, S.F.: Assessing the security risks of a wireless sensor network from a gas compressor station. In: Proceedings of the 2014 6th International Conference on Electronics, Computers and Artificial Intelligence (ECAI), pp. 45–50 (2014)
30. Wadhawan, Y., Neuman, C.: Evaluating resilience of gas pipeline systems under cyber-physical attacks: a function based methodology. In: Proceedings of the 2nd ACM Workshop on Cyber-Physical Systems Security and Privacy, pp. 71–80 (2016)
31. Daniela, T.: Communication security in SCADA pipeline monitoring systems. In: 2011 RoEduNet International Conference 10th Edition: Networking in Education and Research, pp. 1–5 (2011)
32. He, F., Nwafor, J.: Gas pipeline recovery from disruption using multi-objective optimization. In: 2017 IEEE International Symposium on Technologies for Homeland Security (HST), pp. 1–6 (2017)
33. Judson, N.: Interdependence of the electricity generation system and the natural gas system and implications for energy security. Technical report, Lincoln Laboratory, MIT (2013). https://apps.dtic.mil/dtic/tr/fulltext/u2/a584764.pdf
34. Levine, S., Carpenter, P., Thapa, A.: Understanding natural gas market. Technical report, American Petroleum Institute (2014). https://www.api.org/~/media/Files/Oil-and-Natural-Gas/Natural-Gas-primer/Understanding-Natural-Gas-Markets-Primer-High.pdf
35. Natgas: Natural gas-marketing. Technical report (2013). http://naturalgas.org/naturalgas/marketing/
36. Salem, M.B., Hershkop, S., Stolfo, S.J.: A survey of insider attack detection research. In: Insider Attack and Cyber Security, pp 69–90. Springer (2008)
37. Malik, N., Collins, R., Vamburkar, M.: (2018) Cyber attack pings data systems of at least four gas networks. Technical report (2018). https://www.bloomberg.com/news/articles/2018-04-03/day-after-cyber-attack-a-third-gas-pipeline-data-system-shuts

38. Radmand, P., Talevski, A., Petersen, S., Carlsen, S.: Taxonomy of wireless sensor network cyber security attacks in the oil and gas industries. In: 2010 24th IEEE International Conference on Advanced Information Networking and Applications, pp. 949–957. IEEE (2010)
39. Jia, L., Thomas, R.J., Tong, L.: Malicious data attack on real-time electricity market. 2011 IEEE International Conference on Acoustics, pp. 5952–5955. Speech and Signal Processing (ICASSP), IEEE (2011)
40. Shafi, Q.: Cyber physical systems security: a brief survey. In: 2012 12th International Conference on Computational Science and Its Applications, IEEE, pp. 146–150 (2012)
41. Wang, W., Lu, Z.: Cyber security in the smart grid: survey and challenges. Comput. Netw. **57**(5), 1344–1371 (2013)
42. Wang, D., Guan, X., Liu, T., Gu, Y., Sun, Y., Liu, Y.: A survey on bad data injection attack in smart grid. In: 2013 IEEE PES Asia-Pacific Power and Energy Engineering Conference (APPEEC), pp. 1–6. IEEE (2013)
43. Kim, J., Tong, L.: On phasor measurement unit placement against state and topology attacks. In: 2013 IEEE International Conference on Smart Grid Communications (SmartGridComm), pp. 396–401. IEEE (2013)
44. Pricop, E., Mihalache, S.F.: Assessing the security risks of a wireless sensor network from a gas compressor station. In: Proceedings of the 2014 6th International Conference on Electronics, Computers and Artificial Intelligence (ECAI), pp. 45–50. IEEE (2014)
45. Beasley, C., Zhong, X., Deng, J., Brooks, R., Venayagamoorthy, G.K.: A survey of electric power synchrophasor network cyber security. IEEE PES Innovative Smart Grid Technologies, pp. 1–5. IEEE, Europe (2014)
46. He, H., Yan, J.: Cyber-physical attacks and defences in the smart grid: a survey. IET Cyber-Phys. Syst.: Theory Appl. **1**(1), 13–27 (2016)
47. Nazir, S., Patel, S., Patel, D.: Assessing and augmenting SCADA cyber security: a survey of techniques. Comput. Secur. **70**, 436–454 (2017)
48. Liu, X., Li, Z.: Local topology attacks in smart grids. IEEE Trans. Smart Grid **8**(6), 2617–2626 (2017)
49. Liang, G., Zhao, J., Luo, F., Weller, S.R., Dong, Z.Y.: A review of false data injection attacks against modern power systems. IEEE Trans. Smart Grid **8**(4), 1630–1638 (2017)
50. Ding, D., Han, Q.L., Xiang, Y., Ge, X., Zhang, X.M.: A survey on security control and attack detection for industrial cyber-physical systems. Neurocomputing **275**, 1674–1683 (2018)
51. Chung, H.M., Li, W.T., Yuen, C., Chung, W.H., Zhang, Y., Wen, C.K.: Local cyber-physical attack for masking line outage and topology attack in smart grid. IEEE Trans. Smart Grid (2018)
52. Zhang, J., Blum, R.S., Poor, H.V.: Approaches to secure inference in the internet of things: performance bounds, algorithms, and effective attacks on internet of things sensor networks. IEEE Signal Process. Mag. **35**(5), 50–63 (2018)
53. Wang, Z., Blum, R.S.: Topology attack detection in natural gas delivery networks. In: 2019 53rd Annual Conference on Information Sciences and Systems (CISS), pp. 1–6. IEEE (2019)
54. Fillatre, L., Nikiforov, I., Willett, P., et al.: Security of SCADA systems against cyber-physical attacks. IEEE Aerosp. Electron. Syst. Mag. **32**(5), 28–45 (2017)
55. Nagananda, K.G., Kishore, S., Blum, R.S.: A phasor measurement unit scheduling scheme for transmission of synchrophasor data in electric power systems. IEEE Trans. Smart Grid **6**(5), 2519–2528 (2015)
56. Bland, E.: GPS spoofing could threaten national security (2008). http://www.nbcnews.com/id/26992456
57. Basnight, Z., Butts, J., Lopez Jr., J., Dube, T.: Firmware modification attacks on programmable logic controllers. Int. J. Crit. Infrastruct. Prot. **6**(2), 76–84 (2013)
58. Wang, Z., Blum, R.S.: A statistical learning-based algorithm for topology verification in natural gas networks based on noisy sensor measurements. IEEE Trans. Inf. Forensics Secur. **15**, 3653–3666 (2020)
59. Raymond, D.R., Midkiff, S.F.: Denial-of-Service in wireless wensor networks: attacks and defenses. IEEE Pervasive Comput. **7**(1), 74–81 (2008). https://doi.org/10.1109/mprv.2008.6

60. Pelechrinis, K., Iliofotou, M., Krishnamurthy, S.V.: Denial of service attacks in wireless networks: the case of jammers. IEEE Commun. Surv. Tutor. **13**(2), 245–257 (2011). https://doi.org/10.1109/SURV.2011.041110.00022
61. Kailkhura, B., Nadendla, V.S.S., Varshney, P.K.: Distributed inference in the presence of eavesdroppers: a survey. IEEE Commun. Mag. **53**(6), 40–46 (2015)
62. Poor, H.V., Schaefer, R.F.: Wireless physical layer security. Proc. Nat. Acad. Sci. **114**(1), 19–26 (2017)
63. Gao, W., Morris, T., Reaves, B., Richey, D.: On SCADA control system command and response injection and intrusion detection. In: 2010 eCrime Researchers Summit, pp 1–9. IEEE (2010)
64. Kiss, I., Genge, B., Haller, P., Sebestyén, G.: Data clustering-based anomaly detection in industrial control systems. In: 2014 IEEE 10th International Conference on Intelligent Computer Communication and Processing (ICCP), pp. 275–281. IEEE (2014)
65. Arnold, C., Butts, J., Thirunarayan, K.: Detecting integrity attacks on industrial control systems. In: International Conference on Critical Infrastructure Protection, pp. 3–13. Springer (2014)
66. Beaver, J.M., Borges-Hink, R.C., Buckner, M.A.: An evaluation of machine learning methods to detect malicious SCADA communications. In: 2013 12th International Conference on Machine Learning and Applications, vol 2, pp 54–59. IEEE (2013)
67. Feng, C., Li, T., Chana, D.: Multi-level anomaly detection in industrial control systems via package signatures and LSTM networks. In: 2017 47th Annual IEEE/IFIP International Conference on Dependable Systems and Networks (DSN), pp. 261–272. IEEE (2017)
68. Shirazi, S.N., Gouglidis, A., Syeda, K.N., Simpson, S., Mauthe, A., Stephanakis, I.M., Hutchison, D.: Evaluation of anomaly detection techniques for SCADA communication resilience. In: 2016 Resilience Week (RWS), pp. 140–145. IEEE (2016)
69. Nader, P., Honeine, P., Beauseroy, P.: One-class classification framework based on shrinkage methods. J. Signal Process. Syst. **90**(3), 341–356 (2018)
70. Nader, P., Honeine, P., Beauseroy, P.: l_p-norms in one-class classification for intrusion detection in SCADA systems. IEEE Trans. Ind. Inform. **10**(4), 2308–2317 (2014)
71. Cárdenas, A.A., Amin, S., Lin, Z.S., Huang, Y.L., Huang, C.Y., Sastry, S.: Attacks against process control systems: risk assessment, detection, and response. In: Proceedings of the 6th ACM Symposium on Information, Computer and Communications Security, pp. 355–366. ACM (2011)
72. Guan, Y., Ge, X.: Distributed attack detection and secure estimation of networked cyber-physical systems against false data injection attacks and jamming attacks. IEEE Trans. Signal Inf. Process. Over Netw. **4**(1), 48–59 (2017)
73. Van Long, D., Fillatre, L., Nikiforov, I.: Sequential monitoring of SCADA systems against cyber/physical attacks. IFAC-PapersOnLine **48**(21), 746–753 (2015)
74. Housh, M., Ohar, Z.: Model-based approach for cyber-physical attack detection in water distribution systems. Water Res. **139**, 132–143 (2018)
75. Amin, S., Litrico, X., Sastry, S.S., Bayen, A.M.: Cyber security of water SCADA systems part ii: attack detection using enhanced hydrodynamic models. IEEE Trans. Control Syst. Technol. **21**(5), 1679–1693 (2012)
76. Liu, M., Zang, S., Zhou, D.: Fast leak detection and location of gas pipelines based on an adaptive particle filter. Int. J. Appl. Math. Comput. Sci. **15**(4), 541 (2005)
77. Ntalampiras, S.: Detection of integrity attacks in cyber-physical critical infrastructures using ensemble modeling. IEEE Trans. Ind. Inform. **11**(1), 104–111 (2014)
78. Perez, R.L., Adamsky, F., Soua, R., Engel, T.: Machine learning for reliable network attack detection in SCADA systems. 2018 17th IEEE International Conference on Trust, pp. 633–638. Security and Privacy in Computing and Communications, IEEE (2018)
79. Poor, H.V.: An Introduction to Signal Detection and Estimation. Springer Science & Business Media (2013)
80. Fillatre, L., Nikiforov, I., et al.: A statistical method for detecting cyber/physical attacks on SCADA systems. In: 2014 IEEE Conference on Control Applications (CCA), pp 364–369. IEEE (2014)

81. Belsito, S., Lombardi, P., Andreussi, P., Banerjee, S.: Leak detection in liquefied gas pipelines by artificial neural networks. AIChE J. **44**(12), 2675–2688 (1998)
82. Ojha, A., Kekatos, V., Baldick, R.: Solving the natural gas flow problem using semidefinite program relaxation. In: 2017 IEEE Power & Energy Society General Meeting, pp. 1–5. IEEE (2017)
83. Osiadacz, A.J.: Simulation and Analysis of Gas Networks (1987)
84. Spivak, M.: Calculus on Manifolds: A Modern Approach to Classical Theorems of Advanced Calculus. CRC Press (2018)
85. Zhang, X.D.: Matrix Analysis and Applications. Cambridge University Press (2017)

Secure Dynamic Nonlinear Heterogeneous Vehicle Platooning: Denial-of-Service Cyber-Attack Case

Mohammad Hossein Basiri, Nasser L. Azad, and Sebastian Fischmeister

Abstract Connected and Automated Vehicles (CAVs), as a large class of cyber-physical systems, have recently emerged as an effective autonomous driving mechanism in intelligent transportation systems in terms of improvement in safety, fuel economy, road throughput, and driving comfort. This chapter deals with a Secure Distributed Nonlinear Model Predictive Control (Secure–DNMPC) algorithm consisting of (i) detection and (ii) mitigation phases to securely control a string of CAVs, namely *vehicle platoons*. The approach ensures the desired control performance of a nonlinear heterogeneous platoon equipped by different communication topologies under the premise of the existence of Denial-of-Service (DoS) attacks. The proposed method is also capable of providing safe and secure control of dynamic platoons in which arbitrary vehicles might perform cut-in and/or cut-out maneuvers. Convergence time and stability analysis of the system are also investigated in some cases. Furthermore, to handle DoS attacks modeled by an exceeding time delay in inter-vehicular data transmission, we propose the integration of an Unscented Kalman Filter (UKF) design within the controller resulting in a novel Secure–DNMPC–UKF co-design. This, in essence, estimates the delayed system states and feeds the predicted values to the Secure–DNMPC, which efficiently mitigates the attack effects. Simulation results demonstrate the fruitfulness of the proposed method.

1 Introduction

This section introduces the general notion of Cyber-Physical Systems (CPSs) along with one of their subclasses, namely Connected and Automated Vehicles (CAVs), and explains some of the most important and prevalent security-related issues that

M. H. Basiri (✉) · S. Fischmeister
Department of Electrical and Computer Engineering, University of Waterloo, Waterloo, ON, Canada
e-mail: mh.basiri@uwaterloo.ca

S. Fischmeister
e-mail: sfischme@uwaterloo.ca

N. L. Azad
Department of Systems Design Engineering , University of Waterloo, Waterloo, ON, Canada
e-mail: nlashgar@uwaterloo.ca

A. I. Awad et al. (eds.), *Security in Cyber-Physical Systems*, Studies in Systems, Decision and Control 339, https://doi.org/10.1007/978-3-030-67361-1_10

need to be taken into account while dealing with these systems. Then, we will review the related work and explicitly state the contributions of the current chapter.

1.1 State-of-the-Art

Cyber-Physical Systems (CPSs) are among the fast emerging profound infrastructures enabling traditional physical plants to operate in a wide area and a distributed fashion. Networking, computation, communication, and control are tightly interwoven to foster a CPS [1]. These components are categorized in cyber and physical layers, each of which interacts with the other parts to receive external data, process them, and generate appropriate output signals. Never may a CPS perform without the proper and timely functioning of its constituents. Instances of CPS include, but are not limited to, automotive control, medical monitoring, robotic systems, and smart grid [2].

Apart from the physical layer, the cyber one has been broadly shown to be prone to external intelligent cyber-attacks. Data integrity, confidentiality, and availability are the major crucial concerns of cyber-security that an intelligent intruder might target [3, 4]. Various attacks have been introduced to destruct one or more of the aforementioned security aspects of a CPS. False data injection, GPS spoofing, eavesdropping, Denial-of-Service (DoS), and replay attack are some of the paradigms [5–9]. Several well-known attacks on CPS include Stuxnet on a Supervisory Control and Data Acquisition (SCADA) system [10, 11], attacks on the wireless network channels in smart power grid systems [12], and compromising Anti-lock Braking System (ABS) sensors of a vehicle [13]. Hence, the security-related issues, such as attack detection and secure state estimation and control of CPS, have been converted to attracting challenges in the control community.

Constituting an important application of CPS, autonomous driving has greatly emerged during the last decade. Due to the huge growth in the number of vehicles driving in the world, traffic congestion threatens driving safety. This will potentially result in increasing the risk of accidents. Autonomous vehicles and autonomous driving are another aspects which have got a great deal of attention. Through this technological development, driving safety can be highly enhanced as most car accidents are caused by human errors and distractions while driving. Over 90% of all car accidents are caused by human errors [14]. From this point of view, self-driving cars can remove a considerable amount of human errors resulting in safer transportation. In Canada alone, there were close to 111,000 road-related injuries and over 1,800 fatalities reported in 2014 [15]. Autonomous cars have many other advantages, such as getting faster to the destination, reducing governmental costs and car ownership [16, 17].

The degree of autonomy incorporated in autonomous driving is categorized in 6 different levels (levels 0-5). Level 0 (no automation) is the most basic one in which no autonomy is incorporated. The vehicle is fully controlled by a human driver. In level 1 (driver assistance), the vehicle can assist the driver with some functions, such as steering, acceleration, or braking. In level 2 (partial automation), the vehicle lets the driver disinvolve with some of these tasks. The driver still has the main role in monitoring the environment and in taking care of most safety-critical functions. The driver is responsible for taking full control of the vehicle when needed. In level 3 (conditional automation), the vehicle performs all the environment monitoring tasks. In safe conditions, the driver can leave the safety-critical functions like braking to the vehicle; however, his attention is still required. Level 4 (high automation) of autonomy is able to take care of monitoring the environment, steering, acceleration, and braking. In addition, the vehicle is capable of changing lanes, turning and using signals. However, the vehicle can not perform decisions in more complex scenarios, such as traffic jams or merging onto the highway. Level 5 (complete automation) exploits full autonomy, which requires no human intervention, pedals, brakes, or a steering wheel.

Connected and Automated Vehicles (CAVs), as a large class of CPS, have recently emerged as an effective autonomous driving mechanism in intelligent transportation systems in terms of improvement in safety, fuel economy, road throughput, and driving comfort. Vehicles participating in a realistic platoon most likely bear variant nonlinear dynamics forming a heterogeneous platoon. Platoons could be formed based on different spacing policies and governed by different formation control techniques such as traditional linear/nonlinear controllers, optimal control methods, and more advanced consensus algorithms [18–20]. It has been widely proved in the literature that nonlinear control techniques are mandatory to achieve desired formation objectives, such as maintaining a safe gap among consecutive cars while tracking the speed profile of the leader vehicle.

CAVs can communicate with each other and exchange their date through Vehicle-to-Vehicle (V2V) and/or Vehicle-to-Infrastructure (V2I) wireless communications. Getting more developed through using more effective data communication structures, connected vehicles are equipped with different information flow topologies to facilitate and improve the efficacy of data transfers. Predecessor-follower (PF), predecessor-leader follower (PLF), two-predecessors follower (TPF), two-predecessors-leader follower (TPLF), all-predecessors follower (APF), all-predecessors-leader follower (APLF), and h–nearest neighbor are some of the instances [21, 22]. These structures can be exploited either in a unidirectional or a bidirectional data transmit (see Fig. 1).

As was mentioned before, autonomous cars can be equipped with wireless data communication devices such that they can transfer data such as inter-vehicular distance and speed. In this respect, CAVs typically take advantage of V2V and/or V2I communication environments. V2V communications can provide direct data transfer with a much lower delay compared to radars [23]. The V2V communications enable vehicles to drive closely with short inter-vehicular distances. This will increase the amount of road throughput and reduce the need for developing more road networks. The vehicles can exchange data, such as inter-vehicular distance, speed and acceleration. In this context, Cooperative Adaptive Cruise Control (CACC) system has been widely developed, featuring the possibility of coordination between connected vehicles aiming at enhancing fuel efficiency, safety, driving comfort, and road throughput.

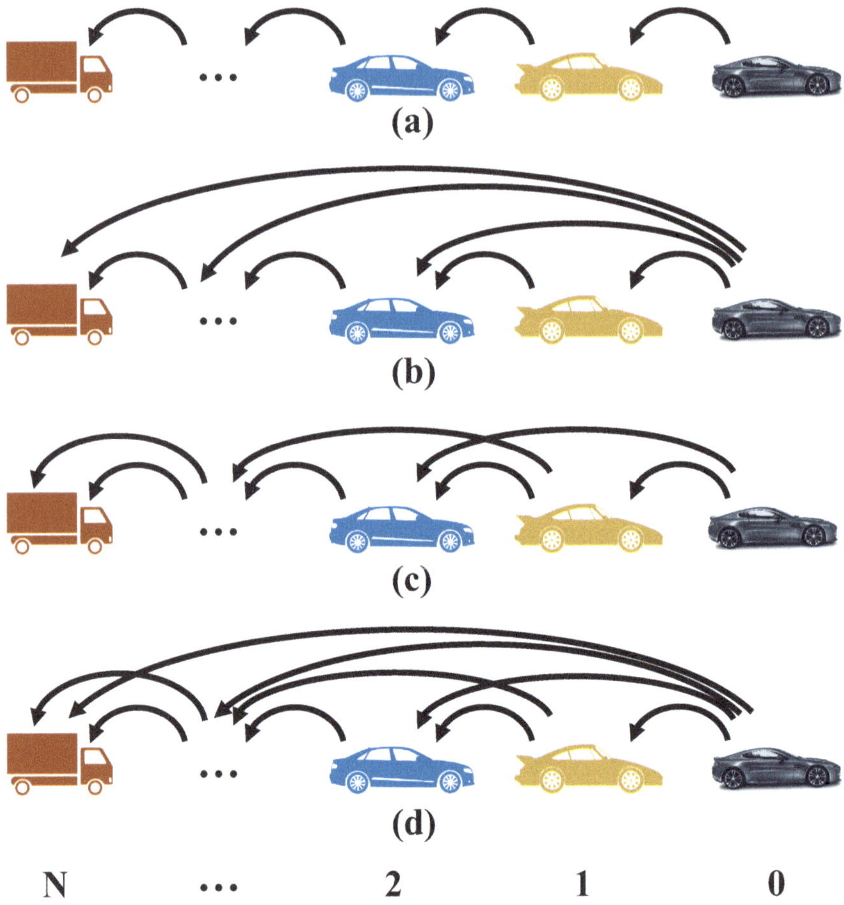

Fig. 1 Unidirectional topology: **a** PF, **b** PLF, **c** TPF, and **d** TPLF, (vehicle 0 is the Leader Vehicle (LV))

This system, which is the advanced version of Adaptive Cruise Control (ACC), lets neighboring vehicles form a platoon, which is a string of vehicles following a common speed profile.

Despite the benefits of wireless connectivity among these vehicles, this makes the whole system susceptible to cyber-attacks. One such a prevalent attack, that has broadly drawn the attention of both cyber-security and control communities, is Denial-of-Service (DoS) attack. A DoS intelligent intruder aims at jamming communication links among cars through overwhelming the beacon node by fake requests, hence, hinders the network from processing legitimate requests. This can result in huge performance degradation and even hazardous collisions.

This chapter concerns with the control problem of a large class of CPS, namely platoon formation, which has had a leading role in autonomous driving systems.

Basically, a platoon is a string of connected and automated vehicles, all driving with a pre-specified safe gap among consecutive cars and following a shared speed profile generated by the leader vehicle. This physical layer, together with the wireless connectivity among the participating vehicles, as the cyber layer, constitutes the system as a whole CPS. Vehicle platooning is well-known due to its advantages such as enhancement of road throughput, fuel economy, driving comfort, and safety; however, it suffers from the vulnerability of wireless connections among the cars to devastating malware [24–26]. In particular, an outsider attacker might compromise inter-vehicular data to fool the on-board sensors and controller of the receiver vehicle resulting in unnecessary acceleration/brake actions. Besides, he may cause a failure in the network by jamming it or injecting a huge amount of delay, which in essence makes the outdated transferred data useless. The latter is the scenario from which a DoS attacker takes advantage and will be the focus of this chapter.

1.2 Related Work

Recently, much research has been done in investigating the security of networked control systems from various perspectives [27–31]. Communication-related protection methods, such as encryption of wireless channels, are techniques to avoid receiving compromised data via the wireless infrastructures [32]. On the other hand, control-oriented concepts, such as game-theoretic methods, are also among the leading methodologies which address the security issue of general cyber-physical systems with a considerable amount of care [33, 34]. Although there has been a large amount of research addressing the security of CPS, those systems still suffer from the lack of secure performance in the presence of possible malicious intruders [31, 35, 36]. This needs to be noted that the introduced techniques for fault-tolerant control might be applied to security problems; however, the majority of those technique cannot handle the devastation imposed by an intelligent intruder [37]. The reason is that an intelligent attacker has some a priori knowledge of the system dynamics and/or the controller which is not the case in a random fault. Furthermore, specific components of a system may be targeted by an intelligent adversary based on his own criteria, such as optimizing the amount of consumed energy or the intended level of devastation. In this regard, different methodologies based on system/graph/game notions have been introduced to address security issues of general control systems [34, 38–40].

In the recent decade, there has been a vast range of studies addressing security issues of vehicle platoons [41–48]. In essence, researchers have been concerned about possible vulnerabilities of vehicle platoons against cyber attacks as well as communication delays [24, 42–53]. Particularly, in [46], a DoS attack detection and estimation scheme based on sliding mode observer has been proposed for a linear homogeneous car following system. Also, authors of [47, 48] study the performance degradation of a linear homogeneous vehicle following controller caused by unreliable wireless communications. Various types of intrusions imposed by either insider and/or outsider adversary on connected vehicles have been investigated in literature, which include but are not limited to DoS, GPS spoofing, masquerading, insider/outsider eavesdropping [54]. Each of these attacks can potentially degrade system performance by violating one or more of the data availability, data confidentiality, and data integrity. A detailed and formal attack classification in a three-dimensional attack space is given in [55]. Network-aware control methods have also been proposed to handle possible communication failures through the platoon. Those approaches mainly consider random communication failures with an emphasis on the control/stability performance of the whole platoon without considering intelligent cyber attacks [51, 56, 57].

However, the lack of a systematic approach adhering to control performance objectives of a dynamic nonlinear heterogeneous platoon while mitigating the DoS attack effects is yet sensible. Thus, in this study, we focus on an attacked dynamic nonlinear heterogeneous platoon in which arbitrary vehicles might perform cut-in/cut-out maneuvers. Variant nonlinear dynamics of the participating cars are considered in the model to form a realistic nonlinear heterogeneous platoon.

1.3 Contributions

Contributions of this chapter are explicitly as follows. Under the premise of the existence of a DoS attacker of either a network blocker or a huge time delay injector, we propose a Secure Distributed Nonlinear Model Predictive Control (Secure–DNMPC) framework to detect and mitigate the attack effects while ensuring fulfillment of the platoon control objectives. The algorithm is flexible to adopt different communication topologies handling inter-vehicular data transfer among the vehicles. Convergence time and stability analysis of the algorithm is proved in some cases. Furthermore, in case of a DoS attacker as an exceeding time delay injector, since the transferred data are still available while the attack is underway, we propose to make use of the outdated system states and take benefit of them to implement the control strategy instead of simply ignoring the data and using the most recent one. This will effectively improve the control performance of the whole system. In essence, we propose to embed a UKF as the state observer within the design of the Secure–DNMPC to adapt the algorithm to the delayed data transmission. This results in a novel Secure–DNMPC–UKF co-design. In addition, this gives the opportunity to either consider non-ideal noisy sensors or take into account the contaminated sent data due to the noisy surrounding environment and road conditions.

1.4 Chapter Organization

The remainder of this chapter is organized as follows. Section 2 presents the system modeling, including the platoon model and different types of DoS attack descriptions. Section 3 details the design of the secure controller together with some stability analysis results. Adaptation of the algorithm to handle dynamic maneuvers together with convergence time analysis are given in Sect. 4. Section 5 demonstrates the simulation results on a Two-Predecessor Follower (TPF) attacked nonlinear dynamic heterogeneous platoon. Finally, Sect. 6 concludes the chapter.

2 System Modeling

In this section, we present the considered platoon model and its control objectives. In addition, we give different DoS attack descriptions on which we focus in this chapter.

2.1 Platoon Model

Let us consider a platoon of vehicles, consisting of a Leader Vehicle (LV) and N Follower Vehicles (FVs) indexed by $\mathcal{N} := \{1, \ldots, N\}$. In this chapter, we consider the longitudinal dynamics and unidirectional communication topologies. Let Δt be the discrete time interval and $p_i(t)$, $v_i(t)$, and $T_i(t)$ denote the position, velocity, and the integrated driving/breaking torque of the i-th FV at time t, respectively. For the i-th FV, we denote the vehicle mass, the coefficient of aerodynamic drag, the coefficient of rolling resistance, the inertial lag of longitudinal dynamics, the tire radius, the mechanical efficiency of the driveline, and the control input by m_i, $C_{A,i}$, $f_{r,i}$, τ_i, r_i, η_i, and $u_i(t) \in \mathbb{R}$, respectively and g is the gravity constant. The dynamics of the i-th FV can be stated via the following discrete-time nonlinear model [58]

$$\begin{cases} \boldsymbol{x}_i(t+1) = \boldsymbol{\phi}_i(\boldsymbol{x}_i(t)) + u_i(t)\,\boldsymbol{\psi}_i \\ \quad\;\; \boldsymbol{y}_i(t) = \boldsymbol{\gamma}\,\boldsymbol{x}_i(t), \end{cases} \tag{1}$$

where $\boldsymbol{x}_i(t) := [p_i(t), v_i(t), T_i(t)]^\top \in \mathbb{R}^3$ and $\boldsymbol{y}(t) := [p_i(t), v_i(t)]^\top \in \mathbb{R}^2$ are the states and outputs of each vehicle, respectively. Also, $\boldsymbol{\psi}_i := [0, 0, (1/\tau_i)\,\Delta t]^\top$, $\boldsymbol{\gamma} := \begin{bmatrix} 1 & 0 & 0 \\ 0 & 1 & 0 \end{bmatrix}$, and

$$\boldsymbol{\phi}_i(\boldsymbol{x}_i(t)) := \begin{bmatrix} p_i(t) + v_i(t)\,\Delta t \\ v_i(t) + \frac{\Delta t}{m_i}\left(\frac{\eta_i}{r_i} T_i(t) - C_{A,i}\, v_i^2(t) - m_i\, g\, f_{r,i}\right) \\ T_i(t) - (1/\tau_i)\, T_i(t)\,\Delta t \end{bmatrix}. \tag{2}$$

Stacking the states, outputs, and the control input signals of all vehicles into vectors yields the platoon dynamics as follows

$$\begin{cases} X(t+1) = \boldsymbol{\Phi}(X(t)) + \boldsymbol{\Psi} U(t), \\ Y(t+1) = \boldsymbol{\Theta} \cdot X(t+1), \end{cases} \tag{3}$$

where $X(t) = [x_1(t)^\top, x_2(t)^\top, \ldots, x_N(t)^\top]^\top \in \mathbb{R}^{3N\times 1}$, $Y(t) = [y_1(t)^\top, y_2(t)^\top, \ldots, y_N(t)^\top]^\top \in \mathbb{R}^{2N\times 1}$, $U(t) = [u_1(t), u_2(t), \ldots, u_N(t)]^\top \in \mathbb{R}^{N\times 1}$. Besides, $\boldsymbol{\Phi} = [\phi_1^\top, \phi_2^\top, \ldots, \phi_N^\top]^\top \in \mathbb{R}^{3N\times 1}$, $\boldsymbol{\Psi} = \mathrm{diag}\{\psi_1, \psi_2, \ldots, \psi_N\} \in \mathbb{R}^{3N\times N}$, and $\boldsymbol{\Theta} = I_N \otimes \gamma \in \mathbb{R}^{2N\times 3N}$.

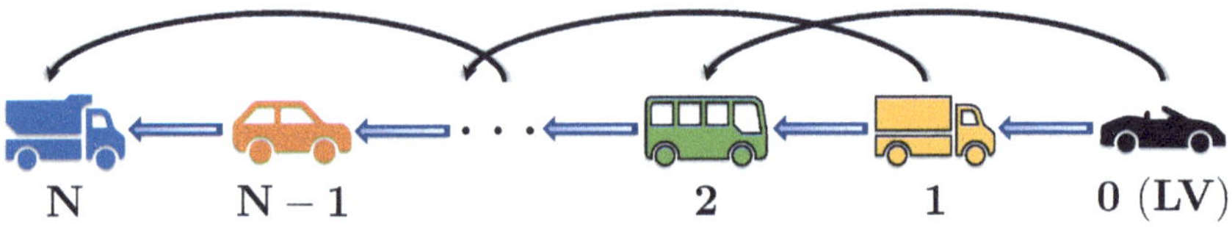

Fig. 2 TPF heterogeneous vehicle platoon with a leader and N followers

Let $\mathcal{A} = [a_{ij}] \in \mathbb{R}^{N\times N}$ be the adjacency matrix of the underlying platoon graph topology where $a_{ij} = 1\,(=0)$ means that the j-th FV can (cannot) send information to the i-th FV, and $\mathcal{D} = \text{diag}\{\text{deg}_1, \text{deg}_2, \ldots, \text{deg}_n\}$ be the degree matrix, where $\text{deg}_i = \Sigma_{j=1}^{n} a_{ij}$. Also, let $p_i = 1\ (=0)$ mean that the i-th FV is (not) pinned to the LV and gets (does not get) information from it. Suppose $\mathbb{P}_i := \{0\}$ if $p_i = 1$ and $\mathbb{P}_i := \varnothing$ if $p_i = 0$. The pinning matrix is then defined by $\mathcal{P} = \text{diag}\{p_1, p_2, \ldots, p_n\}$. We denote $\mathbb{N}_i := \{j | a_{ij} = 1, j \in \mathcal{N}\}$ and $\mathbb{O}_i := \{j | a_{ji} = 1, j \in \mathcal{N}\}$ as the sets of FVs which the i-th FV can get information from and send information to, respectively. The set $\mathbb{I}_i := \mathbb{N}_i \cup \mathbb{P}_i$ is the set of all vehicles sending information to the i-th FV. In this study, for convenience, we consider a dynamic heterogeneous platoon equipped by Two-Predecessor Follower (TPF) communication topology shown in Fig. 2; however, it is straightforward to adapt our algorithm to other communication topologies.

Assumption 1 The directed graph of the platoon topology contains a spanning tree rooted at the LV. This assumption is necessary for stability in both homogeneous [21] and heterogeneous [58] platooning. This ensures that all vehicles get the leader's information either directly or indirectly.

2.2 *Platoon Control Objectives*

The control objectives of the platoon are to track the speed profile generated by the leader while keeping the safe desired distance between the vehicles. Mathematically, we aim at $\lim_{t\to\infty} |v_i(t) - v_0(t)| = 0$ and $\lim_{t\to\infty} |p_{i-1}(t) - p_i(t) - d| = 0$ where d is the desired constant distance between every two consecutive vehicles. We also denote the distance between the i-th and j-th FVs by $d_{i,j}$.

Two types of output are considered here, which are the predicted and assumed outputs. The former is obtained by the calculated control input from optimization, which is fed to the system. The latter is obtained by shifting the optimal output of the last-step optimization problem. Let $\boldsymbol{y}_i^p(k|t)$ and $\boldsymbol{y}_i^a(k|t)$ denote the predicted output and the assumed output, respectively. We explain the details of how to obtain these two outputs in the following sections. The predicted and assumed states are denoted by $\boldsymbol{x}_i^p(k|t)$ and $\boldsymbol{x}_i^a(k|t)$, respectively.

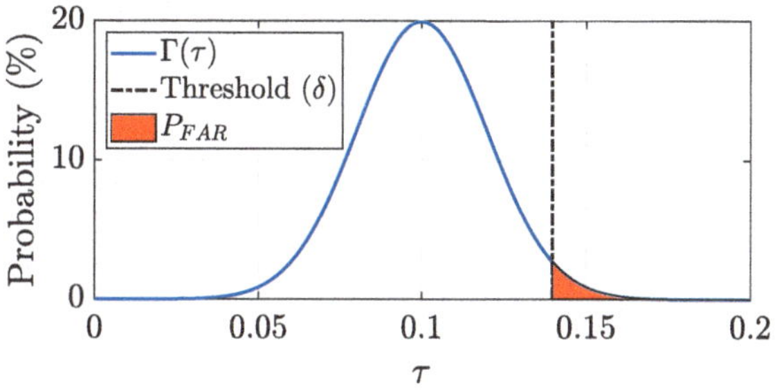

Fig. 3 Probability distribution function of the false alarm rate and the threshold

2.3 *Attack Description*

We mainly focus on a widespread cyber-attack, called the DoS attack. Basically, endangering the security of the system, a DoS attacker jams the network by flooding it with fake requests such that the shared network gets overwhelmed by these demands; hence, becomes too busy to process the legitimate requests sent by the authorized users [27, 59]. This inherently causes packet loss or at least suffering delays in data transfers. In our application, we study two different DoS attack modeling introduced in the literature, i.e.,

- The DoS attacker is able to block the communication link among two nonconsecutive neighboring vehicles, which results in missing inter-vehicular data received by the follower vehicle. In essence, if the communication link among vehicle i and $i-2$ is attacked during $t \in [t_0, t_1]$, the vehicle i is only able to receive the valid data up to $t = t_0$ and has the exact same data until the attack is over, i.e., vehicle i will restart to receive updated data from vehicle $i-2$ at $t > t_1$. In the rest of the chapter, we denote $\tau_a = t_1 - t_0$ as the attack period for notational convenience.
- Another prevalent DoS attack type is to view the intruder as who injects a relatively large delay in the data transmission network. Hence, in this case, during the attack period τ_a, the follower vehicles receive the data with the time delay τ_r. This time delay is much larger compared to a threshold for a practical DSRC network.[1] The threshold can be calculated based on the acceptable probability of false alarm rate P_{FAR}

$$P_{\text{FAR}} = \int_{\delta}^{\infty} \Gamma(\tau)d\tau \leq P_{\text{FAR, acceptable}}, \tag{4}$$

where $\Gamma(\tau)$ is the probability density function of the time delay, and δ is the chosen threshold (shown in Fig. 3) [46]. The threshold δ can also be determined using Monte-Carlo simulations, False Positives and True Negatives [31].

We will propose countermeasures in subsequent sections to face both of the aforementioned attack modelings.

[1] It should be noted that the acceptable time delay heavily depends on the application. Here, we focus on the automotive control application.

3 Secure Controller Design for Dynamic Heterogeneous Platooning

Details of the proposed secure controller are given in this section. In addition, some preliminary concepts needed to develop the method is explained. Furthermore, closed-loop system stability along with the convergence time analysis are presented in this section.

3.1 Overview

To countermeasure the DoS attacker explained in the previous section, we take advantage of a modified version of the Distributed Nonlinear Model Predictive Control (DNMPC) approach proposed in [58], called Secure–DNMPC, which aims at mitigating the effects of the attack while achieving the desired control objectives. The algorithm basically consists of two main phases, namely i) detection and ii) mitigation phase. In the first phase, we seek to detect if a DoS attack is underway. If an attack is detected in which the attacked communication link corresponds to the ego-vehicle with its immediate preceding or following vehicle, then the algorithm ignores the data received through the V2V link (until the attack is over) and switches to the on-board sensors followed by the implementation of the DNMPC. Otherwise, if the blocked link corresponds to the farther neighbors of the ego-vehicle, the victim vehicle makes use of the most recent updated data prior to the attack commence, and the mitigation phase starts by performing Secure–DNMPC. Inherently, in the second phase, each vehicle solves a local optimal control problem detailed as follows to generate its own optimal control input signal, which is used to compute the assumed states. The assumed states are then exchanged with the neighbors. Moreover, if the intruder targets the communication link by injecting a huge amount of time delay in data transmission, denoted by τ_r, the algorithm switches to the Secure–DNMPC–UKF mode to make use of the delayed states as much as possible. Specifically, in this case, the controller employs the observer to estimate the delayed states and provides the controller with the predicted data. The mechanism of the controller in both of the above-mentioned cases are detailed in the following sections.

Assumption 2 As a standard assumption and from a practical point of view, we assume that the attacker has a limited resource of energy preventing him from jamming the network ceaselessly [34, 60, 61].

Remark

It is notable that the DoS attacker never attacks a link between two consecutive vehicles. The reason is that in the algorithm, the positions and velocities are transmitted, which can be reliably measured by on-board sensors mounted on an ego-vehicle such as GPS and radar. Thus, once a follower detects that those quantities are no longer updated, it can switch to its redundant sensors to obtain real-time data.

3.2 Design of the Secure Controller

Consider a predictive horizon N_p for the model predictive control employed to control the platoon. Suppose the predicted control inputs over the horizon are $\mathcal{U}_i^p(t-\tau) := \{u_i^p(0|t-\tau), \ldots, u_i^p(N_p-1|t-\tau)\}$ which need to be calculated by the following optimization problem, which is the local NMPC problem that each vehicle needs to solve at each time instant t

$$\underset{\mathcal{U}_i^p(t-\tau)}{\text{minimize}} \quad J_i(\boldsymbol{y}_i^p, u_i^p, \boldsymbol{y}_i^a, \boldsymbol{y}_{-i}^a) \tag{5a}$$

$$\text{subject to} \quad \boldsymbol{x}_i^p(k+1|t-\tau) = \boldsymbol{\phi}_i(\boldsymbol{x}_i^p(k|t-\tau)) + u_i^p(k|t-\tau)\,\boldsymbol{\psi}_i, \tag{5b}$$

$$\boldsymbol{y}_i^p(k|t-\tau) = \boldsymbol{\gamma}\,\boldsymbol{x}_i^p(k|t-\tau), \tag{5c}$$

$$\boldsymbol{x}_i^p(0|t-\tau) = \boldsymbol{x}_i(t-\tau), \tag{5d}$$

$$u_i^p(k|t-\tau) \in \mathfrak{U}_i, \tag{5e}$$

$$\boldsymbol{y}_i^p(N_p|t-\tau) = \frac{1}{|\mathbb{I}_i|}\sum_{j\in\mathbb{I}_i}\left(\boldsymbol{y}_j^a(N_p|t-\tau) + \tilde{\boldsymbol{d}}_{i,j}\right), \tag{5f}$$

$$T_i^p(N_p|t-\tau) = h_i(v_i^p(N_p|t-\tau)), \tag{5g}$$

where $\boldsymbol{y}_{-i}(t) := [\boldsymbol{y}_{i_1}^\top, \ldots, \boldsymbol{y}_{i_m}^\top]^\top$ (if $\{i_1, \ldots, i_m\} := \mathbb{N}_i$), $\mathfrak{U}_i = \{u_i \mid u_i \in [\underline{u}_i, \bar{u}_i]\}$ defines the feasible bounds on the control input, $|\mathbb{I}_i|$ is the cardinality of $\mathbb{I}_i$, $\tilde{\boldsymbol{d}}_{i,j} := [d_{i,j}, 0]^\top$, and τ is either τ_a or τ_r depending on the attack model. The last two terminal constraints are to make the DNMPC algorithm stable. For a detailed description of the above constraints, the interested reader is referred to [58].

The objective function (5a) is defined as the summation of all local cost functions

$$\begin{aligned} J_i(\boldsymbol{y}_i^p, u_i^p, \boldsymbol{y}_i^a, \boldsymbol{y}_{-i}^a) := \sum_{k=0}^{N_p-1}\Big(&\|\boldsymbol{y}_i^p(k|t-\tau) - \boldsymbol{y}_{\text{des},i}(k|t-\tau)\|_{\boldsymbol{Q}_i} \\ &+ \|u_i^p(k|t-\tau) - h_i(v_i^p)\|_{R_i} + \|\boldsymbol{y}_i^p(k|t-\tau) - \boldsymbol{y}_i^a(k|t-\tau)\|_{\boldsymbol{F}_i} \\ &+ \sum_{j\in\mathbb{N}_i}\|\boldsymbol{y}_i^p(k|t-\tau) - \boldsymbol{y}_j^a(k|t-\tau) - \tilde{\boldsymbol{d}}_{i,j}\|_{\boldsymbol{G}_i}\Big), \end{aligned} \tag{6}$$

in which, for a weight matrix $A \succeq 0$, $\|x\|_A := x^\top A\, x$. In (6), $0 \preceq Q_i, F_i, G_i \in \mathbb{R}^2$ and $0 \leq R_i \in \mathbb{R}$ are the weight matrices which are the NMPC regularization factors. In fact, the matrices Q_i, R_i, F_i, G_i penalize for deviation of the predicted output from the desired output $y_{\text{des},i}(k|t-\tau)$, deviation of the predicted control input from the equilibrium, deviation of the predicted output from the assumed output, and deviation of the predicted output from the neighbors' assumed trajectories, respectively. For the i-th FV, the desired state and control signal are $x_{\text{des},i}(t) := [p_{\text{des},i}(t), v_{\text{des},i}(t), T_{\text{des},i}(t)]^\top$ and $u_{\text{des},i}(t) := T_{\text{des},i}(t)$, respectively, where $p_{\text{des},i}(t) := p_0(t) - i\, d$, $v_{\text{des},i}(t) := v_0$, $T_{\text{des},i}(t) := h_i(v_0)$ where $h_i(v_0) := (r_i/\eta_i)(C_{A,i}\, v_0^2 + m_i\, g\, f_{r,i})$ is the external drag. The desired output is $y_{\text{des},i}(t) := \gamma\, x_{\text{des},i}(t) \in \mathbb{R}^2$.

Having injected the DoS attack on the communication network of two nonconsecutive vehicles, the follower car fails to receive updated data from its neighbor. It should be noted that what distinguishes the intelligent intruder from an intrinsic network time delay is that the data received after a huge time delay is no longer useful to generate the correct control input. To combat this attacker, we propose to integrate an Unscented Kalman Filter (UKF) within our Secure–DNMPC such that the receiver can estimate the missing data and feed the predicted values to the NMPC controller. Consequently, the NMPC controller ignores the delayed states and makes use of the predicted values as long as the attack is running. We refer to this mode of the controller as the Secure–DNMPC–UKF mode. The controller is then switched back to Secure–DNMPC once either the attack is over or the injected time delay falls below the specified threshold.

Embedding the UKF within our design takes us one more step closer to a more realistic vehicle platoon system. In particular, through our proposed co-design, we can take the process and sensor noise into account as well, which is of high importance, especially for measurement sensors. From one side, assuming ideal non-noisy sensors, as done in most of the existing works in the literature, is a contrived assumption. On the other hand, the signals sent through the environment from one vehicle to another will be most likely compromised by some noise due to surrounding weather and road conditions.

The reason for choosing UKF over Extended Kalman Filter (EKF) is to avoid the propagation of the state distribution approximation error through the system dynamics caused by the first-order linearization performed in EKF. This is vital in terms of ensuring the safety of the platoon as the propagated error in the true posterior mean and covariance of the transformed Gaussian random variable may be large and cause unsafe driving behavior. Remarkably, adopting UKF does not impose anymore computational burden compared to EKF. The interested reader is referred to [62] for more details on the superiority of UKF over EKF for nonlinear state estimation. Figure 4 shows a flowchart illustrating the procedure of the Secure–DNMPC–UKF co-design.

Before delving into the details of the Secure–DNMPC–UKF algorithm, a quick overview of the basics of Unscented Kalman Filtering is given in the following.

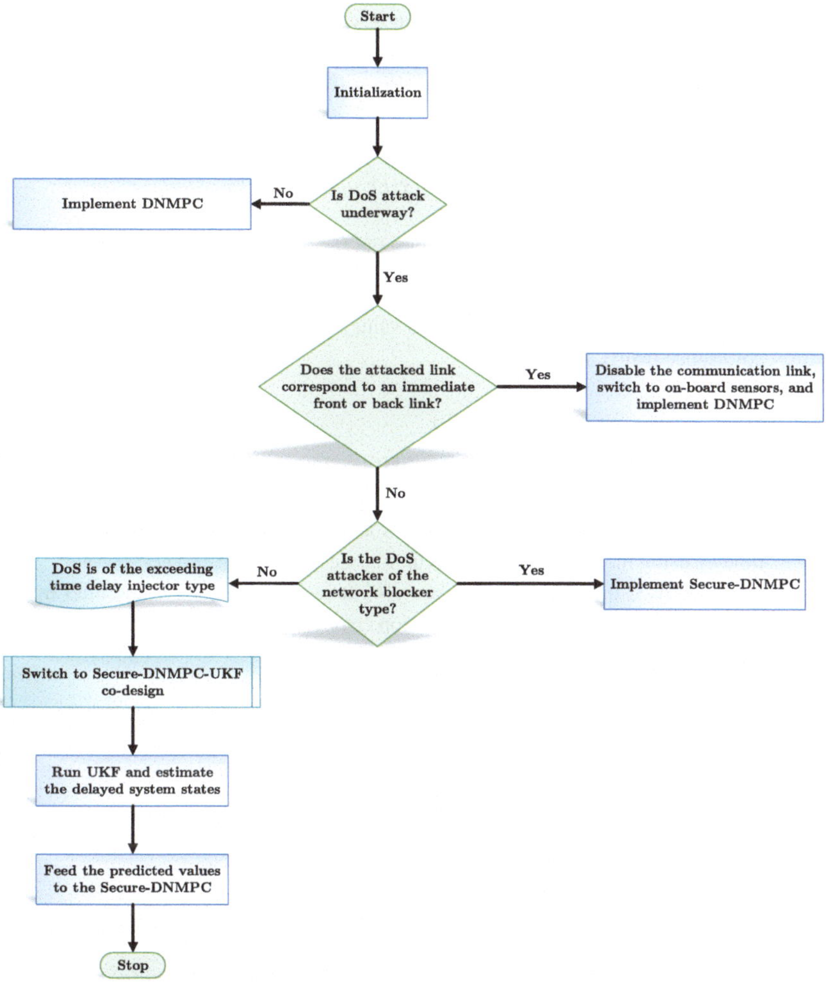

Fig. 4 Procedure of the proposed Secure–DNMPC–UKF co-design

Principles of Unscented Kalman Filtering

Unscented Kalman Filter, as a nonlinear state observer, basically relies on the unscented transformation to capture the statistical properties of state estimates via nonlinear functions. The observer initially captures the mean and covariance of the state estimates through a set of so-called sigma points. The algorithm makes use of those sigma points as the inputs of the process and measurement functions to generate a new set of states. Subsequently, a set of state estimates and state estimation error covariance are obtained using the mean and covariance of the previously transformed points.

Let us consider an n-state nonlinear system described by the following nonlinear state transition and measurement functions comprised by additive zero-mean process noise $\boldsymbol{w}[k] \sim (0, \boldsymbol{Q}[k])$ and measurement noise $\boldsymbol{v}[k] \sim (0, \boldsymbol{R}[k])$

$$\begin{cases} \boldsymbol{x}[k+1] = f(\boldsymbol{x}[k], u_s[k]) + \boldsymbol{w}[k] \\ \boldsymbol{y}[k] = h(\boldsymbol{x}[k], u_m[k]) + \boldsymbol{v}[k] \end{cases} \tag{7}$$

The filter takes the following steps to obtain the state estimates and the state estimation error covariance

(1) The filter is initialized with an initial value for state $x[0]$ and state estimation error covariance matrix P

$$\hat{\boldsymbol{x}}[0|-1] = \mathbb{E}(\boldsymbol{x}[0]) \tag{8}$$

$$P[0|-1] = \mathbb{E}[(\boldsymbol{x}[0] - \hat{\boldsymbol{x}}[0|-1)(\boldsymbol{x}[0] - \hat{\boldsymbol{x}}[0|-1)^\top] \tag{9}$$

where $\hat{\underline{x}}[k]$ is the state estimate at time k and $\hat{\boldsymbol{x}}[k_1|k_0]$ denotes the state estimate at time k_1 using the measurement data up to time k_0.

(2) Having used the measurement data $\boldsymbol{y}[k]$ at each time instant k, the filter updates the state estimate and the state estimation error covariance:

(a) Choose the sigma points $\hat{\boldsymbol{x}}^{(i)}[k|k-1]$ at time k

$$\hat{\boldsymbol{x}}^{(0)} = \hat{\boldsymbol{x}}[k|k-1] \tag{10}$$

$$\hat{\boldsymbol{x}}^{(i)}[k|k-1] = \hat{\boldsymbol{x}}[k|k-1] + \Delta\boldsymbol{x}^{(i)}, \quad i = 1, 2, \ldots, 2n \tag{11}$$

$$\Delta\boldsymbol{x}^{(i)} = (\sqrt{cP[k|k-1]})_i, \quad i = 1, 2, \ldots, n \tag{12}$$

$$\Delta\boldsymbol{x}^{(n+i)} = -(\sqrt{cP[k|k-1]})_i, \quad i = 1, 2, \ldots, n \tag{13}$$

where $c = \alpha^2(n+\kappa)$ is a scaling factor and $(\sqrt{cP})_i$ is the i-th column of the $\sqrt{cP}$ matrix [63].

(b) For each of the sigma points, use the nonlinear measurement function to compute the predicted measurements

$$\hat{\boldsymbol{y}}^{(i)}[k|k-1] = h(\hat{\boldsymbol{x}}^{(i)}[k|k-1], u_m[k]), \quad i = 1, 2, \ldots, 2n \tag{14}$$

(c) In order to obtain the predicted measurement at time k, integrate the predicted measurements

$$\hat{\boldsymbol{y}}[k] = \Sigma_{i=0}^{2n} W_n^{(i)} \hat{\boldsymbol{y}}^{(i)}[k|k-1] \tag{15}$$

$$W_n^{(0)} = 1 - \frac{n}{\alpha^2(n+\kappa)} \tag{16}$$

$$W_n^{(i)} = \frac{1}{2\alpha^2(n+\kappa)}, \quad i = 1, 2, \ldots, 2n \tag{17}$$

(d) By adding the measurement noise $\boldsymbol{R}[k]$, estimate the covariance matrix of the predicted measurement

$$P_y = \Sigma_{i=0}^{2n} W_c^{(i)} (\hat{\boldsymbol{y}}^{(i)}[k|k-1] - \hat{\boldsymbol{y}}[k])(\hat{\boldsymbol{y}}^{(i)}[k|k-1] - \hat{\boldsymbol{y}}[k])^\top + \boldsymbol{R}[k] \tag{18}$$

$$W_c^{(0)} = (2 - \alpha^2 + \beta) - \frac{n}{\alpha^2(n+\kappa)} \tag{19}$$

$$W_c^{(i)} = \frac{1}{2\alpha^2(n+\kappa)}, \quad i = 1, 2, \ldots, 2n \tag{20}$$

For the details on effects of parameters α, β, and κ the reader is referred to [63].

(e) Estimate the cross-covariance between $\hat{\boldsymbol{x}}[k|k-1]$ and $\hat{\boldsymbol{y}}[k]$

$$P_{xy} = \frac{1}{2\alpha^2(n+\kappa)} \Sigma_{i=1}^{2n} (\hat{\boldsymbol{x}}^{(i)}[k|k-1] - \hat{\boldsymbol{x}}[k|k-1])(\hat{\boldsymbol{y}}^{(i)}[k|k-1] - \hat{\boldsymbol{y}}[k|k-1])^\top \tag{21}$$

Note that $\hat{\boldsymbol{x}}^{(0)}[k|k-1] - \hat{\boldsymbol{x}}[k|k-1] = 0$.

(f) Compute the estimated state and state estimation error covariance at time step k

$$K = P_{xy} P_y^{-1} \tag{22}$$

$$\hat{\boldsymbol{x}}[k|k] = \hat{\boldsymbol{x}}[k|k-1] + K(\boldsymbol{y}[k] - \hat{\boldsymbol{y}}[k]) \tag{23}$$

$$P[k|k] = P[k|k-1] - K P_y K_k^\top \tag{24}$$

where K is the Kalman gain.

(3) Now the state and state estimation error covariance can be predicted at time instant $k+1$

(a) Choose the sigma points $\hat{\boldsymbol{x}}^{(i)}[k|k]$ at time instant k.

$$\hat{\boldsymbol{x}}^{(0)}[k|k] = \hat{\boldsymbol{x}}[k|k] \tag{25}$$

$$\hat{\boldsymbol{x}}^{(i)}[k|k] = \hat{\boldsymbol{x}}[k|k] + \Delta\boldsymbol{x}^{(i)}, \quad i = 1, 2, \ldots, 2n \tag{26}$$

$$\Delta\boldsymbol{x}^{(i)} = (\sqrt{cP[k|k]})_i, \quad i = 1, 2, \ldots, n \tag{27}$$

$$\Delta\boldsymbol{x}^{(n+i)} = -(\sqrt{cP[k|k]})_i, \quad i = 1, 2, \ldots, n \tag{28}$$

(b) In order to get the predicted states at time $k+1$, combine the predicted states

$$\hat{\boldsymbol{x}}[k+1|k] = \Sigma_{i=0}^{2n} W_n^{(i)} \hat{\boldsymbol{x}}^{(i)}[k+1|k] \tag{29}$$

$$W_n^{(0)} = 1 - \frac{n}{\alpha^2(n+\kappa)} \tag{30}$$

$$W_n^{(i)} = \frac{1}{2\alpha^2(n+\kappa)}, \quad i = 1, 2, \ldots, 2n \tag{31}$$

(4) To account for the process noise, add $\boldsymbol{Q}[k]$ and compute the covariance of the predicted state

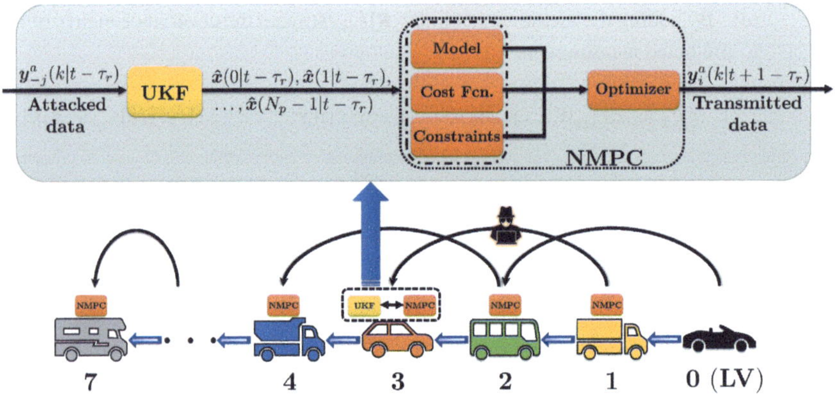

Fig. 5 Schematic of the proposed Secure–DNMPC–UKF co-design

$$P[k+1|k] = \Sigma_{i=0}^{2n} W_c^{(i)} (\hat{\boldsymbol{x}}^{(i)}[k+1|k] - \hat{\boldsymbol{x}}[k+1|k])(\hat{\boldsymbol{x}}^{(i)}[k+1|k] - \hat{\boldsymbol{x}}[k+1|k])^\top + \boldsymbol{Q}[k] \tag{32}$$

$$W_c^{(0)} = (2 - \alpha^2 + \beta) - \frac{n}{\alpha^2(n+\kappa)} \tag{33}$$

$$W_c^{(i)} = \frac{1}{2\alpha^2(n+\kappa)}, \quad i = 1, 2, \ldots, 2n \tag{34}$$

For more details on the observer for the case of non-additive process/measurement noise, please see [63].

The proposed algorithm is summarized in Algorithm 1, which is the extended version of the authors' previous work [64] for static platoons. We further note that $\boldsymbol{y}_i^a(t)$ represents the data sent by the vehicle i to the set $\mathbb{O}_i$ while $\boldsymbol{y}_{-j}^a$ denotes the data received by the vehicle i from its neighbors $j \in \mathbb{N}_i$. Superscript a, p, and $*$ are to distinguish between assumed, predicted, and optimal quantities, respectively. The assumed quantities are the ones transmitted by the vehicles in the platoon. Figure 5 illustrates the Secure–DNMPC–UKF co-design in which $\hat{\boldsymbol{x}}(k|t)$ denotes the estimated state at time instant k using the measured data up to time t.

3.3 *Stability Analysis of Secure–DNMPC*

In this section we study the stability of the Secure–DNMPC algorithm incorporating the time delay τ imposed by the DoS attacker. Prior to stability analysis, let us first introduce the following Lemma.

Lemma 1 ([58]) *For any platoon wherein all the vehicles can receive data (directly/ indirectly) from the leader vehicle, the eigenvalues of* $(\mathcal{D}+\mathcal{P})^{-1}\mathcal{A}$ *lie within the unit circle disk, i.e.*

$$\left|\lambda_i\left\{(\mathcal{D}+\mathcal{P})^{-1}\mathcal{A}\right\}\right| < 1. \tag{35}$$

Now, we can prove the stability of the Secure–DNMPC algorithm.

Algorithm 1 SECURE- DNMPC- UKF FOR DYNAMIC NONLINEAR HETEROGENEOUS VEHICLE PLATOONING UNDER DOS ATTACK

1: **Initialization:**
Assumed values for vehicle i are set at time $t = 0$,
$u_i^a(k|0) = h_i(v_i(0)),\ \boldsymbol{y}_i^a(k|0) = \boldsymbol{y}_i^p(k|0),\quad k = 0, 1, \ldots, N_p - 1$
2: **while** $t \le t_{\text{final}}$ **do**
3: Cut-in/cut-out CHECK ▷ Check to see if cut-in/cut-out occurred
4: Adjust data send-to/receive-from vehicles based on the occurred cut-in/cut-out
5: **if** $p_{-j}^a(t) = p_{-j}^a(t-1),\ j \in \mathbb{N}_i$ **then** ▷ Check to see if a DoS is underway
6: **if** $j = i - 1$ or $j = i + 1$ **then** ▷ Check to see if the attacked link
7: corresponds to a predecessor or a follower
8: Disable communication link, switch to on-board sensors, $\tau \leftarrow 0$, and **Go to:** 13
9: **else**
10: **if** Attacker blocks the communication link **then** ▷ Check to see if the attacker is of
11: the blockage type
12: $\tau \leftarrow \tau_a$
13: **for** Each vehicle i **do** ▷ Implement Secure–DNMPC
14: Solve Problem 5 at time $t > 0$ and yield $u_i^*(k|t-\tau),\ k = 0, 1, \ldots, N_p - 1$
15: Compute: $\begin{cases} \boldsymbol{x}_i^*(k+1|t-\tau) = \boldsymbol{\phi}_i(\boldsymbol{x}_i^*(k|t-\tau)) + \boldsymbol{\psi}_i u_i^*(k|t-\tau), \\ \boldsymbol{x}_i^*(0|t-\tau) = \boldsymbol{x}_i(t-\tau),\quad k = 0, 1, \ldots, N_p - 1 \end{cases}$
16: Compute: $u_i^a(k|t-\tau+1) = \begin{cases} u_i^*(k+1|t-\tau), & k = 0, 1, \ldots, N_p - 2 \\ h_i\left(v_i^*(N_p|t-\tau)\right), & k = N_p - 1 \end{cases}$
17: Compute: $\begin{cases} \boldsymbol{x}_i^a(k+1|t-\tau+1) = \boldsymbol{\phi}_i\left(\boldsymbol{x}_i^a(k|t-\tau+1)\right) + \boldsymbol{\psi}_i u_i^a(k|t-\tau+1) \\ \boldsymbol{x}_i^a(0|t-\tau+1) = \boldsymbol{x}_i^*(1|t-\tau),\quad k = 0, 1, \ldots, N_p - 1 \end{cases}$
18: Compute: $\boldsymbol{y}_i^a(k|t-\tau+1) = \boldsymbol{\gamma}\boldsymbol{x}_i^a(k|t-\tau+1),\quad k = 0, 1, \ldots, N_p - 1$
19: Send $\boldsymbol{y}_i^a(k|t-\tau+1)$ to the vehicles lie in the set $\mathbb{O}_i$, and receive $\boldsymbol{y}_{-j}^a(k|t-\tau+1)$ from neighboring vehicles $j \in \mathbb{N}_i$ and compute $\boldsymbol{y}_{\text{des},i}(k|t-\tau+1)$
20: Exert the first element of the optimal control signal $u_i(t-\tau) = u_i^*(0|t-\tau)$
21: **end for**
22: **else if** Attacker injects exceeding delay $\tau_r > \delta$ **then** ▷ Check to see if the attacker
23: is of the exceeding time delay injector type
24: $\tau \leftarrow \tau_r$
25: Switch to Secure–DNMPC–UKF mode
26: Estimate the delayed states via UKF
27: Implement the Secure–DNMPC using the predicted states coming from the UKF
28: **end if**
29: **end if**
30: **end if**
31: **end while**

Theorem 1 ([64]) *If a platoon which is under a DoS attack satisfies the condition in Lemma 1, then the terminal output of the system controlled by the Secure–DNMPC proposed in Algorithm 1 asymptotically converges to the desired state, i.e.*

$$\lim_{t\to\infty} \left| \mathbf{y}_i^p(N_p|t-\tau) - \mathbf{y}_{des,i}(N_p|t-\tau) \right| = 0. \tag{36}$$

4 Dynamic Platoon Control: Handling Cut-in/Cut-out Maneuvers

In this section, we consider a dynamic heterogeneous platoon wherein arbitrary vehicle(s) might perform cut-in/cut-out maneuvers. Here, we demonstrate the ability of the proposed algorithm to handle dynamic maneuvers while the platoon is subject to the cyber-attack.

First, we consider a secure dynamic heterogeneous platoon and prove some results based on which we extend the results to an insecure platoon. Assume there exist N_{ci} cut-in and N_{co} cut-out maneuvers in total while the number of initial FVs in the platoon is N. Let $\mathcal{N}_{ci} := \{1, \ldots, N_{ci}\}$ and $\mathcal{N}_{co} := \{1, \ldots, N_{co}\}$. We denote the time of the i-th cut-in and the j-th cut-out maneuvers by $t_{ci,i}$ and $t_{co,j}$, respectively. The following theorem determines the time of convergence of a dynamic platoon including possible cut-in and cut-out maneuvers.

Lemma 2 ([58, Theorem 2]) *If Assumption 1 is satisfied, then Problem (5) guarantees convergence of the output to the desired output in at most N time steps, i.e.,* $\mathbf{y}_i^p(N_p|t) = \mathbf{y}_{des,i}(N_p|t), \forall t \geq N$, *for a static platoon (without any dynamic maneuvers).*

Theorem 2 *When having cut-in and/or cut-out maneuvers in a secured dynamic platoon, if Assumption 1 is satisfied, the Problem (5) guarantees convergence of the output to the desired output in at most*

$$\begin{aligned} t_{conv,\,secure} := &\max_{i,j}\left[t_{ci,i}, t_{co,j} \mid \forall i \in \mathcal{N}_{ci}, \forall j \in \mathcal{N}_{cj}\right] \\ &+ N + N_{ci} - N_{co}, \end{aligned} \tag{37}$$

time steps, i.e., $\mathbf{y}_i^p(N_p|t) = \mathbf{y}_{des,i}(N_p|t), \forall t \geq t_{conv,\,secure}$.

Proof Let $\mathcal{L} := \mathcal{D} - \mathcal{A}$ be the Laplacian matrix of the underlying platoon graph topology. When a new cut-in or cut-out occurs, some new chaos is introduced to the system so we can consider the latest cut-in/cut-out maneuver. Considering the latest cut-in, one vehicle is added to the number of existing vehicles, let it be N. If the platoon graph is unidirectional and satisfies Assumption 1, the new $\mathcal{A} \in \mathbb{R}^{(N+1)\times(N+1)}$ is a lower-triangular matrix. Moreover, according to [58, Lemma 4], we have $\mathcal{D} + \mathcal{P} > 0$, yielding the eigenvalues of $(\mathcal{D} + \mathcal{P})^{-1}\mathcal{A}$ to be zero and this matrix to be nilpotent with degree at most $N + 1$. Based on [58, Lemma 1] and [58, Theorem 1], $\mathbf{y}_i^p(N_p|t)$ converges to the desired output in at most $N + 1$

steps. Extending this to N_{ci} cut-in maneuvers requires $N + N_{ci}$ time steps after the latest cut-in. Similar analysis can be performed for the cut-out maneuvers, resulting in $N - N_{co}$ time steps after the latest cut-out because the number of vehicles has been reduced. In general, having N_{ci} cut-in and N_{co} cut-out maneuvers will need $N + N_{ci} - N_{co}$ time steps after the latest maneuver which can be formulated as $\max_{i,j}[t_{ci,i}, t_{co,j} \mid \forall i \in N_{ci}, \forall j \in N_{cj}]$.

Corollary 1 *Lemma 2, for the static platoon, is a special case of Theorem 2 which is for a dynamic platoon.*

Proof When neither cut-in nor cut-out happen, the time of convergence is $t_{\text{conv, secure}} = 0 + N + 0 + 0 = N$ according to Theorem 2.

Special Cases

Four special cases of the dynamic platoon are as follows:

E.g. (1) One cut-in happens at $t = 0$ and one cut-out happens at $t = N$: According to Theorem 2, the platoon converges in $t = N + N + 1 - 1 = 2N$. It is correct because before the cut-out, the platoon contains $N + 1$ vehicles until time N. When cut-out happens, the platoon is changed to a platoon with N vehicles which converges in N time steps according to Lemma 2.

E.g. (2) One cut-out happens at $t = 0$ and one cut-in happens at $t = N$: According to Theorem 2, the platoon converges in $t = N + N + 1 - 1 = 2N$, which is correct because in $t \in [0, N]$, the platoon includes $N - 1$ vehicles until time N. When cut-in happens, the platoon is modified to a platoon with N vehicles which converges in N time steps according to Lemma 2.

E.g. (3) Both cut-in and cut-out happen at $t = 0$: According to Theorem 2, the platoon converges in $t = 0 + N + 1 - 1 = N$, which is correct because the platoon includes N vehicles which converges in N time steps according to Lemma 2.

E.g. (4) Both cut-in and cut-out happen at $t = N$: According to Theorem 2, the platoon converges in $t = N + N + 1 - 1 = 2N$, which is correct because in $t \in [0, N]$, the platoon includes N vehicles. After the cut-in/cut-out actions the platoon still includes N vehicles which converges in N time steps according to Lemma 2.

Corollary 2 *When having cut-in and/or cut-out maneuvers in an insecure dynamic platoon, if Assumption 1 is satisfied, the convergence time of the output to the desired output is upper bounded by* $t_{conv,\, secure} + \max\{\tau_r, \tau_a\}$, *i.e.*,[2]

$$\begin{aligned} t_{conv,\, insecure} \leq & \max_{i,j}\left[t_{ci,i}, t_{co,j} \mid \forall i \in N_{ci}, \forall j \in N_{cj}\right] \\ & + N + N_{ci} - N_{co} + \max\{\tau_r, \tau_a\}, \end{aligned} \tag{38}$$

time steps, i.e., $y_i^p(N_p|t-\tau) = y_{des,i}(N_p|t-\tau), \forall t \geq t_{conv,\, insecure}$.

[2]Although, this upper bound might be conservative in some cases (such as in the scenario studied in Sect. 5.2), it provides a safe margin for the convergence time of the controller.

5 Simulation Results

A heterogeneous platoon consisted of seven different vehicles is considered where they can exchange inter-vehicular data among each other through the TPF communication topology. It is assumed that the communication link connecting the vehicle 1 and 3 is subject to a DoS attack. Therefore, vehicle 3 cannot receive real-time data, including the position and velocity of vehicle 1 while the attack is performing (see Fig. 6). Remarkably, to emulate a practical scenario, based on Assumption 2 the external intruder is only able to cause communication degradation among the vehicles for a finite time period. In the following simulations, the DoS attacker starts jamming the communication link from vehicle 1 to 3. Seven different vehicles with realistic parameters form the platoon wherein the leader vehicle starts driving at $v_0(0) = 20\text{m/s}$ for one second, then it accelerates to reach $v_0(2) = 22\text{m/s}$ and continues with this velocity until the end of the simulation. The prediction horizon and desired spacing among consecutive vehicles have been chosen as $N_p = 20$, and $d = 10$ meters, respectively. The parameters of the participating vehicles in the platoon are listed in Table 1, which is in accordance with [65]. We have extended the code in [66] for our security analysis.

Remark

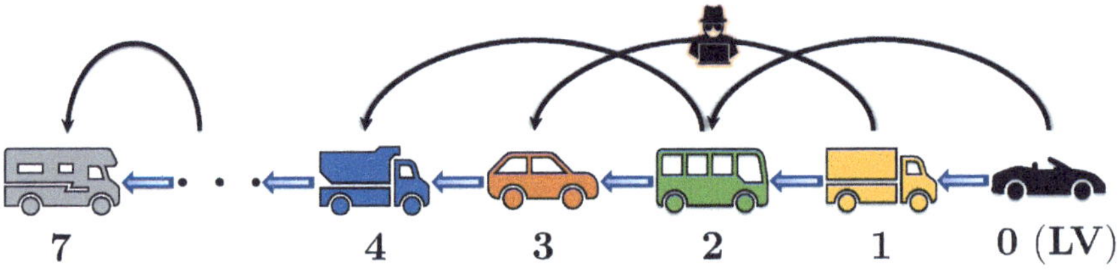

Fig. 6 TPF heterogeneous attacked vehicle platoon with a leader and 7 followers

Table 1 Parameters of the participating vehicles in the platoon

Vehicle index	m_i (kg)	τ_i (s)	$C_{A,i}$ (N s^2 m^{-2})	r_i (m)
1	1035.7	0.51	0.99	0.30
Cut-in	1305.9	0.63	1.00	0.40
2	1849.1	0.75	1.15	0.38
3	1934.0	0.78	1.17	0.39
4	1678.7	0.70	1.12	0.37
5	1757.7	0.73	1.13	0.38
6	1743.1	0.72	1.13	0.37
7	1392.2	0.62	1.06	0.34

To select an appropriate value for the prediction horizon, one has to notice as τ increases, N_p needs to be decreased in order to let the vehicles have enough time to exchange and update their data prior to the attack occurrence. On the other hand, too small values for N_p results in frequent rapid oscillations in the control input which makes the controller unimplementable in practice.

5.1 DoS Attack Modeled as a Network Blocker

In this part, we take one more step to effectively control the dynamic heterogeneous platoon endangered by an intelligent DoS intruder. As was previously described, the attacker could jam the communication network among any two nonconsecutive vehicle to prohibits a follower vehicle from receiving updated data. Having made an expressive scenario incorporating both cut-in and cut-out actions while taking into account a DoS attack, we consider a same setting for the attacked platoon presented in the previous section except assuming a vehicle merges with the platoon at $t = 2$ s to be placed in front of the second FV. Furthermore, we let the fourth FV to perform a cut-out action at $t = 4$ s (see Fig. 7). We note that the desired distance among the vehicles ($d = 10$ m) provides enough space for a regular vehicle to cut-in. The attack happens on the communication environment among the first and third FVs in the time interval $t \in [3, 6]$. Although these tight actions might not seem to happen in practice, they are chosen to challenge the algorithm largely. Figure 8 demonstrate the driving quantities of the respective platoon.

From Fig. 8a, one can see that despite the blockage of the data transfer link from vehicle 1 to 3, there is no collision in the platoon, and the safety has been ensured. Besides, Fig. 8b reveals that the Secure–DNMPC algorithm effectively mitigates the DoS attack and the followers begin to keep tracking the leader's speed profile shortly after the attack is over. Convergence of torque and acceleration are also demonstrated

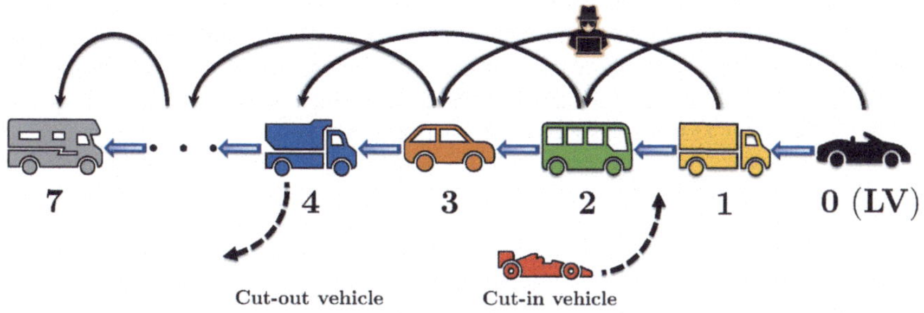

Fig. 7 TPF dynamic heterogeneous attacked vehicle platoon with cut-in and cut-out vehicles

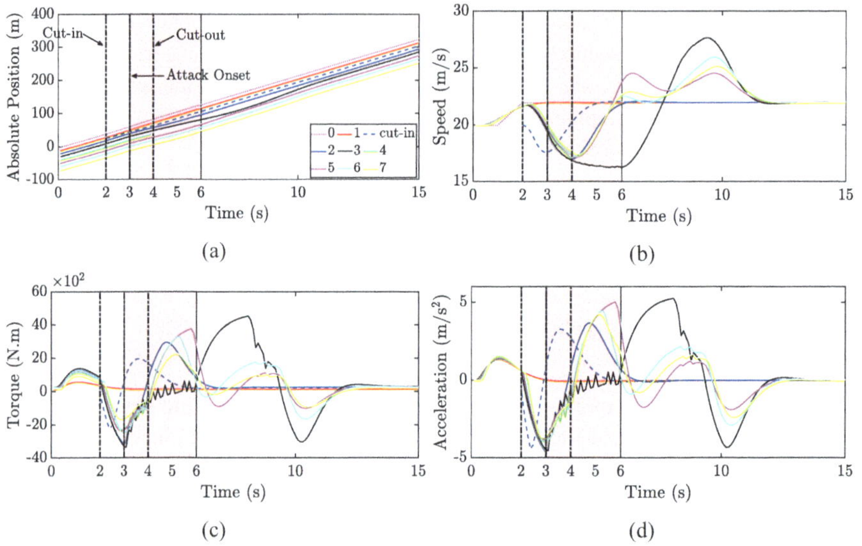

Fig. 8 **a** Absolute position, **b** speed, **c** torque, and **d** acceleration of the TPF dynamic heterogeneous DoS attacked platoon with cut-in/cut-out maneuvers equipped by secure–DNMPC

in Fig. 8c, d. It is worth mentioning that by reducing its speed, the second FV has increased its gap with the first FV to make the desired distance of 10m for the cut-in vehicle. Consequently, the following vehicles have lessened their velocity to keep the desired distance. Figure 8b verifies this fact. A similar analysis exists for the cut-out maneuver where the following vehicles have increased their velocity to reach the desired distance from the vehicles in front.

As one can see, the spacing and speed tracking objectives have been safely fulfilled. To have a clearer look at the spacing objective, Fig. 9 shows the magnified absolute positions and the spacing error of consecutive vehicles. Since all the spacing errors in Fig. 9b are greater than -10 meters, no collision has occurred. Moreover, the relative spacing error shows jumps in the distance error (blue and purple curves) because of the cut-in/cut-out maneuvers.[3] As expected, the spacing error for the cut-out maneuver (purple curve in Fig. 9b) has an opposite sign with respect to the cut-in error (blue curve in Fig. 9b). Furthermore, we see that convergence has been reached in less than 14s which coincides with Corollary 2 because $t_{\text{conv, insecure}} \leq \max(2, 4) + 7 + 1 - 1 + 3 = 14$s.

[3]Note that the jump in the relative spacing error of the third FV (black curve) is due to the DoS attack.

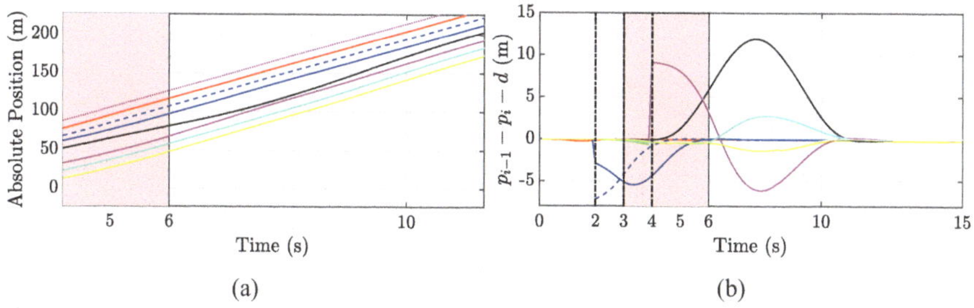

Fig. 9 **a** Magnified absolute position and **b** spacing error of the TPF dynamic heterogeneous DoS attacked platoon with cut-in/cut-out maneuvers equipped by Secure–DNMPC

5.2 DoS Attack Modeled as an Exceeding Time Delay Injection in the Data Transmission

Inherent communication delay of standard 802.11p-based DSRC network ranges from tens to hundreds of milliseconds [67–69]. Here, to ensure modeling a highly devastating attacker, we assume the time delay imposed by the intruder is $\tau_r = 2.5\,\text{s}$. In addition, non-ideal sensors are assumed in the simulations, i.e., an additive zero-mean white Gaussian noise with variance $\sigma^2 = 0.01$ is considered on both the position and velocity sensors.[4] To challenge more the algorithm we introduce a severer attack which happens for a longer period of time, i.e., in the time range $t \in [3, 10]$. It is worth noting that cut-in and cut-out maneuvers are still in effect at $t = 2$sec and $t = 4$sec, respectively. Figure 10 shows the performance of the proposed co-design controller on the attacked platoon with cut-in and cut-out actions. As is demonstrated by the driving quantities, safe distance and velocity tracking requirements have been fulfilled. Furthermore, the convergence has been reached in less than 18s which again verifies Corollary 2 because $t_{\text{conv, insecure}} \leq \max(2, 4) + 7 + 1 - 1 + \max(2.5, 7) = 18\text{s}$. It would also be insightful to compare the results to the case where UKF is not embedded in the design. Figure 11 demonstrates the resulting driving behavior when only relying on the controller leaving out the estimation phase. Occurring collision and violating the control objectives clearly prove the critical role of the observer design.

It is noticeable that by comparing the previous scenarios (Figs. 8, 10, and 11), it reveals that embedding the UKF within our controller design, also has the advantage of reducing the oscillations in the control input caused by the cyber attack. This generation of a smoother control input enhances the driving comfort in practice.

We highlight that the proposed algorithm has also been successfully tested on different platoon formations such as Two-Predecessor Leader Follower (TPLF), with

[4]This could also be considered as the environment effects on the transmitted signals. Modeling the environment effect with white Gaussian noise in V2V communications is widely used in literature [70, 71].

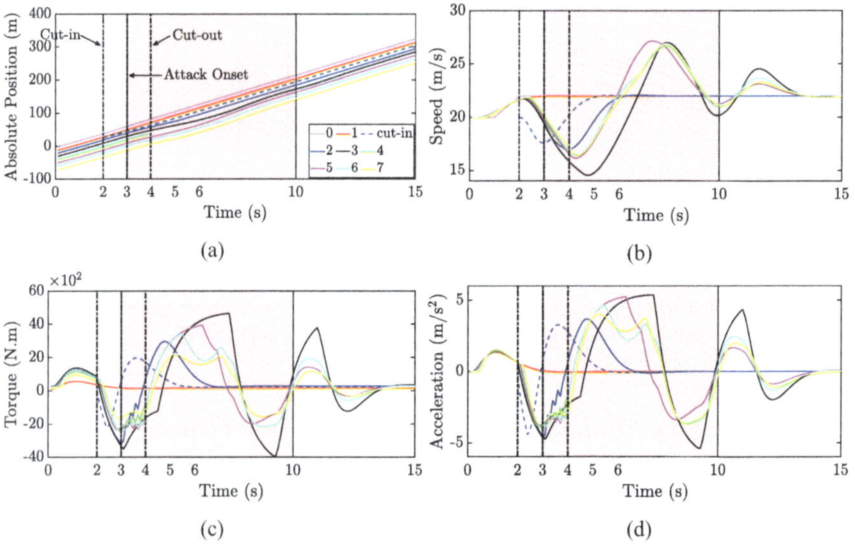

Fig. 10 **a** Absolute position, **b** speed, **c** torque, and **d** acceleration of the TPF dynamic heterogeneous DoS attacked platoon with cut-in/cut-out maneuvers equipped by Secure–DNMPC–UKF co-design

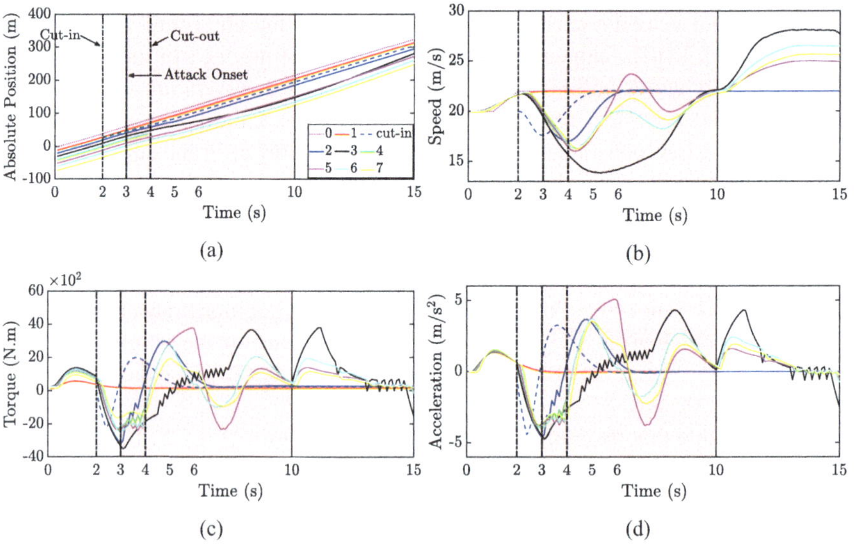

Fig. 11 **a** Absolute position, **b** speed, **c** torque, and **d** acceleration of the TPF dynamic heterogeneous DoS attacked platoon with cut-in/cut-out maneuvers without UKF design

different spacing policies such as Constant Time Headway (CTH) policy, and also on Federal Test Procedure (FTP) drive cycle to emulate urban driving.

6 Conclusion and Future Directions

This chapter dealt with a broadly concerned control problem, namely the dynamic heterogeneous platoon control. A platoon mainly consists of networking and data transmission among the vehicles, forming the cyber layer, and the physical environment composed of the participant cars, forming the physical layer. This cyber-physical system is highly prone to cyber-attacks endangering the wireless connectivity among the vehicles. This vulnerability to external attackers needs to be fully addressed as an insecure communication layer in a platoon can cause manipulated and/or missing data received by the followers resulting in dangerous hazards. In this chapter, we focused on the widespread so-called DoS attack in which the intelligent intruder targets the wireless links by overwhelming the node by invalid requests, hence, either blocks the network or prevents it from timely data transfer. We proposed a Secure–Distributed Nonlinear Model Predictive Control (Secure–DNMPC) framework to ensure a safe and secure dynamic platooning which fulfills both the safe distancing between the cars and speed tracking requirements. The method is capable of handling cut-in/cut-out maneuvers under the premise of the existence of a cyber DoS attack. The algorithm is basically comprised of detection and mitigation phases.

Furthermore, we introduced a novel Secure–DNMPC–UKF co-design for the case when the DoS attacker injects a huge amount of time delay in the network compared to the intrinsic practical DSRC time delay. This makes use of the available but outdated data to estimate and predict future states. The proposed approach also provides the opportunity to consider non-ideal sensors which contaminate the measured data. In addition, compromised signals sent through a realistic noisy environment can be considered as well. Simulation results demonstrated the efficacy of the introduced technique. As a future direction, one can think of generalizing the given algorithm to a multi-platooning scenario in which two or more attacked platoons drive in parallel, and arbitrary vehicles wish to exit their own platoon and merge with an adjacent one. Also, other types of attacks and the corresponding countermeasures could be considered.

Acknowledgments The authors would like to thank Dr. Yang Zheng (SEAS and CGBC at Harvard University, Cambridge, MA, USA) for the fruitful discussions during this work. Also, the authors would like to thank the support from Natural Sciences and Engineering Research Council (NSERC) of Canada.

References

1. Wang, X.: Cyber-Physical Systems: A Reference. Springer, Berlin, Heidelberg (2021)
2. Song, H., Fink, G., Jeschke, S.: Security and Privacy in Cyber-Physical Systems. Wiley Online Library (2017)
3. Guo, S., Zeng, D.: Cyber-Physical Systems: Architecture, Security and Application. Springer (2019)
4. Koç, Ç.K.: Cyber-Physical Systems Security. Springer (2018)
5. Liu, H., Niu, B., Li, Y.: False-data-injection attacks on remote distributed consensus estimation. IEEE Trans. Cybern. (2020)
6. Humayed, A., Lin, J., Li, F., Luo, B.: Cyber-physical systems security—a survey. IEEE Int. Things J. **4**(6), 1802–1831 (2017)
7. Ashibani, Y., Mahmoud, Q.H.: Cyber physical systems security: analysis, challenges and solutions. Comput. Secur. **68**, 81–97 (2017)
8. Ali, S., Al Balushi, T., Nadir, Z., Hussain, O.K.: Cyber Security for Cyber Physical Systems, vol. 768. Springer (2018)
9. Basiri, M.H., Azad, N.L., Fischmeister, S.: Distributed time-varying kalman filter design and estimation over wireless sensor networks using OWA sensor fusion technique. In: 2020 28th Mediterranean Conference on Control and Automation (MED), pp. 325–330. IEEE (2020)
10. Langner, R.: Stuxnet: Dissecting a cyberwarfare weapon. IEEE Secur. Privacy **9**(3), 49–51 (2011)
11. Farwell, J.P., Rohozinski, R.: Stuxnet and the future of cyber war. Survival **53**(1), 23–40 (2011)
12. Liu, Y., Ning, P., Reiter, M.K.: False data injection attacks against state estimation in electric power grids. ACM Trans. Inf. Syst. Secur. (TISSEC) **14**(1), 13 (2011)
13. Shoukry, Y., Martin, P., Tabuada, P., Srivastava, M.: Non-invasive spoofing attacks for anti-lock braking systems. In: International Workshop on Cryptographic Hardware and Embedded Systems, pp. 55–72. Springer (2013)
14. Singh, S.: Critical reasons for crashes investigated in the national motor vehicle crash causation survey. Technical report (2015)
15. Transport Canada: Canadian motor vehicle traffic collision statistics (2014)
16. Road Safety in Canada (2011). http://www.tc.gc.ca/eng/motorvehiclesafety/tp-tp15145-1201.htm
17. Godsmark, P., Kirk, B., Gill, V., Flemming, B.: Automated vehicles: the coming of the next disruptive technology (2015)
18. Gao, F., Hu, X., Li, S.E., Li, K., Sun, Q.: Distributed adaptive sliding mode control of vehicular platoon with uncertain interaction topology. IEEE Trans. Ind. Electron. **65**(8), 6352–6361 (2018)
19. Li, Y., Tang, C., Peeta, S., Wang, Y.: Nonlinear consensus-based connected vehicle platoon control incorporating car-following interactions and heterogeneous time delays. IEEE Transactions on Intelligent Transportation Systems (2018)
20. Liu, P., Kurt, A., Ozguner, U.: Distributed model predictive control for cooperative and flexible vehicle platooning. IEEE Trans. Control Syst. Technol. (99), 1–14 (2018)
21. Zheng, Y., Li, S.E., Wang, J., Cao, D., Li, K.: Stability and scalability of homogeneous vehicular platoon: Study on the influence of information flow topologies. IEEE Trans. Intell. Transp. Syst. **17**(1), 14–26 (2016)
22. Pirani, M., Hashemi, E., Simpson-Porco, J.W., Fidan, B., Khajepour, A.: Graph theoretic approach to the robustness of k-nearest neighbor vehicle platoons. IEEE Trans. Intell. Transp. Syst. **18**(11), 3218–3224 (2017)
23. Van Arem, B., Van Driel, C.J., Visser, R.: The impact of cooperative adaptive cruise control on traffic-flow characteristics. IEEE Trans. Intell. Transp. Syst. **7**(4), 429–436 (2006)
24. Rawat, D.B., Bajracharya, C.: Vehicular Cyber Physical Systems. Springer, Technical report (2017)

25. Kargl, F., Papadimitratos, P., Buttyan, L., Müter, M., Schoch, E., Wiedersheim, B., Thong, T.V., Calandriello, G., Held, A., Kung, A., et al.: Secure vehicular communication systems: implementation, performance, and research challenges. IEEE Commun. Mag. **46**(11), 110–118 (2008)
26. Koopman, P., Wagner, M.: Autonomous vehicle safety: an interdisciplinary challenge. IEEE Intell. Transp. Syst. Mag. **9**(1), 90–96 (2017)
27. Amin, S., Cárdenas, A.A., Sastry, S.S.: Safe and secure networked control systems under denial-of-service attacks. In: International Workshop on Hybrid Systems: Computation and Control, pp. 31–45. Springer (2009)
28. Mo, Y., Sinopoli, B.: Secure control against replay attacks. In: 47th Annual Allerton Conference on Communication, Control, and Computing, 2009. Allerton 2009, pp. 911–918. IEEE (2009)
29. Mo, Y., Garone, E., Casavola, A., Sinopoli, B.: False data injection attacks against state estimation in wireless sensor networks. In: 2010 49th IEEE Conference on Decision and Control (CDC), pp. 5967–5972. IEEE (2010)
30. Chabukswar, R., Sinopoli, B., Karsai, G., Giani, A., Neema, H., Davis, A.: Simulation of network attacks on SCADA systems. In: First Workshop on Secure Control Systems, pp. 587–592 (2010)
31. Basiri, M.H., Thistle, J.G., Simpson-Porco, J.W., Fischmeister, S.: Kalman filter based secure state estimation and individual attacked sensor detection in cyber-physical systems. In: American Control Conference (ACC), vol. 2019, pp. 3841–3848. IEEE (2019)
32. Siegel, J.E., Erb, D.C., Sarma, S.E.: A survey of the connected vehicle landscape-architectures, enabling technologies, applications, and development areas. IEEE Trans. Intell. Transp. Syst. **19**(8), 2391–2406 (2018)
33. Zhu, Q., Basar, T.: Game-theoretic methods for robustness, security, and resilience of cyber-physical control systems: games-in-games principle for optimal cross-layer resilient control systems. IEEE Control Syst. **35**, 45–65 (2015)
34. Li, Y., Shi, L., Cheng, P., Chen, J., Quevedo, D.E.: Jamming attacks on remote state estimation in cyber-physical systems: a game-theoretic approach. IEEE Trans. Autom. Control **60**(10), 2831–2836 (2015)
35. Cardenas, A., Amin, S., Sinopoli, B., Giani, A., Perrig, A., Sastry, S., et al.: Challenges for securing cyber physical systems. In: Workshop on Future Directions in Cyber-Physical Systems Security, vol. 5 (2009)
36. Fawzi, H., Tabuada, P., Diggavi, S.: Secure estimation and control for cyber-physical systems under adversarial attacks. IEEE Trans. Autom. Control **59**(6), 1454–1467 (2014)
37. Cardenas, A.A., Amin, S., Sastry, S.: Secure control: Towards survivable cyber-physical systems. In: 2008 The 28th International Conference on Distributed Computing Systems Workshops, pp. 495–500. IEEE (2008)
38. Pasqualetti, F., Dörfler, F., Bullo, F.: Attack detection and identification in cyber-physical systems. IEEE Trans. Autom. Control **58**(11), 2715–2729 (2013)
39. Liu, Y.C., Bianchin, G., Pasqualetti, F.: Secure trajectory planning against undetectable spoofing attacks. Automatica **112**, 108655 (2020)
40. Weerakkody, S., Liu, X., Son, S.H., Sinopoli, B.: A graph-theoretic characterization of perfect attackability for secure design of distributed control systems. IEEE Trans. Control of Netw. Syst. **4**(1), 60–70 (2016)
41. Zhang, T., Zou, Y., Zhang, X., Guo, N., Wang, W.: Data-driven based cruise control of connected and automated vehicles under cyber-physical system framework. IEEE Trans. Intell. Trans. Syst. (2020)
42. Basiri, M.H., Pirani, M., Azad, N.L., Fischmeister, S.: Security of vehicle platooning: a game-theoretic approach. IEEE Access **7**(1), 185565–185579 (2019)
43. Parkinson, S., Ward, P., Wilson, K., Miller, J.: Cyber threats facing autonomous and connected vehicles: Future challenges. IEEE Trans. Intell. Transp. Syst. **18**(11), 2898–2915 (2017)
44. Petit, J., Shladover, S.E.: Potential cyberattacks on automated vehicles. IEEE Trans. Intell. Transp. Syst. **16**(2), 546–556 (2014)

45. Zhang, T., Antunes, H., Aggarwal, S.: Defending connected vehicles against malware: Challenges and a solution framework. IEEE Int. Things J. **1**(1), 10–21 (2014)
46. Biron, Z.A., Dey, S., Pisu, P.: Real-time detection and estimation of denial of service attack in connected vehicle systems. IEEE Trans. Intell. Transp. Syst. **19**(12), 3893–3902 (2018)
47. Lei, C., Van Eenennaam, E., Wolterink, W.K., Karagiannis, G., Heijenk, G., Ploeg, J.: Impact of packet loss on cacc string stability performance. In: 2011 11th International Conference on ITS Telecommunications, pp. 381–386. IEEE (2011)
48. Ploeg, J., Semsar-Kazerooni, E., Lijster, G., van de Wouw, N., Nijmeijer, H.: Graceful degradation of CACC performance subject to unreliable wireless communication. In: 16th International IEEE Conference on Intelligent Transportation Systems (ITSC 2013), pp. 1210–1216. IEEE (2013)
49. Azees, M., Vijayakumar, P., Deborah, L.J.: Comprehensive survey on security services in vehicular ad-hoc networks. IET Intell. Transp. Syst. **10**(6), 379–388 (2016)
50. Dadras, S., Gerdes, R.M., Sharma, R.: Vehicular platooning in an adversarial environment. In: Proceedings of the 10th ACM Symposium on Information, Computer and Communications Security, pp. 167–178. ACM (2015)
51. Öncü, S., Ploeg, J., Van de Wouw, N., Nijmeijer, H.: Cooperative adaptive cruise control: Network-aware analysis of string stability. IEEE Trans. Intell. Transp. Syst. **15**(4), 1527–1537 (2014)
52. Qin, W.B., Orosz, G.: Experimental validation of string stability for connected vehicles subject to information delay. IEEE Trans. Control Syst. Technol. (2019)
53. Qin, W.B., Gomez, M.M., Orosz, G.: Stability analysis of connected cruise control with stochastic delays. In: American Control Conference, vol. 2014, pp. 4624–4629. IEEE (2014)
54. Laurendeau, C., Barbeau, M.: Threats to security in DSRC/WAVE. In: International Conference on Ad-Hoc Networks and Wireless, pp. 266–279. Springer (2006)
55. Teixeira, A., Shames, I., Sandberg, H., Johansson, K.H.: A secure control framework for resource-limited adversaries. Automatica **51**, 135–148 (2015)
56. Harfouch, Y.A., Yuan, S., Baldi, S.: An adaptive switched control approach to heterogeneous platooning with intervehicle communication losses. IEEE Trans. Control of Netw. Syst. **5**(3), 1434–1444 (2017)
57. Dolk, V.S., Ploeg, J., Heemels, W.M.H.: Event-triggered control for string-stable vehicle platooning. IEEE Trans. Intell. Transp. Syst. **18**(12), 3486–3500 (2017)
58. Zheng, Y., Li, S.E., Li, K., Borrelli, F., Hedrick, J.K.: Distributed model predictive control for heterogeneous vehicle platoons under unidirectional topologies. IEEE Trans. Control Syst. Technol. **25**(3), 899–910 (2017)
59. Yuan, Y., Zhu, Q., Sun, F., Wang, Q., Başar, T.: Resilient control of cyber-physical systems against denial-of-service attacks. In: 6th International Symposium on Resilient Control Systems (ISRCS), vol. 2013, pp. 54–59. IEEE (2013)
60. Zhang, H., Cheng, P., Shi, L., Chen, J.: Optimal denial-of-service attack scheduling with energy constraint. IEEE Trans. Autom. Control **60**(11), 3023–3028 (2015)
61. Sun, Q., Zhang, K., Shi, Y.: Resilient model predictive control of cyber-physical systems under dos attacks. IEEE Trans. Ind. Inform. (2019)
62. Wan, E.A., Van Der Merwe, R.: The unscented kalman filter for nonlinear estimation. In: Proceedings of the IEEE 2000 Adaptive Systems for Signal Processing, Communications, and Control Symposium (Cat. No. 00EX373), pp. 153–158. IEEE (2000)
63. Simon, D.: Optimal State Estimation: Kalman, H Infinity, and Nonlinear Approaches. Wiley (2006)
64. Basiri, M.H., Azad, N.L., Fischmeister, S.: Attack resilient heterogeneous vehicle platooning using secure distributed nonlinear model predictive control. In: 2020 28th Mediterranean Conference on Control and Automation (MED), pp. 307–312. IEEE (2020)
65. Wang, J.Q., Li, S.E., Zheng, Y., Lu, X.Y.: Longitudinal collision mitigation via coordinated braking of multiple vehicles using model predictive control. Integr. Comput.-Aided Eng. **22**(2), 171–185 (2015)
66. Zheng, Y.: DMPC for platoons (2019). https://github.com/zhengy09/DMPC_for_platoons

67. Yao, Y., Rao, L., Liu, X., Zhou, X.: Delay analysis and study of IEEE 802.11-p based DSRC safety communication in a highway environment. In: Proceedings IEEE INFOCOM, vol. 2013, pp. 1591–1599. IEEE (2013)
68. Wang, Y., Duan, X., Tian, D., Lu, G., Yu, H.: Throughput and delay limits of 802.11-p and its influence on highway capacity. Procedia-Soc. Behav. Sci. **96**, 2096–2104 (2013)
69. Ma, X., Chen, X., Refai, H.H.: Performance and reliability of DSRC vehicular safety communication: a formal analysis. EURASIP J. Wirel. Commun. Netw. **2009**, 1–13 (2009)
70. Kukshya, V., Krishnan, H.: Experimental measurements and modeling for vehicle-to-vehicle dedicated short range communication (DSRC) wireless channels. In: IEEE Vehicular Technology Conference, pp. 1–5. IEEE (2006)
71. Sabouni, R., Hafez, R.M.: Performance of DSRC for V2V communications in urban and highway environments. In: 2012 25th IEEE Canadian Conference on Electrical and Computer Engineering (CCECE), pp. 1–5. IEEE (2012)

The manufacturer's authorised representative in the EU is Springer Nature Customer Service Centre GmbH, Europaplatz 3, 69115 Heidelberg, Germany. If you have any concerns regarding our products, please contact ProductSafety@springernature.com

Printed and bound by CPI Group (UK) Ltd, Croydon, CR0 4YY
30/09/2026
02225692-0009